Progress in Analytical Methods for the Characterization, Quality and Safety of the Beehive Products

Editors

Gavino Sanna
Marco Ciulu
Yolanda Picò
Nadia Spano
Carlo I.G. Tuberoso

MDPI • Basel • Beijing • Wuhan • Barcelona • Belgrade • Manchester • Tokyo • Cluj • Tianjin

Editors

Gavino Sanna
Department of Chemical,
Physical, Mathematical and
Natural Sciences
Sassari University
Sassari
Italy

Marco Ciulu
Department of Chemical,
Physical, Mathematical and
Natural Sciences
Sassari University
Sassari
Italy

Yolanda Picò
Centre d'Investigacions sobre
Desertificació (CIDE),
Moncada (València), Spain
Valencia University
Moncada (Valencia)
Spain

Nadia Spano
Department of Chemical,
Physical, Mathematical and
Natural Sciences
Sassari University
Sassari
Italy

Carlo I.G. Tuberoso
Department of Life and
Environmental Sciences
Cagliari University
Monserrato (Cagliari)
Italy

Editorial Office
MDPI
St. Alban-Anlage 66
4052 Basel, Switzerland

This is a reprint of articles from the Special Issue published online in the open access journal *Molecules* (ISSN 1420-3049) (available at: www.mdpi.com/journal/molecules/special_issues/analytical_bee_product).

For citation purposes, cite each article independently as indicated on the article page online and as indicated below:

LastName, A.A.; LastName, B.B.; LastName, C.C. Article Title. *Journal Name* **Year**, *Volume Number*, Page Range.

ISBN 978-3-0365-4134-1 (Hbk)
ISBN 978-3-0365-4133-4 (PDF)

Cover image courtesy of Ignazio Floris

© 2022 by the authors. Articles in this book are Open Access and distributed under the Creative Commons Attribution (CC BY) license, which allows users to download, copy and build upon published articles, as long as the author and publisher are properly credited, which ensures maximum dissemination and a wider impact of our publications.

The book as a whole is distributed by MDPI under the terms and conditions of the Creative Commons license CC BY-NC-ND.

Contents

About the Editors vii

Preface to "Progress in Analytical Methods for the Characterization, Quality and Safety of the Beehive Products" ix

Piotr M. Kuś
Honey as Source of Nitrogen Compounds: Aromatic Amino Acids, Free Nucleosides and Their Derivatives
Reprinted from: *Molecules* **2020**, *25*, 847, doi:10.3390/molecules25040847 1

Aleksandra Wilczyńska and Natalia Żak
The Use of Fluorescence Spectrometry to Determine the Botanical Origin of Filtered Honeys
Reprinted from: *Molecules* **2020**, *25*, 1350, doi:10.3390/molecules25061350 11

Marco Ciulu, Elisa Oertel, Rosanna Serra, Roberta Farre, Nadia Spano and Marco Caredda et al.
Classification of Unifloral Honeys from SARDINIA (Italy) by ATR-FTIR Spectroscopy and Random Forest
Reprinted from: *Molecules* **2020**, *26*, 88, doi:10.3390/molecules26010088 29

Michał Miłek, Aleksandra Bocian, Ewelina Kleczyńska, Patrycja Sowa and Małgorzata Dżugan
The Comparison of Physicochemical Parameters, Antioxidant Activity and Proteins for the Raw Local Polish Honeys and Imported Honey Blends
Reprinted from: *Molecules* **2021**, *26*, 2423, doi:10.3390/molecules26092423 39

Paola Montoro, Gilda D'Urso, Adam Kowalczyk and Carlo Ignazio Giovanni Tuberoso
LC-ESI/LTQ-Orbitrap-MS Based Metabolomics in Evaluation of Bitter Taste of *Arbutus unedo* Honey
Reprinted from: *Molecules* **2021**, *26*, 2765, doi:10.3390/molecules26092765 55

Severyn Salis, Nadia Spano, Marco Ciulu, Ignazio Floris, Maria I. Pilo and Gavino Sanna
Electrochemical Determination of the "Furanic Index" in Honey
Reprinted from: *Molecules* **2021**, *26*, 4115, doi:10.3390/molecules26144115 69

Andrea Mara, Sara Deidda, Marco Caredda, Marco Ciulu, Mario Deroma and Emanuele Farinini et al.
Multi-Elemental Analysis as a Tool to Ascertain the Safety and the Origin of Beehive Products: Development, Validation, and Application of an ICP-MS Method on Four Unifloral Honeys Produced in Sardinia, Italy
Reprinted from: *Molecules* **2022**, *27*, 2009, doi:10.3390/molecules27062009 83

Radmila Pavlovic, Gigliola Borgonovo, Valeria Leoni, Luca Giupponi, Giulia Ceciliani and Stefano Sala et al.
Effectiveness of Different Analytical Methods for the Characterization of Propolis: A Case of Study in Northern Italy
Reprinted from: *Molecules* **2020**, *25*, 504, doi:10.3390/molecules25030504 99

Xiasen Jiang, Linchen Tao, Chunguang Li, Mengmeng You, George Q. Li and Cuiping Zhang et al.
Grouping, Spectrum–Effect Relationship and Antioxidant Compounds of Chinese Propolis from Different Regions Using Multivariate Analyses and Off-Line Anti-DPPH Assay
Reprinted from: *Molecules* **2020**, *25*, 3243, doi:10.3390/molecules25143243 125

Meifei Zhu, Jian Tang, Xijuan Tu and Wenbin Chen
Determination of Ascorbic Acid, Total Ascorbic Acid, and Dehydroascorbic Acid in Bee Pollen Using Hydrophilic Interaction Liquid Chromatography-Ultraviolet Detection
Reprinted from: *Molecules* **2020**, *25*, 5696, doi:10.3390/molecules25235696 137

Eliza Matuszewska, Joanna Matysiak, Grzegorz Rosiński, Elżbieta Kedzia, Weronika Zabek and Jarosław Zawadziński et al.
Mining the Royal Jelly Proteins: Combinatorial Hexapeptide Ligand Library Significantly Improves the MS-Based Proteomic Identification in Complex Biological Samples
Reprinted from: *Molecules* **2021**, *26*, 2762, doi:10.3390/molecules26092762 147

Maria Luisa Astolfi, Marcelo Enrique Conti, Elisabetta Marconi, Lorenzo Massimi and Silvia Canepari
Effectiveness of Different Sample Treatments for the Elemental Characterization of Bees and Beehive Products
Reprinted from: *Molecules* **2020**, *25*, 4263, doi:10.3390/molecules25184263 163

Eliza Matuszewska, Agnieszka Klupczynska, Krzysztof Maciołek, Zenon J. Kokot and Jan Matysiak
Multielemental Analysis of Bee Pollen, Propolis, and Royal Jelly Collected in West-Central Poland
Reprinted from: *Molecules* **2021**, *26*, 2415, doi:10.3390/molecules26092415 183

Małgorzata Starowicz, Paweł Hanus, Grzegorz Lamparski and Tomasz Sawicki
Characterizing the Volatile and Sensory Profiles, and Sugar Content of Beeswax, Beebread, Bee Pollen, and Honey
Reprinted from: *Molecules* **2021**, *26*, 3410, doi:10.3390/molecules26113410 201

Maria Luisa Astolfi, Marcelo Enrique Conti, Martina Ristorini, Maria Agostina Frezzini, Marco Papi and Lorenzo Massimi et al.
An Analytical Method for the Biomonitoring of Mercury in Bees and Beehive Products by Cold Vapor Atomic Fluorescence Spectrometry
Reprinted from: *Molecules* **2021**, *26*, 4878, doi:10.3390/molecules26164878 217

About the Editors

Gavino Sanna

Gavino Sanna is an Associate Professor of Analytical Chemistry in the Department of Chemical, Physical, Mathematical and Natural Sciences of the University of Sassari, Italy. His general research interest is focused on the assessment, the validation, and the application to real samples of instrumental methods for the determination of organic and inorganic analytes in foods, environmental and biotical matrices. He is also active in the fields of the science of the materials and in the study of the interactions between metal ions and biological ligands. He is author of more than 100 international contributions including papers, book chapters and patents, and he is member of the Editorial Board of international Journals indexed by Scopus and/or WoS of Analytical Chemistry, Environmental Sciences and Food Science and Technology, for which he has also served as Guest Editor for many Special Issues. Finally, he also obtained the scientific habilitation as a full professor in Analytical Chemistry.

Marco Ciulu

Marco Ciulu is a food and analytical chemist. He obtained his PhD in Chemical Sciences and Technologies at the University of Sassari (Italy) with a Thesis focused on the development of innovative chromatographic methods for the assessment of quality, freshness and authenticity of beehive products. After his PhD, he continued to work as a post-doctoral researcher in Italy, Spain, France and Germany in various research projects mostly concerning the evaluation and improvement of the quality of food products of animal and plant origin and the promotion of sustainable production. In his latest position at the University of Göttingen (Germany) he was lecturer of "Chromatographic analysis of animal products"and was involved in several research projects concerning the nutritional evaluation of meat, fish and eggs obtained from alternative protein sources, local breeds and animal-welfare friendlier productions. He recently obtained the Italian scientific habilitation as associate professor of Food Chemistry.

Yolanda Picò

Yolanda Picò is a Full Professor of Food Chemistry and Nutrition of the University of Valencia, Spain. Her line of priority research was the development and validation of methods of analysis for the determination of emerging organic contaminants in food and the environment. Over her career, she has published over 300 articles included in SCI, is editor of 5 books on the analysis of food contaminants and their determination by LC-MS and several Special Issues of prestigious journals

Nadia Spano

Nadia Spano is an Associate Professor of Analytical Chemistry in the Department of Chemical, Physical, Mathematical and Natural Sciences of the University of Sassari, Italy. She has a long experience in research areas related to development, optimization, validation and application of quali-quantitative methods for determination of organic and inorganic analytes in real samples, employing different instrumental techiniques. These research interests relate to and concern mainly, but not only, the food, health and environmental sectors, and have allowed her numerous international publications.

Carlo I.G. Tuberoso

Carlo I.G. Tuberoso is an Associate Professor of Food Chemistry in the Department of Life and Environmental Sciences of the University of Cagliari (Italy), and he also obtained the scientific habilitation as a full professor in Food Chemistry. His general research interest is focused on food quality, studying the nutritional, toxicological and process innovation aspects especially of food products typical of the Mediterranean diet. He is also active in the field of the exploitation of agri-food by-products with low-cost and environmentally friendly green-extraction methods. In particular, he is interested in secondary metabolites that show antioxidant and antiradical properties, mainly polyphenols. He has been involved in numerous national and international project, and he is author of more than 100 international contributions in international Journals indexed by Scopus and/or WoS of Food Science and Technology, Analytical Chemistry, and Environmental Sciences.

Preface to "Progress in Analytical Methods for the Characterization, Quality and Safety of the Beehive Products"

The honeybee is one of the oldest forms of animal life, already present since the Neolithic. The origin of the bond between man and honeybees dates back almost 10,000 years. Primeval humans gathered and ate the honey and honeycombs of wild bees, the only available sweetener at that time, up to 7000 BC. Honey is a food consisting of a very complex mixture of nutrients and bioactive compounds, endowed with outstanding nutritional value and many recognized biological activities. Nonetheless, bees also produce and store in hives many other products of enormous interest to humans. Although honey is the most famous of hive products, propolis, bee bread, royal jelly, beeswax, and bee venom are also of great value. Most of them have been objects of increasing interest to the global scientific community, as a growing number of studies have discovered their nutraceutical or pharmaceutical effects on human health. For many of these products, nutrition and/or biological activity is closely linked to their origin. Whereas the botanical origin is of the highest importance in defining the quality of honey and pollen, the assignment of geographical origin is decisive for that of propolis, beeswax and royal jelly. In addition, adulteration processes or the fraudulent assignment of a specific geographic origin to a hive product can result in an unfair increase in market shares and prices to the detriment of authentic products. In addition, bees and hive products can act as effective biomonitoring matrices for the ascertainment of the level of environmental quality. Beyond the definition of the quality of hive products, identification of traces of toxic elements, of persistent organic pollutants and of residues of drugs or phytosanitary products may provide a reliable map of environmental conditions in the ecosystems close to the hive. Of course, the characterization of these matrices in terms of reliable determination of minority or of trace amounts of organic or inorganic analytes is an increasingly challenging task. New techniques and methods of analysis are needed to achieve this, and only the continuous updating of ever more powerful and multivariate approaches to data processing is necessary to ensure reliable responses to today's global consumers.

Gavino Sanna, Marco Ciulu, Yolanda Picò, Nadia Spano, and Carlo I.G. Tuberoso
Editors

Article

Honey as Source of Nitrogen Compounds: Aromatic Amino Acids, Free Nucleosides and Their Derivatives

Piotr M. Kuś

Department of Pharmacognosy and Herbal Medicines, Faculty of Pharmacy, Wroclaw Medical University, ul. Borowska 211a, 50-556 Wrocław, Poland; kus.piotrek@gmail.com or piotr.kus@umed.wroc.pl

Received: 19 January 2020; Accepted: 12 February 2020; Published: 14 February 2020

Abstract: The content of selected major nitrogen compounds including nucleosides and their derivatives was evaluated in 75 samples of seven varieties of honey (heather, buckwheat, black locust, goldenrod, canola, fir, linden) by targeted ultra-high performance liquid chromatography-diode array detector - high-resolution quadrupole time-of-flight mass spectrometry (UHPLC-DAD-QqTOF-MS) and determined by UHPLC-DAD. The honey samples contained nucleosides, nucleobases and their derivatives (adenine: 8.9 to 18.4 mg/kg, xanthine: 1.2 to 3.3 mg/kg, uridine: 17.5 to 51.2 mg/kg, guanosine: 2.0 to 4.1 mg/kg; mean amounts), aromatic amino acids (tyrosine: 7.8 to 263.9 mg/kg, phenylalanine: 9.5 to 64.1 mg/kg; mean amounts). The amounts of compounds significantly differed between some honey types. For example, canola honey contained a much lower amount of uridine (17.5 ± 3.9 mg/kg) than black locust where it was most abundant (51.2 ± 7.8 mg/kg). The presence of free nucleosides and nucleobases in different honey varieties is reported first time and supports previous findings on medicinal activities of honey reported in the literature as well as traditional therapy and may contribute for their explanation. This applies, e.g., to the topical application of honey in herpes infections, as well as its beneficial activity on cognitive functions as nootropic and neuroprotective, in neuralgia and is also important for the understanding of nutritional values of honey.

Keywords: essential and non-essential nutrients; nucleosides; honey composition; uridine; neuropharmacological activities

1. Introduction

Nutrient and medicinal properties of honey have been appreciated since prehistoric times and the knowledge about its values is currently deepened and rediscovered. Honey is not only a source of simple carbohydrates, but also microelements, vitamins, antioxidants, prebiotics, and probiotic bacteria [1–3]. Recently, Begum et al. reported the presence of essential elements like K, Ca, Mn, and Fe in honey along with *Gluconobacter oxydans* that was found to possess probiotic properties with siderophorogenic potential [2]. Other research demonstrated that buckwheat honey increases antioxidant potential of serum, thus possibly could exert antioxidant-related health benefits [4]. Honey is also a good source of carbohydrates supporting during physical exercise and characterized by low glycemic index [3]. One of the specific groups of compounds in honey are those containing nitrogen, which is an essential component of basic compounds in plants such as nucleic acids, proteins, coenzymes, hormones, some vitamins, and chlorophylls. In honey, nitrogen compounds constitute a versatile group present as a minor compounds. The amino acids are found in all honey types in various proportions, depending on their floral source. The differences in their levels were useful to determine the botanical origin in combination with chemometrics [5]. Another, common groups of nitrogen compounds found in honey were water-soluble vitamins such as riboflavin, niacin, nicotinic acid, pantothenic acid or folic acid [6]. Some of the honey varieties contain relevant amounts of different classes of nitrogen compounds that may be useful as a specific or non-specific markers of botanical origin, for example purine alkaloids in *Coffea* spp. honey [7] or high amounts of kynurenic acid in *Castanea sativa* Mill. honey [8]. The latter was found to possess anti-scarring activity as well as downregulating IL-17/1L-23 axis *in vitro*, and this could reduce inflammation in autoimmune diseases such as psoriasis or alopecia [9,10]. Oelschlaegel et al. beside kynurenic acid, reported the presence of its metabolite 4-hydroxyquinoline in cornflower (*Centaurea cyanus* L.) honey [11]. Recently, we reported the presence of 5-*epi*-lithospermoside, a noncyanogenic cyanoglucoside as a marker of *Phacelia tanacetifolia* Benth. honey as well as the occurrence of nucleoside uridine and nucleobase—adenine as well as other purine—xanthine [12]. Nucleobases that are part of nucleosides and further—nucleotides, play fundamental role in construction of DNA and RNA. Those compounds may also have importance for nutritional values of honey, especially uridine that was found to play relevant role, e.g., in cognitive functions [13] as well as guanosine, that exhibits neurotrophic and neuroprotective effects interacting with glutamatergic and adenosinergic systems and calcium-activated potassium channels [14]. Not much is known on occurrence and content of nucleosides in different honey varieties. Therefore, the scope of the current study was targeted ultra-high performance liquid chromatography-diode array detector-high-resolution quadrupole time-of-flight mass spectrometry (UHPLC-DAD-QqTOF-MS) and UHPLC-DAD analyses of the most common types of Polish honeys to evaluate the content of selected nitrogen compounds, including nucleobases and nucleosides.

2. Results and Discussion

Chemical profiles of 75 samples of seven unifloral honey varieties were investigated by targeted UHPLC-DAD and UHPLC-DAD-QqTOF-MS analyses. Six major nitrogen compounds (Adenine Xanthine, Uridine, Tyrosine, Phenylalanine, Guanosine) were identified (Table 1) and quantified (Table 2). The comparison of chromatographic profiles obtained for honey varieties is presented on Figure 1 and demonstrates visible differences. All the mentioned compounds were identified by retention time, UV-Vis spectra and comparison with commercially available standard compounds as well as HRMS. For uridine, beside the main pseudomolecular ion 245.0764 $[M + H]^+$, also 113.0346 $[M - 132 + H]^+$, corresponding to uracil was present. The 132 amu loss corresponds to an unmodified ribose from the ribonucleoside, thus confirming nitrogen-carbon glyosidic bond. Similarly, guanosine was accompanied by 152.0570 $[M - 132 + H]^+$ fragment, corresponding to a pseudomolecular ion of guanine.

Table 1. Nitrogen compounds identified in the investigated honey samples.

	Compound	t_R (min)	UV_{max} (nm)	$[M + H]^+$	Formula	Error (ppm)
1	Adenine	1.17	263	136.0621	$C_5H_5N_5$	1.62
2	Uridine	1.72	262	245.0764	$C_9H_{12}N_2O_6$	3.93
3	Xanthine	1.78	268	153.0405	$C_5H_4N_4O_2$	4.91
4	Tyrosine	2.21	224, 275	182.0812	$C_9H_{11}NO_3$	2.75
5	Guanosine	3.10	254, 274	284.0989	$C_{10}H_{13}N_5O_5$	2.09
6	Phenylalanine	3.41	258	166.0864	$C_9H_{11}NO_2$	2.41

The mean amount of adenine in different honey varieties ranged from about 8.9 to 18.4 mg/kg and was highest in linden (*Tilia* spp.) and lowest in canola (*Brassica napus* L.) honey. Xanthine and guanosine levels were quite low, and ranged from 1.2 in goldenrod (*Solidago* spp.) to 3.3 mg/kg in black locust and 2.0 to 4.1 mg/kg, respectively. Xanthine was not detected by UHPLC-DAD in heather [*Calluna vulgaris* (L.) Hull], buckwheat (*Fagopyrum esculentum* Moench) and fir (*Abies alba* Mill.) honeys. Guanosine in buckwheat and fir honey as well as adenine in the latter honey were not detected as well, due to the presence of overlapping peaks. Mean uridine content ranged from about 17.5 to 51.2 mg/kg and was lowest in canola and highest in black locust (*Robinia pseudoacacia* L.) honeys. The mean content of aromatic amino acids ranged from 7.8 to 263.9 and from 9.5 to 64.1 mg/kg for tyrosine and phenylalanine, respectively. The lowest amounts of tyrosine and phenylalanine were found in canola honey and the highest in buckwheat honey. Previously analyzed *P. tanacetifolia* honey contained higher amounts of adenine, with mean value 18.5 mg/kg almost equal to that found in linden honey. It contained also relatively high amount of uridine (42.8 mg/kg) similar to that found in buckwheat honey but it was also much richer in xanthine (mean value 10.5 mg/kg) than any other investigated variety [1].

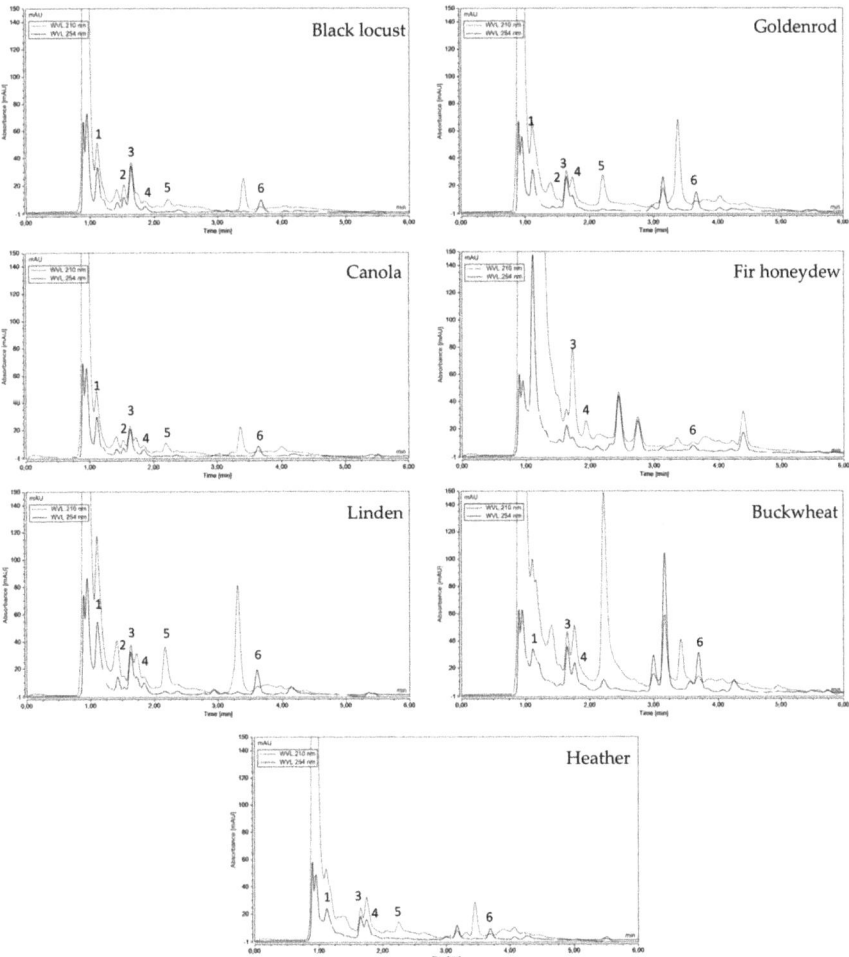

Column: Phenomenex Kinetex EVO C18 110 Å column (150 mm × 2.10 mm, 2.6 µm) eluted with mixtures of 0.2 mol/L phosphoric acid (solvent A) and acetonitrile (solvent B) at a flow rate of 0.4 mL/min. 1—adenine; 2—xanthine; 3—uridine; 4—tyrosine; 5—guanosine; 6—phenylalanine.

Figure 1. Representative UHPLC chromatographic profiles of different Polish unifloral honeys at $\lambda = 210$ nm and $\lambda = 254$ nm.

Table 2. Amounts of the identified nitrogen compounds in varietal honey samples.

	Adenine (mg/kg)	Xanthine (mg/kg)	Uridine (mg/kg)	Tyrosine (mg/kg)	Guanosine (mg/kg)	Phenylalanine (mg/kg)
Heather (n = 10)	$11.8^a \pm 2.9$	nd	$30.7^{bc} \pm 9.9$	$35.8^a \pm 28.9$	$3.4^{ab} \pm 1.1$	$17.1^{abc} \pm 4.0$
Buckwheat (n = 10)	$11.6^a \pm 4.4$	nd	$40.2^{cd} \pm 13.6$	$263.9^b \pm 91.6$	nd*	$34.6^c \pm 17.1$
Black locust (n = 10)	$11.1^a \pm 1.5$	$3.3^b \pm 0.8$	$51.2^d \pm 7.8$	$12.1^a \pm 5.2$	$2.6^a \pm 0.3$	$11.3^{ab} \pm 5.5$
Goldenrod (n = 10)	$14.0^{ab} \pm 2.7$	$1.2^a \pm 0.6$	$24.6^{ab} \pm 2.7$	$44.9^a \pm 25.1$	$2.5^a \pm 1.3$	$64.1^d \pm 20.3$
Canola (n = 10)	$8.9^a \pm 1.4$	$3.0^b \pm 0.6$	$17.5^a \pm 3.9$	$7.8^a \pm 3.3$	$2.0^a \pm 0.7$	$9.5^a \pm 3.6$
Fir (n = 10)	nd*	nd	$32.5^{bc} \pm 12.8$	$31.6^a \pm 17.9$	nd*	$18.1^{abc} \pm 13.3$
Linden (n = 15)	$18.4^b \pm 6.5$	$1.7^a \pm 0.9$	$28.6^{ab} \pm 10.6$	$21.8^a \pm 12.3$	$4.1^b \pm 1.5$	$28.4^{bc} \pm 20.6$
LOD	0.1	0.2	0.3	0.4	0.1	0.3
LOQ	0.4	0.6	1.0	1.1	0.3	0.8

The amount of compounds is expressed in milligrams per kilogram of honey as average (AV) ± standard deviation (SD); nd—not detected; *—overlapping peak; the mean values with different letters within the same column differ significantly at $p < 0.05$; n—number of samples; LOD—limit of detection; LOQ—limit of quantification.

The content of quantified compounds significantly differed between some of the varieties, suggesting strong link to the botanical origin. This indicates that some varieties such as black locust or buckwheat honeys may be more valuable as sources of nutrients.

Adenine is a nucleobase (Figure 2) building nucleotides of the nucleic acids, as well as adenosine triphosphate (ATP), adenosine monophosphate (AMP), deoxy AMP, cyclic AMP, nicotinamide adenine dinucleotide (NAD) playing important roles in metabolism and signaling. Its derivatives possess also significant antiviral and cytostatic activity [2]. Xanthine commonly found in some plants is an adenosine degradation intermediate formed by oxidation of hypoxanthine [3]. Uridine and guanosine are RNA-specific nucleosides consisting of uracil (pyrimidine base) or guanine (purine base) and ribofuranose bound by β-N_1 or β-N_9 glycosidic bond, respectively. Uridine was found to act as nootropic, improving memory and learning ability as well as positively affecting mood. Dietary uridine was found to enhance the improvement in learning and memory of docosahexaenoic acid (DHA) fed gerbils [4] and enhanced synapse formation [5]. In combination with choline it improved cognitive deficits in rats [6]. Uridine, together with other key nutrients—omega-3 fatty acid DHA, and choline—accelerated the formation of a synaptic membrane and affected numbers of synapses formed initiated by neuronal activity [7]. Uridine decreased also depressive symptoms in patients with bipolar disorder and was patented as a treatment in doses starting from 1 g/day (Kondo et al., 2011; Renshaw, 2010). More recently, uridine at doses between 10 mg and 15 g per day was patented as prophylaxis and therapy of some tumors [8]. Interestingly, in double-blind, randomized, comparative, controlled trial, uridine in form of uridine triphosphate trisodium (1.5 mg, corresponding to 0.76 mg of uridine) in combination with cytidine monophosphate disodium (2.5 mg) and vitamin B_{12} (1 mg) administered 3 times a day showed superior improvement in pain reduction efficacy than vitamin B_{12} monotherapy of compressive neuralgia [9]. The amount of uridine corresponds to that present in 15–25 g of most honeys, thus moderate honey consumption (that contain free uridine) may be beneficial, also in prevention of this condition.

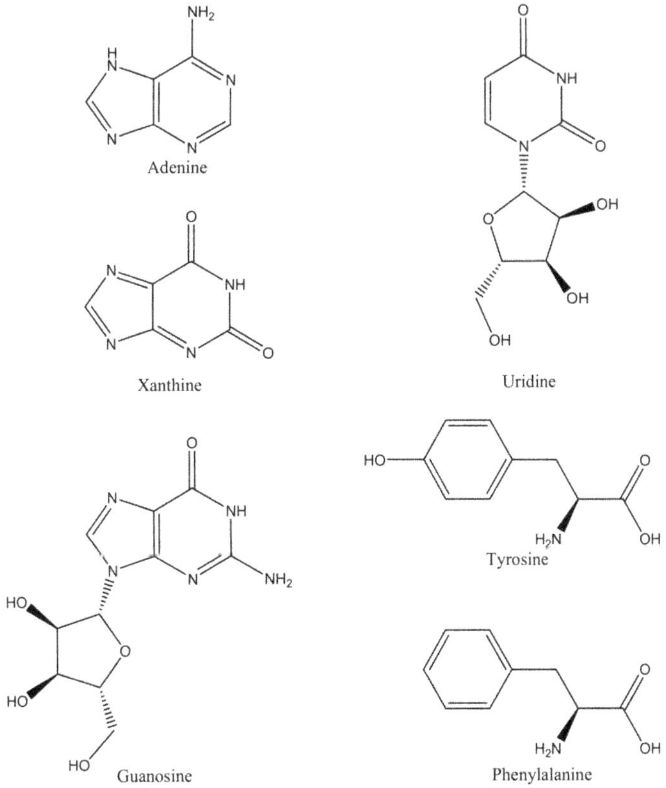

Figure 2. Structures of the nitrogen compounds determined in different honey varieties.

Guanosine is an extracellular signaling molecule exerting anti-inflammatory and antioxidative effects in several *in vivo* and *in vitro* injury models, modulates inflammation, downregulating of NFκB-mediated signaling [10]. It exhibits neurotrophic and neuroprotective effects interacting with glutamatergic, adenosinergic systems and calcium-activated potassium channels [11,12]. Guanosine provides also antidepressant-like effects [13] and attenuates behavioral deficits after traumatic brain injury by modulation of adenosinergic receptors [14]. Considering this, honey consumption may contribute to dietary prevention of different diseases. It may also partially explain and support previous reports on nootropic, memory-enhancing effects as well as neuropharmacological activities, such as antinociceptive, antidepressant but also neuroprotective potential of honey [15]. Guanosine is also known for antiviral activity against HSV-1, EC_{50} determined in infected Vero cells was 0.03 µM [16] and its analogues are widely used in treatment in herpesvirus infections [17]. Knowledge on concentration of guanosine in honey (approximately 5–10 µM), together with known wound-healing properties of honey [18], could support rationale of traditional treatment method of labial herpes by topical application of honey. Such beneficial activity was also confirmed in a small, non-blinded, cross-over study [19].

In general, the available literature data on content of phenylalanine and tyrosine, that are considered respectively essential and conditionally essential amino acids, in varietal honey is not very consistent [20–23] The mean contents of phenylalanine in Serbian black locust (acacia) honey and linden honey (8.5 and 9.2 mg/kg, respectively) as well as Polish black locust and heather honeys (6.4 and 16.0 mg/kg, respectively) were very similar to those found in current study. The values reported for canola (rapeseed), buckwheat and goldenrod honeys were much different (80.8, 10.4, 13.7 mg/kg, respectively). In case of tyrosine content, the results were similar to those reported for Serbian canola honey (8.9 mg/kg), but not to other data on Serbian or Polish varietal honeys [21,22]. Shen et al. found in buckwheat honey also a much higher amount of tyrosine than in other varieties which is consistent with our findings [23]. In fact, this compound was even proposed as marker of buckwheat honey [24]. From a nutritional point of view, the content of aromatic amino acids in honey is low.

Some correlations between amounts of different components were found. The amount of adenine and guanosine was significantly, positively correlated within all dataset ($R^2 = 0.4818$, $p < 0.05$). Amounts of tyrosine and phenylalanine were significantly correlated ($p < 0.05$) within the same some honey types (buckwheat, $R^2 = 0.7223$; black locust, $R^2 = 0.6327$; fir, $R^2 = 0.4607$; linden, $R^2 = 0.5396$; goldenrod, $R^2 = 0.9374$). The levels of uridine and adenine were significantly correlated in goldenrod, $R^2 = 0.7421$ and canola, $R^2 = 0.4651$. The correlations between complementary nucleosides adenine and uridine could be explained by originating from RNA. The nucleobases as well as nucleosides may originate from plant, bee saliva or honey microbiome [25]. The significant differences found between various honey varieties suggest that plant origin may have the most important impact on their content. This is partially visibly also on a dendrogram demonstrating natural clustering of the samples, based on the major compounds (Figure 3). The linkage distance within and between groups indicates tendency for natural clustering and separation from other groups. It is particularly clear for buckwheat, black locust, goldenrod and canola honeys. The cluster containing buckwheat honeys demonstrates the biggest differences from the other samples. The balance between salvage and degradation of nucleotides helps to optimize the plant energy economy while maintaining levels of key elements. This includes nucleotide degradation, that involves removal of the phosphates from nucleotides to form nucleosides as well as cleavage of the nucleobase from the sugar mediated by nucleosidase allowing the base to be recycled as monophosphate [26]. The reason why these compounds would be excreted together with nectar is unknown. However, it is known that the components of nectar often protect against pathogens and modulate the behavior of nectar feeders, thus maximizing benefits for the plant [27].

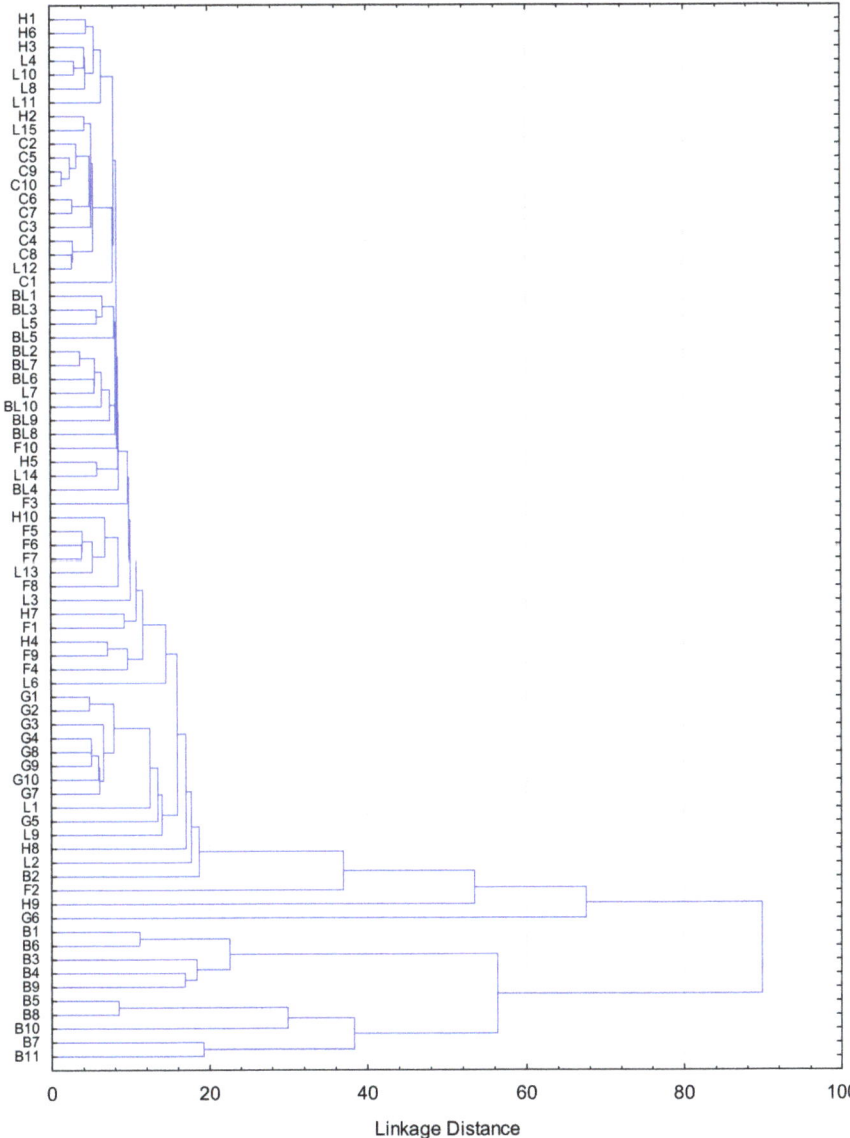

Figure 3. Dendrogram of different Polish unifloral honeys according to cluster analysis of similarity on the basis of the major compounds content (uridine, tyrosine, phenylalanine). H1-10—heather; B1-10—buckwheat; BL1-10—black locust; G1-10—goldenrod; C1-10—canola; F1-10—fir; L1-15—linden.

3. Materials and Methods

3.1. Honey Samples

Seventy five samples of unifloral honey samples—heather [*Calluna vulgaris* (L.) Hull]–10, buckwheat (*Fagopyrum esculentum* Moench)–10, black locust (acacia) (*Robinia pseudoacacia* L.)–10, goldenrod (*Solidago* spp.)–10, canola (rapeseed) (*Brassica napus* L.)–10, fir honeydew (*Abies alba* Mill.)–10, linden (lime-tree) (*Tilia* spp.)–15. Honey samples were obtained from professional beekeepers in different parts of Poland in 2016–2018. The samples were stored in closed glass jars in dark, at 4 °C. Melissopalynological analyses were done according the International Commission for Bee Botany [28].

3.2. Reagents

Acetonitrile (gradient and LC-MS grade), LC-MS grade water, formic acid and phosphoric acid of analytical grade (Fluka™) were from Honeywell (Seelze, Germany), while adenine, guanosine and uridine standards were from Sigma-Aldrich (Steinheim, Germany). Tyrosine and phenylalanine were from Merck (Darmstadt, Germany) and xanthine from Reanal (Budapest, Hungary). Ultrapure H_2O (<0.06 µS/cm) was obtained from Hydrolab HLP20UV (Hydrolab, Straszyn, Poland) device.

3.3. UHPLC-DAD and UHPLC-QqTOF-MS Analysis

The analyses were performed similarly as described previously [1]. In short, UHPLC-DAD Thermo Scientific™-Dionex™ system UltiMate™ 3000 fitted with a pump module, autosampler module, column thermostate and a diode array detector (Thermo Scientific™ Dionex™, Sunnyvale, CA, USA) was used. The system was set to record 3D data as well as set at 210 and 254 nm. The chromatographic separation was obtained on Phenomenex Kinetex® EVO C18 110 Å column (150 mm × 2.10 mm, 2.6 µm, Phenomenex, Torrence, CA, USA) at 35 °C. The gradient was constructed using 0.2 M phosphoric acid in water (solvent A) and acetonitrile (solvent B) as a mobile phase at 0.4 mL/min. The gradient started with 100% of solvent A, decreased to 95% in 4 min, to 5% within 0.5 min, and remained at this concentration for 3 min. The system was washed and stabilized, before each injection (5 µL). The results were elaborated with a Chromeleon v7.2 SR5 software (Thermo Scientific™ Dionex™, Sunnyvale, CA, USA, 2017). Standard solutions were dissolved in methanol or water (xanthine), and diluted in ultrapure water and the calibrations curve concentrations ranged from approximately 0.2 to 30 mg/L (correlation values 0.9999–1.0000). Before analysis, the honey samples were dissolved in ultrapure water (1:5, *w/v*) and filtered through H-PTFE membrane (0.2 µm, Ø 25 mm, Macherey Nagel™, Düren, Germany). The limits of detection (LOD) and quantification (LOQ) were determined according the International Conference on Harmonization of Technical Requirements for Registration of Pharmaceuticals for Human Use (ICH) guidance note [29]. LOD and LOQ were calculated as following: LOD = 3.3 σ/S and LOQ = 10 σ/S, where σ is the standard deviation of blank and S is the slope of proper calibration plot.

UHPLC-QqTOF-MS analyses were performed in similar setting and conditions, except of solvent A, which was 0.1% formic acid in water. ESI-HRMS analysis was performed with Compact™ QqTOF mass spectrometer (Bruker Daltonic, Bremen, Germany). The following settings were applied as previously [1]: positive mode, the ion source temperature 100 °C, nebulizer gas pressure at 2.0 bar, dry gas flow 8.0 L/min and temperature 210 °C. The capillary voltage was set at 4.50 kV. The collision energy was set on 8.0 eV and for MS/MS: 35 eV. Sodium formate clusters at concentration 10 mM were used for internal calibration.

3.4. Statistical Analysis

Statistical analyses were performed using the Statistica 64 v13.1 software (StatSoft Inc., Tulsa, OK, USA) and R for Windows, version 3.6.2 (R-Cran project, http://cran.r-project.org/). The results were expressed as the mean ± SD and analyzed by ANOVA followed by the Tukey test. The Pearson's product-moment correlation was used to assess relationships between the parameters and their significance was assessed in two-tailed test at $p < 0.05$. The variations of parameters among different honey samples were evaluated by cluster analysis (hierarchical-tree clustering). A dendrogram was used as an output of hierarchical clustering to show the hierarchical relationship between different samples and to find the best way to allocate them to clusters.

4. Conclusions

The targeted analysis revealed presence of adenine, xanthine, tyrosine, phenylalanine, and guanosine that varied between different honey types. The occurrence of nucleosides and nucleobases was found not to be specific only for a particular honey type. However, in some cases it significantly varied, indicating those more and less valuable as sources of nutrients. It may be useful for distinguishing or quality evaluation of some honey types. Honey may be considered as moderate dietary source of free uridine and black locust honey contains significantly higher amount of uridine (except buckwheat honey) than most of the investigated varieties. On the other hand, mean levels of adenine and guanosine were significantly more abundant in linden honey than in the most other varieties. The presence of nucleosides and nucleobases may partially explain the beneficial activities, properties reported in scientific literature and traditionally attributed to honey. This includes its topical application in herpes infections as well as its beneficial neurological activities: antinociceptive, nootropic, and neuroprotective among others.

Funding: This work has been financed by the Polish National Science Centre funding granted under the decision DEC-2014/15/N/NZ9/04058.

Acknowledgments: The research was partially performed in Laboratory of Elemental Analysis and Structural Research as well as in the Screening Laboratory of Biological Activity Test and Collection of Biological Material, Faculty of Pharmacy and the Division of Laboratory Diagnostics, Wroclaw Medical University, supported by the ERDF Project within the Innovation Economy Operational Programme POIG.02.01.00-14-122/09.

Conflicts of Interest: The author declare no conflicts of interest.

References

1. Kuś, P.M.; Włodarczyk, M.; Tuberoso, C.I.G. Nitrogen compounds in *Phacelia tanacetifolia* Benth. honey: First time report on occurrence of (−)-5-*epi*-lithospermoside, uridine, adenine and xanthine in honey. *Food Chem.* **2018**, *255*, 332–339. [CrossRef]
2. Wang, C.; Song, Z.; Yu, H.; Liu, K.; Ma, X. Adenine: An important drug scaffold for the design of antiviral agents. *Acta Pharm. Sin. B* **2015**, *5*, 431–441. [CrossRef]
3. Dorland, W.A.N. *Dorland's Medical Dictionary*, 32nd ed.; Elsevier: Philadelphia, PA, USA, 2007; ISBN 0721631428.
4. Holguin, S.; Martinez, J.; Chow, C.; Wurtman, R. Dietary uridine enhances the improvement in learning and memory produced by administering DHA to gerbils. *FASEB J.* **2008**, *22*, 3938–3946. [CrossRef]
5. Wurtman, R.J.; Cansev, M.; Ulus, I.H. Synapse formation is enhanced by oral administration of uridine and DHA, the circulating precursors of brain phosphatides. *J. Nutr. Heal. Aging* **2009**, *13*, 189–197. [CrossRef]
6. de Bruin, N.M.W.J.; Kiliaan, A.J.; de Wilde, M.C.; Broersen, L.M. Combined uridine and choline administration improves cognitive deficits in spontaneously hypertensive rats. *Neurobiol. Learn. Mem.* **2003**, *80*, 63–79. [CrossRef]
7. Wurtman, R.J. A Nutrient combination that can affect synapse formation. *Nutrients* **2014**, *6*, 1701–1710. [CrossRef]
8. Stover, P.J.; Field, M.S. Use of uridine and deoxyuridine to treat folate-responsive pathologies. U.S. Patent US9579337B2, 28 February 2017.
9. Goldberg, H.; Mibielli, M.A.; Nunes, C.P.; Goldberg, S.W.; Buchman, L.; Mezitis, S.G.; Rzetelna, H.; Oliveira, L.; Geller, M.; Wajnsztajn, F. A double-blind, randomized, comparative study of the use of a combination of uridine triphosphate trisodium, cytidine monophosphate disodium, and hydroxocobalamin, versus isolated treatment with hydroxocobalamin, in patients presenting with compress. *J. Pain Res.* **2017**, *10*, 397–404. [CrossRef]
10. Zizzo, M.G.; Caldara, G.; Bellanca, A.; Nuzzo, D.; Di, M.; Rosa, C. Preventive effects of guanosine on intestinal inflammation in 2,4-dinitrobenzene sulfonic acid (DNBS)-induced colitis in rats. *Inflammopharmacology* **2019**, *27*, 349–359. [CrossRef]
11. Lanznaster, D.; Dal-Cim, T.; Tetsadê, C.B.; Piermartiri, C.I.T. Guanosine: A Neuromodulator with therapeutic potential in brain disorders. *Aging Dis.* **2016**, *7*, 657–679. [CrossRef]

12. Dal-cim, T.; Poluceno, G.G.; Lanznaster, D.; de Oliveira, K.A.; Nedel, C.B.; Tasca, C.I. Guanosine prevents oxidative damage and glutamate uptake impairment induced by oxygen / glucose deprivation in cortical astrocyte cultures: Involvement of A 1 and A 2A adenosine receptors and PI3K, MEK, and PKC pathways. *Purinergic Signal.* **2019**, *15*, 465–476. [CrossRef]
13. Rosa, P.B.; Bettio, L.E.B.; Neis, V.B.; Moretti, M.; Werle, I.; Leal, R.B.; Rodrigues, A.L.S. The antidepressant-like effect of guanosine is dependent on GSK-3 β inhibition and activation of MAPK / ERK and Nrf2 / heme oxygenase-1 signaling pathways. *Purinergic Signal.* **2019**, *2*, 491–504. [CrossRef]
14. Dobrachinski, F.; Gerbatin, R.R.; Sartori, G.; Golombieski, R.M.; Antoniazzi, A.; Nogueira, C.W.; Royes, L.F.; Fighera, M.R.; Porciúncula, L.O.; Cunha, R.A.; et al. Guanosine attenuates behavioral deficits after traumatic brain injury by modulation of adenosinergic receptors. *Mol. Neurobiol.* **2019**, *56*, 3145–3158. [CrossRef]
15. Rahman, M.M.; Gan, S.H.; Khalil, I. Neurological effects of honey: Current and future prospects. *Evidence-Based Complement. Altern. Med.* **2014**, *2014*, 958721. [CrossRef]
16. Goudgaon, N.M.; Schinazi, R.F. Activity of acyclic 6-(phenylselenenyl)pyrimidine nucleosides against human immunodeficiency viruses in primary lymphocytes. *J. Med. Chem.* **1991**, *34*, 3305–3309. [CrossRef]
17. Clercq, E. De Nucleosides, nucleotides and nucleic acids guanosine analogues as anti-herpesvirus agents. *Nucleotides and Nucleic Acids* **2000**, *19*, 1531–1541. [CrossRef]
18. Mayer, A.; Slezak, V.; Takac, P.; Olejnik, J.; Majtan, J. Treatment of non-healing leg ulcers with honeydew honey. *J. Tissue Viability* **2014**, *23*, 94–97. [CrossRef]
19. Al-Waili, N.S. Topical honey application vs. acyclovir for the treatment of recurrent herpes simplex lesions. *Med. Sci. Monit.* **2004**, *10*, MT94–MT98.
20. Chen, H.; Jin, L.; Chang, Q.; Peng, T.; Hu, X.; Fan, C.; Pang, G.; Wang, W. Discrimination of botanical origins for Chinese honey according to free amino acids content by high-performance liquid chromatography with fluorescence detection with chemometric approaches. *J. Sci. Food Agric.* **2017**, *97*, 2042–2049. [CrossRef]
21. Janiszewska, K.; Aniołowska, M.; Nowakowski, P. Free amino acids content of honeys from Poland. *Polish J. Food Nutr. Sci.* **2012**, *62*, 85–89. [CrossRef]
22. Kečkeš, J.; Trifković, J.; Andrić, F.; Jovetić, M.; Tešić, Z.; Milojković-Opsenica, D. Amino acids profile of Serbian unifloral honeys. *J. Sci. Food Agric.* **2013**, *93*, 3368–3376. [CrossRef]
23. Shen, S.; Wang, J.; Chen, X.; Liu, T.; Zhuo, Q.; Zhang, S. Evaluation of cellular antioxidant components of honeys using UPLC-MS / MS and HPLC-FLD based on the quantitative composition-activity relationship. *Food Chem.* **2019**, *293*, 169–177. [CrossRef]
24. Zieliński, Ł.; Deja, S.; Jasicka-Misiak, I.; Kafarski, P. Chemometrics as a tool of origin determination of Polish monofloral and multifloral honeys. *J. Agric. Food Chem.* **2014**, *62*, 2973–2981. [CrossRef]
25. Kňazovická, V.; Gábor, M.; Miluchová, M.; Bobko, M.; Medo, J. Diversity of bacteria in Slovak and foreign honey, with assessment of its physico-chemical quality and counts of cultivable microorganisms. *J. Microbiol. Biotechnol. Food Sci.* **2019**, *9*, 414–421. [CrossRef]
26. Mach, J. Uridine ribohydrolase and the balance between nucleotide degradation and salvage. *Plant Cell* **2009**, *21*, 699. [CrossRef]
27. Nepi, M. New perspectives in nectar evolution and ecology: Simple alimentary reward or a complex multiorganism interaction? *Acta Agrobot.* **2017**, *70*, 1704. [CrossRef]
28. Louveaux, J.; Maurizio, A.; Vorwohl, G. Methods of melissopalynology by International Commission for Bee Botany of IUBS. *Bee World* **1970**, *51*, 125–138. [CrossRef]
29. EMEA ICH Topic Q2 (R1) Validation of analytical procedures: Text and methodology. Available online: http://www.ema.europa.eu/docs/en_GB/document_library/Scientific_guideline/2009/09/WC500002662.pdf. (accessed on 14 February 2020).

Sample Availability: Samples of the compounds are not available from the authors.

© 2020 by the author. Licensee MDPI, Basel, Switzerland. This article is an open access article distributed under the terms and conditions of the Creative Commons Attribution (CC BY) license (http://creativecommons.org/licenses/by/4.0/).

Article

The Use of Fluorescence Spectrometry to Determine the Botanical Origin of Filtered Honeys

Aleksandra Wilczyńska * and Natalia Żak

Gdynia Maritime University, Department of Commodity Science and Quality Management, ul. Morska 81-87, 81-225 Gdynia, Poland; n.zak@wpit.umg.edu.pl
* Correspondence: a.wilczynska@wpit.umg.edu.pl

Academic Editors: Gavino Sanna, Marco Ciulu, Yolanda Picò, Nadia Spano and Carlo I.G. Tuberoso
Received: 3 February 2020; Accepted: 8 March 2020; Published: 16 March 2020

Abstract: The aim of this study was to determine whether fluorescence spectrometry can be used to identify the botanical origin of filtered honeys. Sixty-two honey samples with different botanical origins, both filtered and unfiltered, were investigated in order to examine their fluorescence spectra. The results showed that individual honey varieties have different fluorescence spectra, and the filtration process had no impact on these spectra. The results suggest that fluorescence spectroscopy may be a useful method to identify the botanical origin of filtered honeys.

Keywords: filtered honey; botanical origin; fluorescence spectrometry

1. Introduction

Honey is a valuable, natural food product produced by *Apis mellifera* from plant juices (honeydew) and flower nectar. Both are processed by bees and enriched with their scent-gland secretions. It is a product of the highest quality, not obtainable by any other means. The quality of honey depends on environmental conditions, the climate, the beekeeper's interference in the production process, and the method of its collection and storage. However, the chemical composition and sensory characteristics of honey depend mainly on its botanical origin—the type and species of the plant that it comes from. Additionally, the specific chemical composition of different honey types determines their therapeutic properties [1–4].

Due to its high price, natural honey is often adulterated [5]. The results of the honey quality control indicate that many producers use misleading practices regarding the expected nutritional and health properties of honey. The most common adulteration techniques involve feeding sugar syrup to the bees during the nectar flow, or adding sugar to honey to increase the final product volume. [6]. However, the most common practice, recently, is the falsification of the botanical origin of honey, consisting in declaring honey as unifloral, when it is the cheapest multi-flower honey or honey imported from outside the European Union, e.g., from China.

The honey filtration process also helps to falsify it. According to Council Directive 2001/110/EC relating to honey, filtered honey is: "honey obtained by removing foreign inorganic or organic matter in such a way as to result in the significant removal of pollen". As a result of honey filtration, glucose crystals, impurities, yeast cells, wax fragments, and dyes are removed. However, the most important fact is that most pollen grains are lost as a result of filtration. Therefore the filtration process changes slightly the organoleptic properties of honey (color, taste, smell), as well as its chemical composition and biological properties [7,8]. This prevents the honey variety from being correctly identified.

The identification of the biological origin of honey is very important, because it determines the properties of individual honeys. The most common reference method for identifying polyfloral honey types, recommended by Codex Alimentarius and other honey standards, is the time-consuming palynological analysis. However, in some cases, the percentage of pollen is not always decisive, because the production of pollen and nectar by flowers is not always simultaneous, varying between countries and even within the same country, according to the geographical area [9]. Moreover, a palynological analysis is protracted and involves dedicated experts. The authenticity and health benefits of different honey types can also be determined based on their sensory parameters and chemical composition: amino acids, sugars, polyphenols, volatile compounds, etc. [10–12]. The combination of the multi-component analysis and chemometric techniques is also increasingly used [13–20]. However, due to the fact that the filtration process affects the chemical composition of honey, it is almost impossible to use all of these methods to properly identify the botanical origin of filtered honeys.

The authenticity of different honey types may also be ascertained by infrared spectroscopy and fluorescence spectroscopy [21,22]. The advantage of fluorescence spectroscopy is the high sensitivity and specificity of the classification. Fluorescence spectroscopy requires only a minimal sample preparation. The results of the above-mentioned studies confirmed that single synchronous fluorescence spectra of different honeys differ significantly because of their distinct physical and chemical characteristics, and provide sufficient data for the clear differentiation among honey groups. The studies demonstrated that this method is a valuable and promising technique for honey authentication. According to Ruoff [23], honeys are well known to contain numerous fluorophores such as polyphenols and amino acids. Some of them have been proposed as tracers for unifloral honeys—ellagic acid for heather honey, hesperetin for citrus honey, phenylalanine and tyrosine, which have been found to be characteristic for lavender honey and allowed a differentiation from eucalyptus honey, and tryptophan and glutamic acid, which have been shown to be useful for the differentiation between honeydew and blossom honeys. Due to the presence of such strong fluorophores, fluorescence spectroscopy may be helpful for authenticating the botanical origin of honey. Therefore, the aim of our study was to determine whether fluorescence spectra can also be used to identify the botanical origin of filtered honeys.

2. Results and Discussion

The emission spectra (excitation: 200–450 nm; emission: 260–560 nm) considered in this investigation allowed the study of the fluorescence of honey samples and the variation of their botanical origins and filtration. The EEM spectra were measured for the tested honeys. Figure 1 shows the results as contour maps, after removing Rayleigh scattering for 62 samples of different honey types: multifloral, honeydew, acacia, goldenrods, rape, phacelia, lime, buckwheat, and dandelion.

In Figure 1, it appears that each honey type exhibited a specific fluorescence spectrum. It was concluded that emission spectra (260–560 nm) are fingerprints allowing a good identification of the botanical origin of honeys. It can be seen that the results were consistent with the results of other researchers [22,24–29].

It could be seen that dandelion and goldenrods fluoresced very strongly over the emission range of 380–540 nm and the excitation range of 300–435 nm. On the other hand, rape honey samples had an almost negligible emission in this spectral region. But the intensity of the fluorescence was very strong over the emission range of 320–410 nm and the excitation range of 250–310 nm. It could also be seen that acacia, buckwheat, and honeydew honeys were characterized by two very intense emission sources (acacia: 320–385 nm and 390–500 nm, buckwheat: 310–360 nm and 380–450 nm, and honeydew: 320–385 nm and 380–525 nm).

a

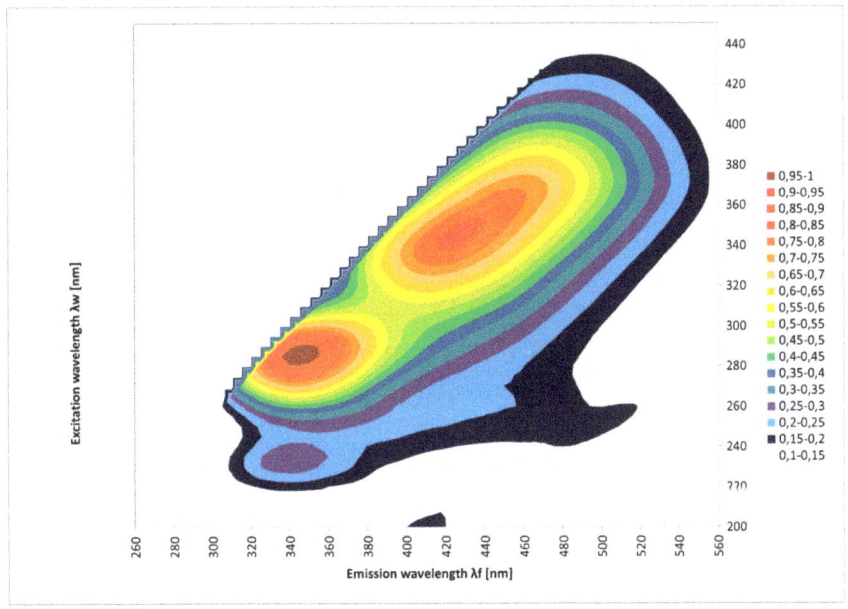

b

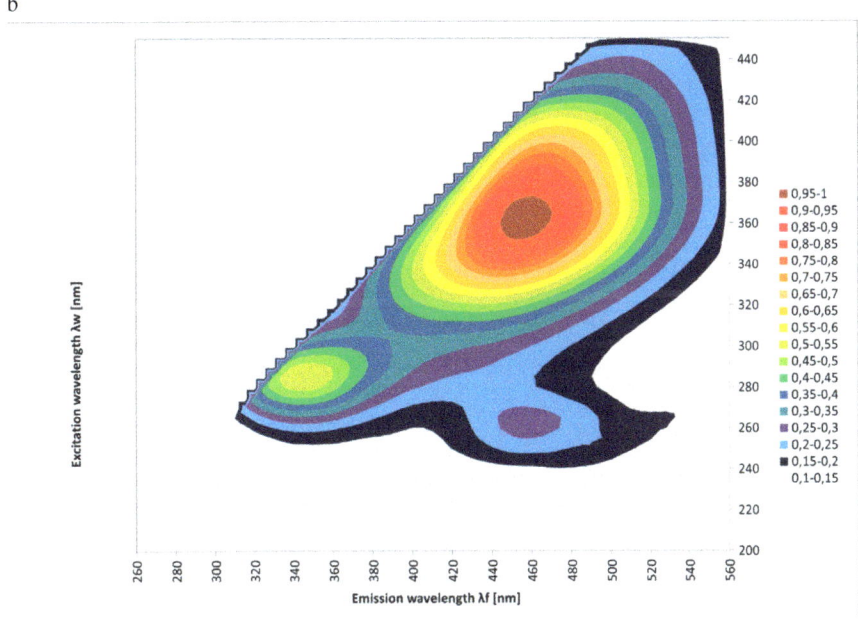

Figure 1. *Cont.*

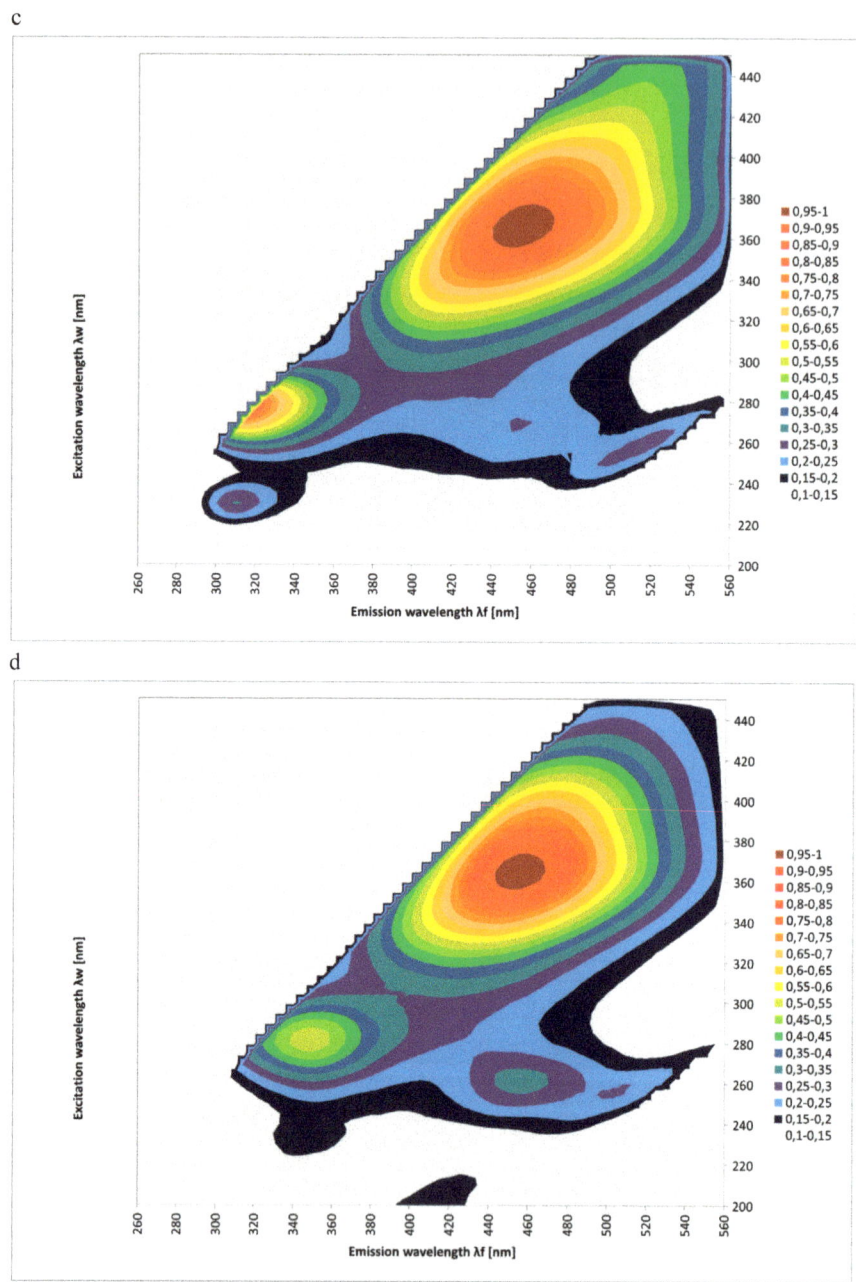

Figure 1. *Cont.*

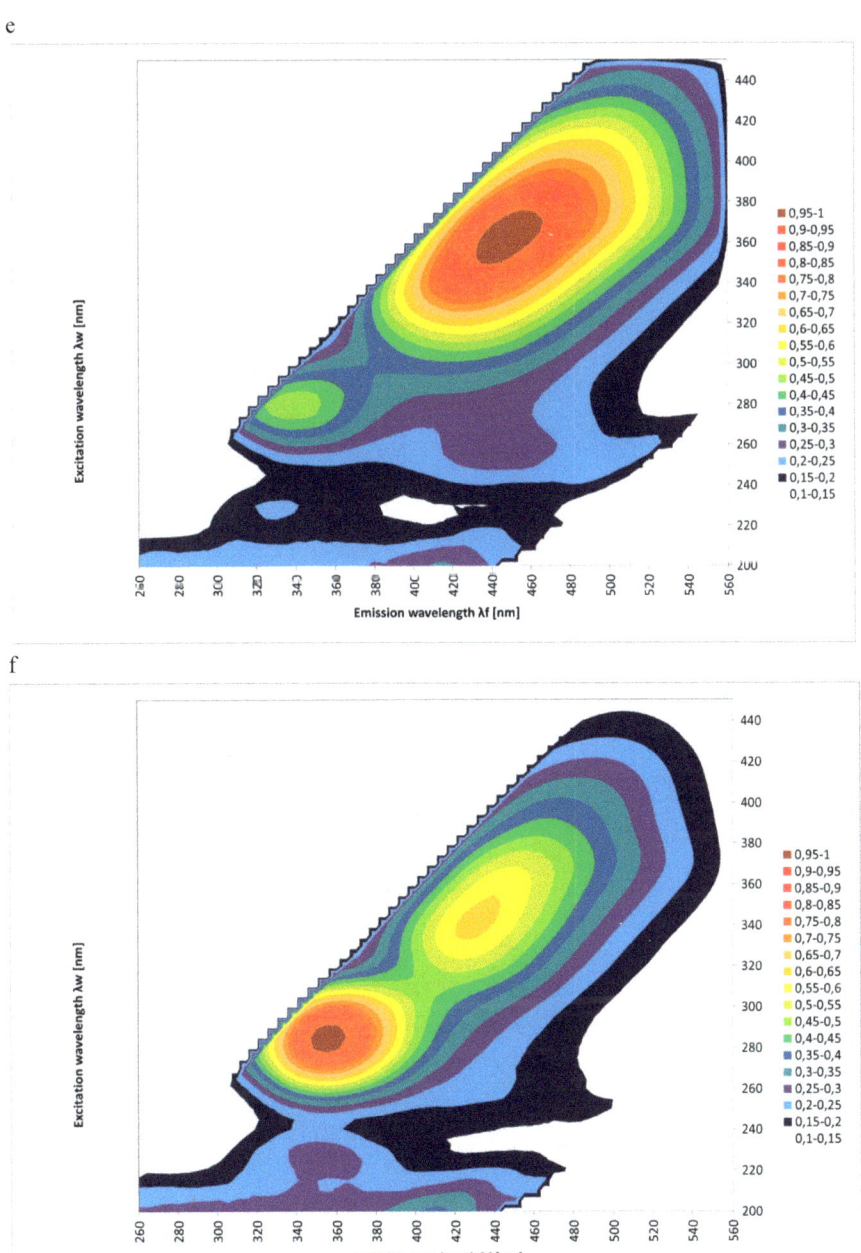

Figure 1. *Cont.*

g

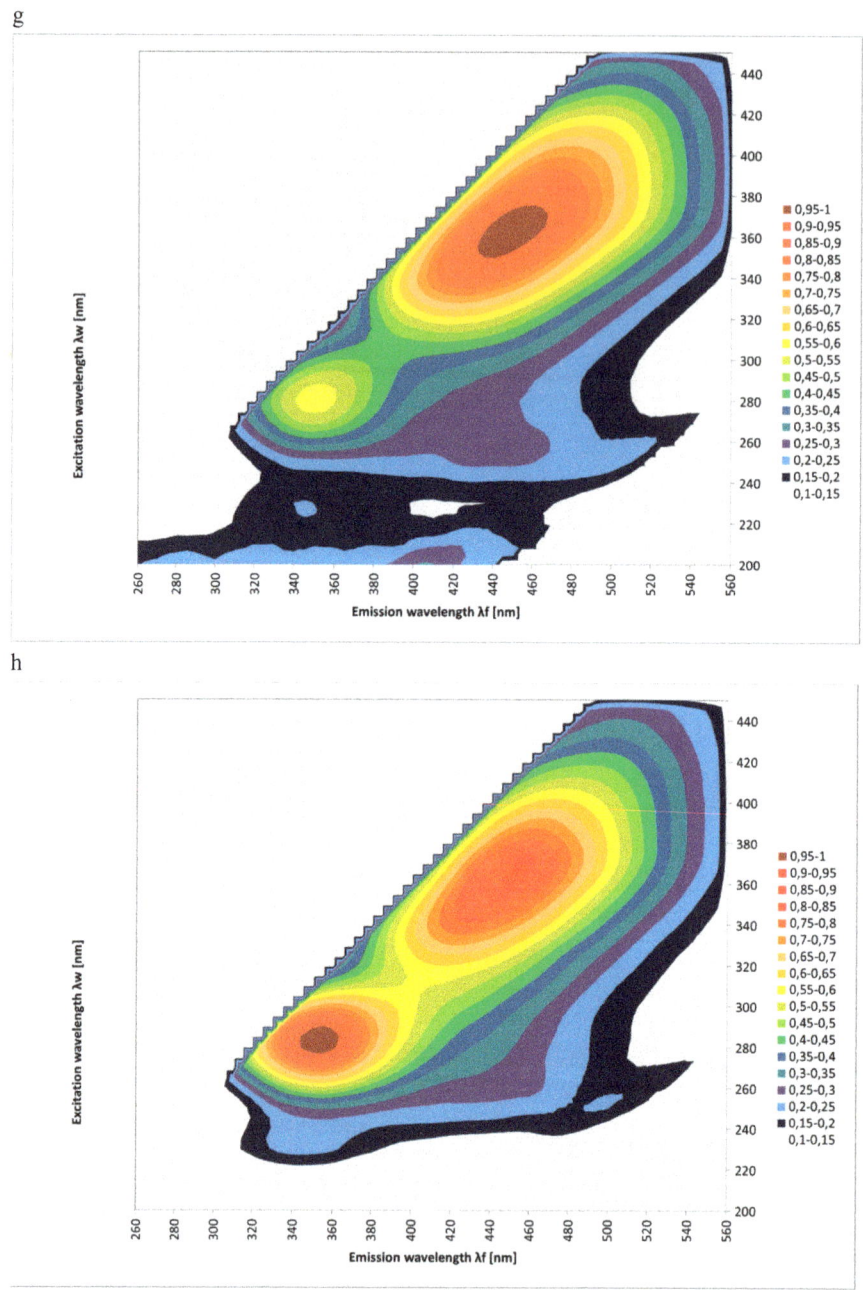

h

Figure 1. *Cont.*

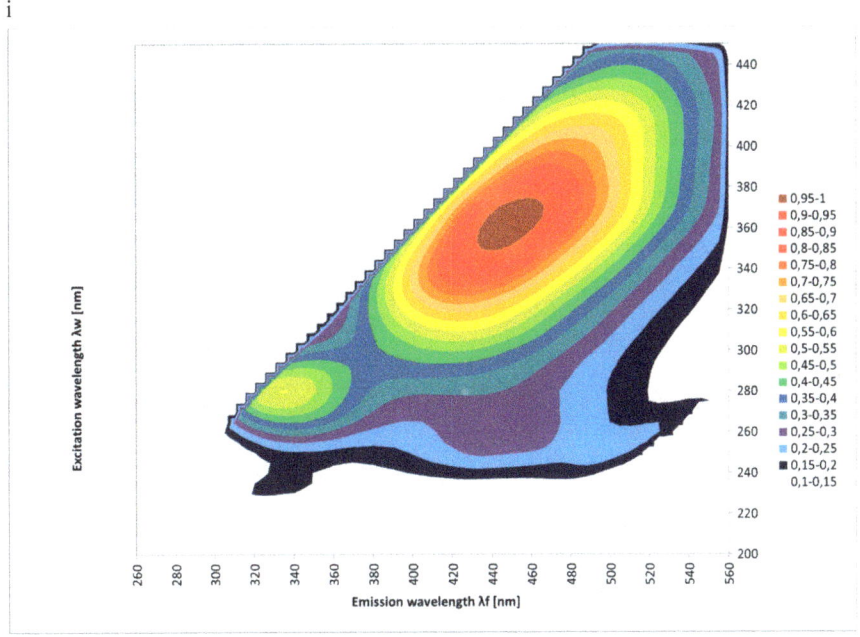

Figure 1. Excitation emission (EEM) spectra of different botanical origins of honey (**a**–acacia, **b**–phacelia, **c**–buckwheat, **d**–lime, **e**–multifloral, **f**–rape, **g**–dandelion, **h**–honeydew, and **i**–goldenrods). Source: own research.

Figure 2 shows the synchronous cross-sections of these spectra obtained at $\Delta\lambda = 100$ nm for honey samples and the variation of their botanical origins. Similar results were presented by Gębala [30] and Lenhardt et al. [22].

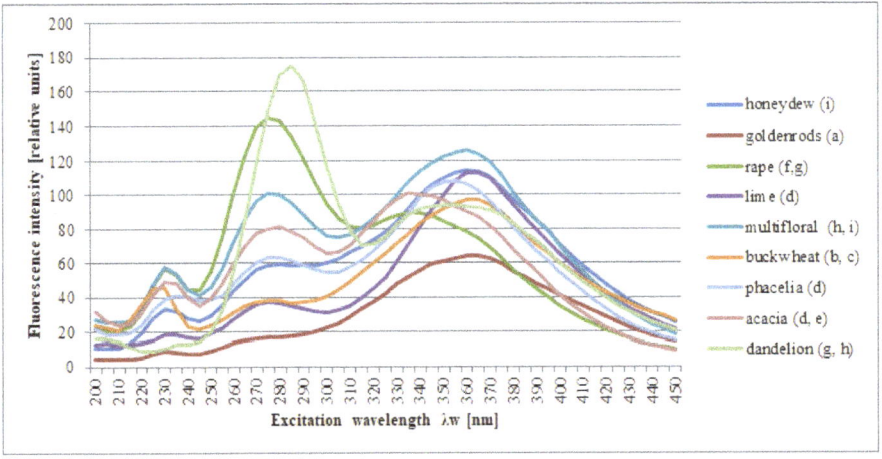

Figure 2. Normalized fluorescence emission spectra of different botanical origins of unfiltered honeys; the letters (**a**–**i**) indicate homogeneous groups, separated as a result of post-hoc analysis (Duncan's test). Source: own research.

Figure 3 shows the fluorescence spectra of the same honey samples, but subjected to a filtration process. Statistical analysis confirmed that the filtration process does not affect the shape of the spectra ($F = 3.65$, $p = 0.056$), while the differences in the spectra of honeys of different botanical origins were statistically significant ($F = 8.59$, $p = 0.00$).

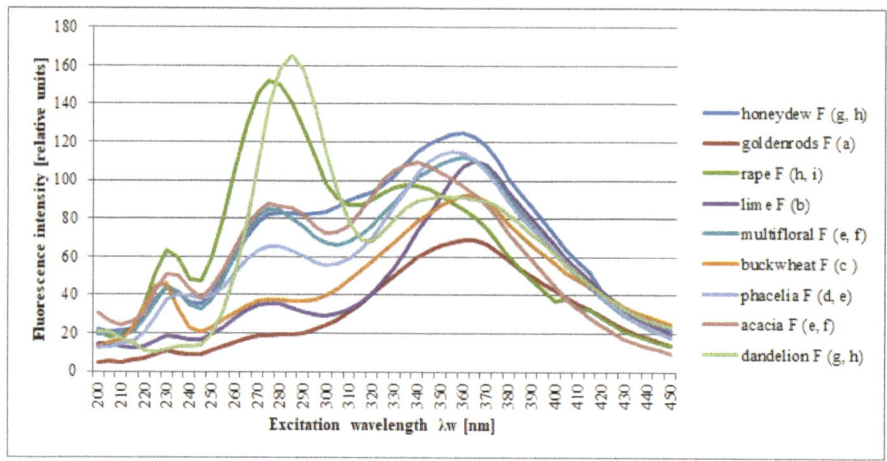

Figure 3. Normalized fluorescence emission spectra of different botanical origins of filtered honeys; the letters (a–i) indicate homogeneous groups, separated as a result of post-hoc analysis (Duncan's test). Source: own research.

The spectra of all tested honeys in connection with the filtration process are characterized by the presence of the same characteristic emission strip of variable intensity, as it can be seen in Figure 4a–i. In Table 1, the mean emission intensities over characteristic spectral regions for filtered and unfiltered honeys of different types are summarized.

Filtered honeys exhibited higher fluorescence intensities for acacia honey. It was possible to observe characteristic changes in the synchronic spectrum of the filtered acacia honey in relation to the spectrum of the unfiltered acacia honey ($\Delta\lambda = 100$ nm): increased intensities in the short-wave strip, increased intensity in the strip of intermediate range, and no change in the long-wave spectrum. There was a noticeable tendency for a rise to three emission excitances: 230 nm for unfiltered and filtered honeys, 275 nm and 340 nm for the filtered honey, and 280 nm and 335 nm for the unfiltered honey. Considering the filtration, an increase in the intermediate strip emission intensity (from 235 to 430 nm) was observed (Figure 4a and Table 1).

In the case of the synchronic spectrum of the phacelia honey ($\Delta\lambda = 100$ nm) there were three emission excitations: 230 nm, 275 nm, and 355 nm. Considering the filtration, an increase in the intensity of the emission line was observed from 250 to 410 nm (Figure 4b and Table 1).

Figure 4c shows the set of synchronic spectra for buckwheat honeys obtained at $\Delta\lambda = 100$ nm. It can be observed that the spectra of all studied buckwheat honeys were characterized by the presence of the same emission strips. Their intensity is different only in the range of 325 to 415 nm. In this range, the intensity of unfiltered honeys increased significantly (Figure 4c and Table 1).

The same results were obtained for the set of synchronic spectra for the lime honeys obtained at $\Delta\lambda = 100$ nm. It could be observed that the spectra of all studied lime honeys were characterized by the presence of the same emission strips and intensity. The only exception was the intensity of the emission in the range of 275 to 365 nm. In this range, the intensity of unfiltered honeys increased slowly (Figure 4d and Table 1).

a

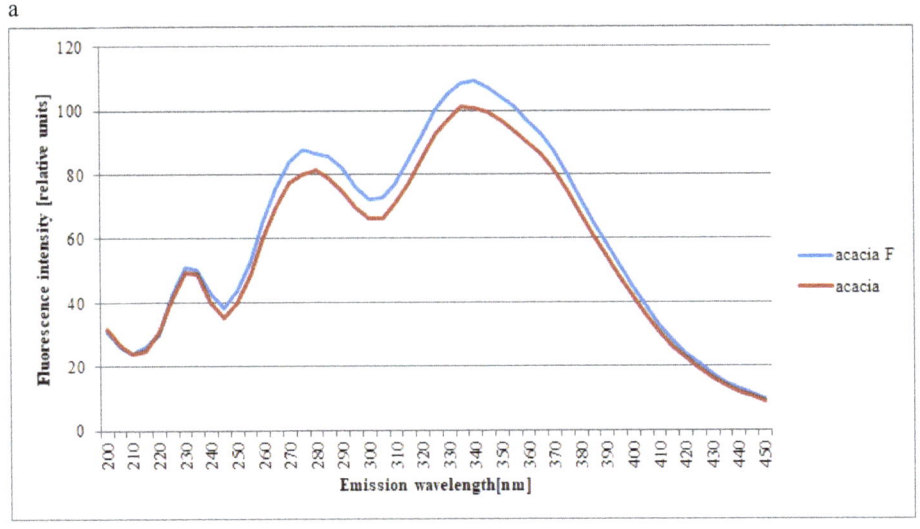

b

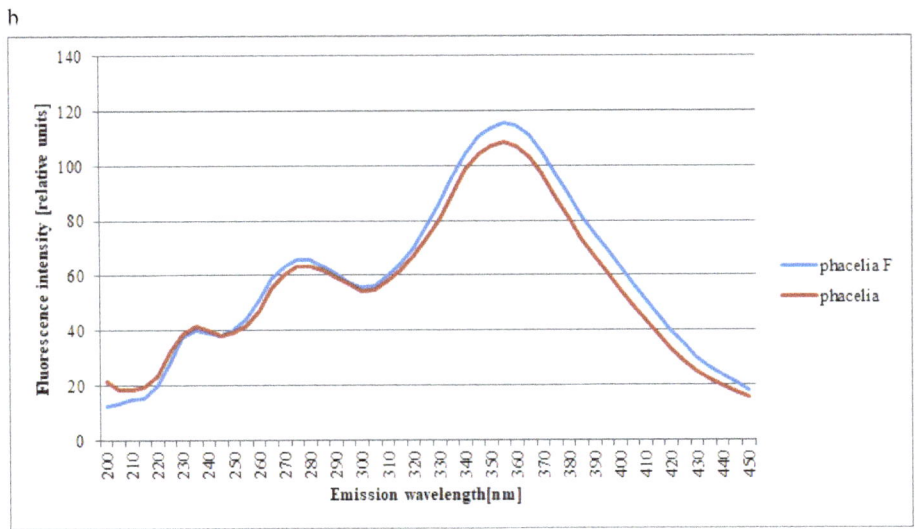

Figure 4. *Cont.*

c

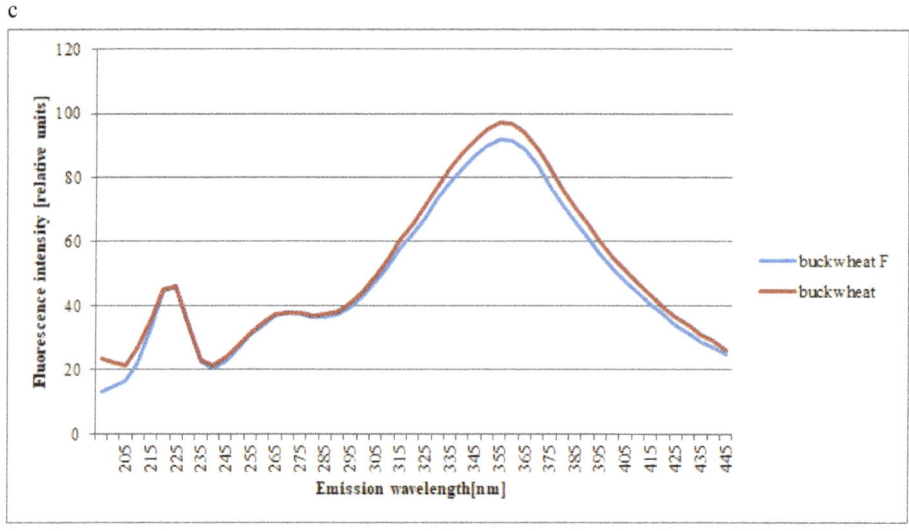

d

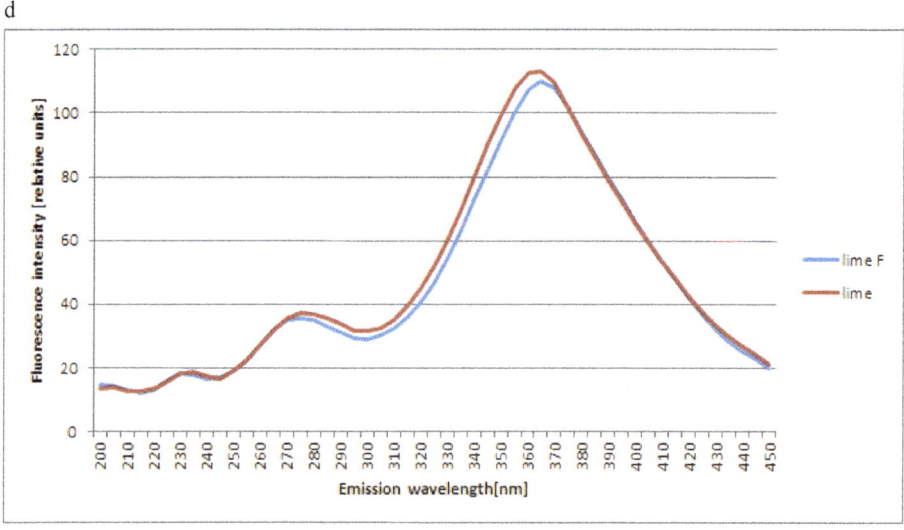

Figure 4. *Cont.*

e

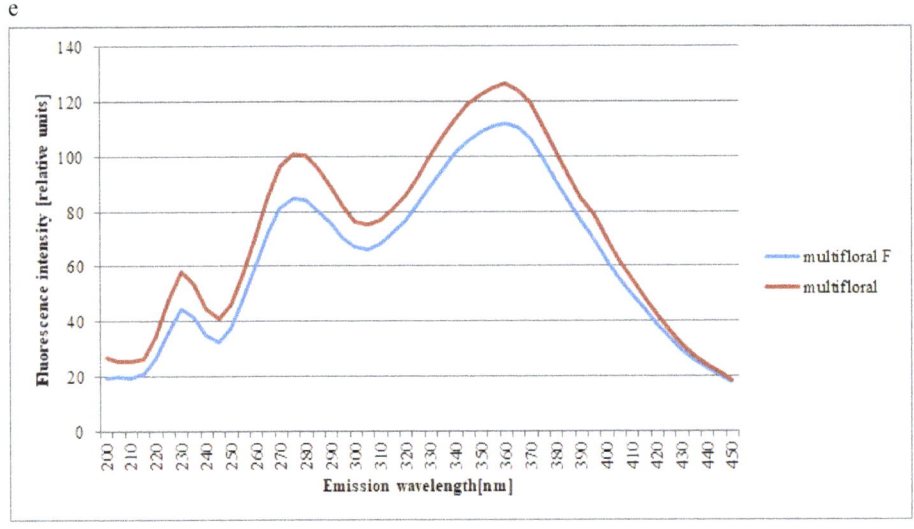

f

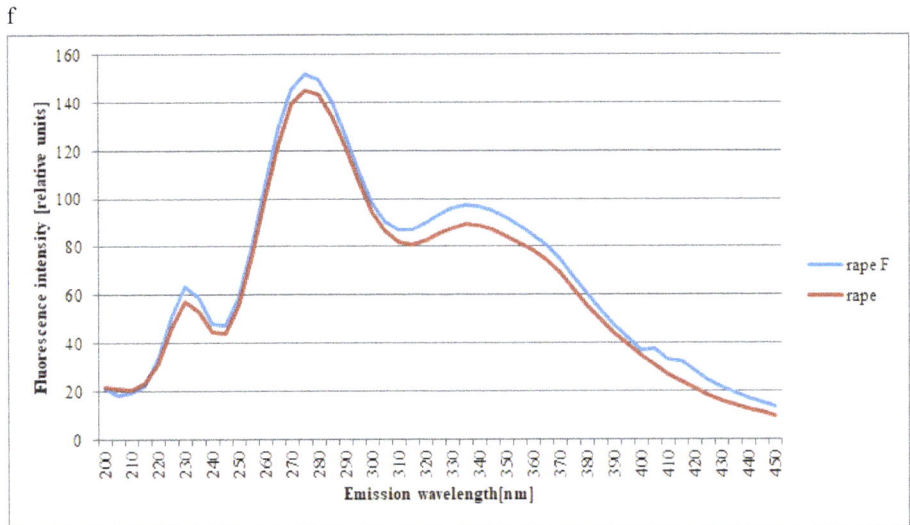

Figure 4. *Cont.*

g

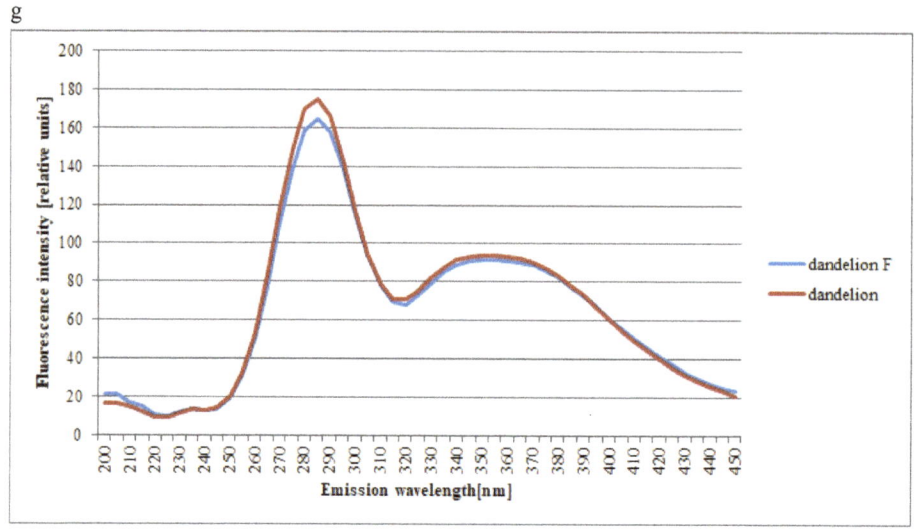

h

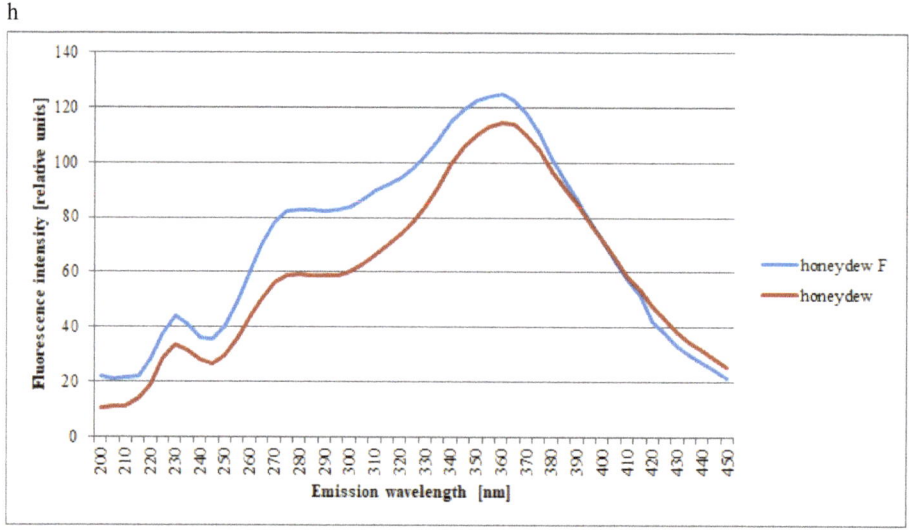

Figure 4. *Cont.*

i

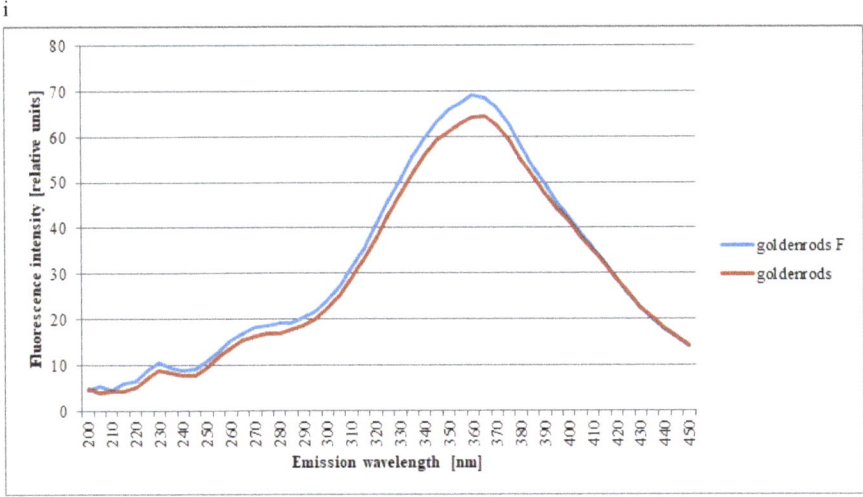

Figure 4. Normalized fluorescence emission spectra of different botanical origins of filtered and unfiltered honeys (a–acacia, b–phacelia, c buckwheat, d–lime, e–multifloral, f–rape, g–dandelion, h–honeydew, and i–goldenrods) F-filtered and unfiltered. Source: own research.

Filtered honeys included lower fluorescence intensities for multifloral honeys. It was possible to observe characteristic changes in the synchronic spectrum of the filtered multifloral honey in relation to the spectrum of the unfiltered acacia honey ($\Delta\lambda$ = 100 nm): increased intensities in the short-wave strip, increased intensity in the strip in the intermediate range, and no change in the long-wave spectrum. There was a noticeable tendency for up to three emission excitances: 230 nm, 275 nm, and 360 nm. The fluorescence intensity of filtered and unfiltered honeys was the same from 420 to 450 nm (Figure 4e and Table 1).

Filtered honeys included higher fluorescence intensities for rape honeys. It was possible to observe characteristic changes in the synchronic spectrum of filtered rape honeys in relation to the spectrum of unfiltered acacia honey ($\Delta\lambda$ = 100 nm): no change in the short-wave strip (225 nm), increased intensity in the strip in the intermediate range, and in the long-wave spectrum. There was a noticeable tendency for up to two emission excitances: 230 nm, and 275 nm. The third maximum (335 nm) was observed to be flatter (Figure 4f and Table 1).

Figure 4g and Table 1 show the set of synchronic spectra for dandelion filtered and unfiltered honeys obtained at $\Delta\lambda$ = 100 nm. It can be observed that the spectra of all studied dandelion honeys were characterized by the presence of the same emission strips. Their intensity was different only in the range of 200 to 285 nm. The third maximum (350 nm for unfiltered honeys and 355 nm for filtered honeys) was observed. In this range, the intensity of unfiltered honeys gently increased.

It was possible to observe characteristic changes in the synchronic spectrum of the filtered honeydew honey in relation to the spectrum of the unfiltered acacia honey ($\Delta\lambda$ = 100 nm). The short-wave and intermediate range strip spectrum (from 200 to 380 nm) of the filtered honeys included higher fluorescence intensities than unfiltered honeys. There was a noticeable tendency for up to three emission excitances: 230 nm, 280 nm, and 360 nm. In addition, no changes could be observed in the long-wave spectrum including filtered and unfiltered honeys, from 380 to 450 nm (Figure 4h and Table 1).

Figure 4i and Table 1 show the set of synchronic spectra for goldenrod honeys obtained at $\Delta\lambda$ = 100 nm. It can be observed that the spectra of all studied goldenrods honeys were characterized by the presence of the same emission strips. Their intensity was different only in the range of 320 to 415 nm. In this range, the intensity of unfiltered honeys increased significantly. The fluorescence intensity of filtered and unfiltered honeys was the same from 420 to 450 nm—no changes in the long-wave spectrum.

Table 1. Mean emission intensities over characteristic spectral regions for filtered and unfiltered honeys of different types.

Spectral Region	Excitation	Type of Honey	Kind of Processing	Mean Intensities of Fluorescence	SD
1st	200	dandelion		21.19	1.69
1st	230	honeydew		33.34	3.43
1st	230	goldenrods		8.96	2.62
1st	230	rape		45.78	2.82
1st	230	multifloral		58.15	3.98
1st	230	acacia		51.04	3.50
1st	235	phacelia		41.46	3.62
2nd	275	rape		37.59	2.26
2nd	275	lime		77.52	9.15
2nd	275	multifloral		100.96	6.17
2nd	275	buckwheat		37.59	2.21
2nd	275	phacelia	unfiltered	63.43	4.26
2nd	280	honeydew		59.42	6.32
2nd	280	acacia		86.65	5.83
2nd	285	dandelion		110.27	6.28
3rd	335	rape		67.70	2.46
3rd	335	acacia		108.37	4.63
3rd	355	dandelion		106.82	2.46
3rd	355	phacelia		108.45	3.96
3rd	360	honeydew		114.42	4.69
3rd	360	multifloral		126.41	4.19
3rd	360	buckwheat		91.98	2.90
3rd	365	goldenrods		64.49	4.13
3rd	365	lime		107.44	4.13
1st	200	dandelion		25.41	1.69
1st	230	honeydew		43.98	2.96
1st	230	goldenrods		10.66	4.94
1st	230	rape		45.59	3.06
1st	230	multifloral		44.56	3.79
1st	230	acacia		37.89	3.78
1st	235	phacelia		40.16	3.05
2nd	275	rape		38.65	2.21
2nd	275	lime		35.35	1.74
2nd	275	multifloral		84.70	6.76
2nd	275	buckwheat		37.91	2.20
2nd	275	phacelia	filtered	65.57	3.96
2nd	275	acacia		80.11	5.39
2nd	280	honeydew		82.98	5.58
2nd	285	dandelion		105.81	6.07
3rd	335	rape		72.68	2.78
3rd	340	acacia		100.66	3.69
3rd	350	dandelion		103.96	2.71
3rd	355	phacelia		115.21	3.68
3rd	360	honeydew		124.82	4.69
3rd	360	goldenrods		69.06	6.26
3rd	360	multifloral		111.97	3.89
3rd	360	buckwheat		97.24	3.58
3rd	365	lime		110.20	4.75

SD–standard deviation. Source: own research.

An appropriate classification and confirmation of the honey's authenticity is extremely important, because the health-promoting effect of honey is related to its chemical composition, and hence, to its botanical origin. The obtained results have once again confirmed that fluorescence spectrometry is an excellent method for the fast and nondestructive identification of the botanical origin of honey, and these results were consistent with those of other researchers [26–31]. However, it has been demonstrated for the first time that fluorescence spectrometry can be used also to determine the botanical origin of filtered honeys. As mentioned in the introduction, the main fluorophores in honey are phenolic compounds and aromatic amino acids. The EEMs of honey are characterized by a spectral region characterized by high emission intensities, and these emissions come from these compounds. The results obtained in this study confirm our previous observations that the filtration process does not change the phenolic content of honey [8]. Therefore, the spectra of filtered honeys differed slightly in the fluorescence intensity, while the shape of the spectra did not change.

3. Materials and Methods

3.1. Samples

Sixty-two honey samples of 9 different botanical origins (multifloral: 12, honeydew: 12, acacia: 3, goldenrods: 3, rape: 9, phacelia: 5, lime: 5, buckwheat: 8, dandelion: 5) were analyzed to evaluate the effect of filtration on the fluorescence spectra. The honeys, provided by a local beekeepers' association from the Pomeranian province, were harvested in 2017. Each sample was available as unfiltered and originally filtered forms, to compare the honeys directly.

The samples (100–150 g) were subjected to a filtration process—for this purpose they were rapidly heated to 45 °C for 5–10 min, to reduce the honey viscosity and dissolve any crystals. The honey was mixed thoroughly, then filtered through Schott filters (Duran, Mainz, Germany) with a pore size <100 μm, under a pressure of 0.3–0.4 MPa. The honey was then cooled down. The filtered and unfiltered samples were stored without light at room temperature until the analysis (no longer than 48 h).

3.2. Methods

Fluorescence spectra were determined by a method patented by Gębala and Przybyłowski [30,31]. The studies were carried out using a set-up based on the Fluorescence Spectrophotometer F-7000 Hitachi, Japan. A special adapter was built for it, in order to change its traditional measurement range (Figure 5.). During the measurement, the fluorescence intensity was measured from the surface of the sample, where the excitation radiation falls on. A reflective geometry enable to eliminate the effects of the internal filter associated with high absorbance of the sample—weakening of the fluorescence intensity due to the absorbance of excitation and the radial radiation emitted [30,31].

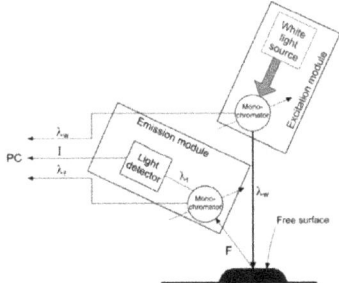

Figure 5. Scheme of surface fluorescence measurement. Source: [30,31].

The dimensional fluorescence spectra were measured at room temperature and daylight. Honey samples were liquefied at 40 °C and pipetted into 0.5 mL quartz cuvettes before measurements. The fluorescence spectra were obtained by recording the emission spectra (from 220 to 560 nm with a 10-nm step) corresponding to excitation wavelengths ranging between 200 and 450 nm (with a 5-nm step), and automatically normalized to the excitation intensity by the instrument. The sensitivity of the excitation and emission measurements was stated at a voltage equal 600 V. The difference between the fluorescent light wavelength (γ_F) and the excitation light wavelength (γ_W) was preferably 100 nm [30,31].

The fluorescence spectra were normalized by reducing the area under each spectrum to a value of 1 [24]. This was to reduce scattering effects and compare the investigated honey samples.

All analyses were done in triplicate. The final results are presented as a set of numerical data in the form of contour maps (excitation emission (EEM)) and the synchronous cross-sections of these spectra were obtained at $\Delta\lambda = 100$ nm for honey samples.

To determine the effect of the filtration or botanical origin on the shape of the spectra, one- and two-way analysis of variance (ANOVA) were used. Statistical hypotheses were verified at the significance level of $p = 0.05$.

4. Conclusions

1. The methodology proposed here allowed honey samples to be distinguished based on their different botanical origins, by using the simple and fast analysis of their fluorescence spectra.

2. The fluorescence spectra were the same for the filtered and unfiltered honeys, but the intensity of the fluorescence was different. This means that fluorescence spectra can also be used to identify filtered honeys.

Author Contributions: A.W.: developed the concept of this work, performed the literature review, and developed the final version of this article. N.Ż.: performed the researches and analyzed the results. All authors have read and agreed to the published version of the manuscript.

Funding: This research received no external funding.

Acknowledgments: The authors acknowledge the financial support of Gdynia Maritime University (projects: WPiT/2019/PZ/04 and WPiT/2019/PI/06).

Conflicts of Interest: The authors declare no conflict of interest.

References

1. Bogdanov, S.; Jurendic, T.; Sieber, R.; Gallmann, P. Honey for nutrition and health: A review. *JACN* **2008**, *27*, 677–689. [CrossRef] [PubMed]
2. De Silva, P.M.; Gauche, C.; Gonzaga, L.V.; Costa, A.C.O. Honey: Chemical composition, stability and authenticity. *Food Chem.* **2016**, *196*, 309–323. [CrossRef] [PubMed]
3. Escuredo, O.; Dobre, I.; Fernández-González, M.; Seijo, M.C. Contribution of botanical origin and sugar composition of honeys on the crystallization phenomenon. *Food Chem.* **2014**, *149*, 84–90. [CrossRef]
4. Kaskoniene, V.; Venskutonis, P.R.; Ceksteryte, V. Carbohydrate composition and electrical conductivity of different origin honeys from Lithuania. *LWT-Food Sci. Technol.* **2010**, *43*, 801–807. [CrossRef]
5. Soares, S.; Amaral, J.S.; Oliveira, M.B.P.P.; Mafra, I. A comprehensive review on the main honey authentication issues: Production and origin, Comp. *Rev. Food Sci. and Food Safety.* **2017**, *16*, 1072–1100. [CrossRef]
6. Matysiak-Żurowska, D.; Borowicz, A. A comparison of spectrophotometric Winkler method and HPLC technique for determination of 5-hydroxymethylfurfural in natural honey. *Chem. Anal.* **2009**, *54*, 939–947.
7. Beckmann, K.; Beckh, G.; Luellmann, C.; Speer, K. Characterization of filtered honey by electrophoresis of enzyme fractions. *Apidologie* **2010**, *42*, 59. [CrossRef]
8. Wilczyńska, A. Effect of filtration on colour, antioxidant activity and total phenolics of honey. *LWT-Food Sci. Technol.* **2014**, *57*, 767–774. [CrossRef]
9. Juan-Borrás, M.; Domenech, E.; Hellebrandova, M.; Escriche, I. Effect of country origin on physicochemical, sugar and volatile composition of acacia, sunflower and tilia honeys. *Food Res. Int.* **2014**, *60*, 86–94. [CrossRef]
10. Escriche, I.; Kadar, M.; Juan-Borrás, M.; Domenech, E. Using flavonoids, phenolic compounds and headspace volatile profile for botanical authentication of lemon and orange honeys. *Food Res. Int.* **2011**, *44*, 1504–1513. [CrossRef]
11. Persano-Oddo, L.; Bogdanov, S. Determination of honey botanical origin: Problems and issues. *Apidologie* **2004**, *35*, 2–3. [CrossRef]
12. Aliferis, K.A.; Ttarantilis, P.A.; Harizanis, P.C.; Alissandrakis, E. Botanical discrimination and classification of honey samples applying gas chromatography/mass spectrometry fingerprinting of headspace volatile compounds. *Food Chem.* **2010**, *121*, 856–862. [CrossRef]

13. Ruoff, K.; Luginbuehl, W.; Bogdanov, S.; Bosset, J.O.; Esterman, B.; Ziorko, T.; Kheradmandan, S.; Amad, R. Quantitative determination of physical and chemical measurands in honey by near-infrared spectrometry. *Eur. Food Res. Technol.* **2007**, *225*, 415–423. [CrossRef]
14. Terrab, A.; Gustavo-González, A.; Díez, M.J.; Heredia, F.J. Characterization of Moroccan unifloral honeys using multivariate analysis. *Eur. Food Res. Technol.* **2003**, *218*, 88–95. [CrossRef]
15. Oroian, M.; Ropciuc, S.; Buculei, A. Romanian honey authentication based on physico-chemical parameters and chemometrics. *J. Food Measur. Charac.* **2017**, *11*, 719–725. [CrossRef]
16. Popek, S.; Halagarda, M.; Kursa, K. A new model to identify botanical origin of Polish honeys based on the physicochemical parameters and chemometric analysis. *LWT–Food Sci. Technol.* **2017**, *77*, 482–487. [CrossRef]
17. Kuś, P.M.; van Ruth, S. Discrimination of Polish unifloral honeys using overall PTR-MS and HPLC fingerprints combined with chemometrics. *LWT–Food Sci. Technol.* **2015**, *62*, 69–75. [CrossRef]
18. Kaczmarek, A.; Muzolf-Panek, M.; Tomaszewska-Gras, J.; Konieczny, P. Predicting the botanical origin of honeys with chemometric analysis according to their antioxidant and physicochemical properties. *Pol. J. Food Nutr. Sci.* **2019**, *69*, 191–201. [CrossRef]
19. Dżugan, M.; Tomczyk, M.; Sowa, P.; Grabek-Lejko, D. Antioxidant Activity as Biomarker of Honey Variety. *Molecules* **2018**, *23*, 2069. [CrossRef]
20. Kropf, U.; Korosec, M.; Bertoncelj, J.; Ogrinc, N.; Necemer, M.; Kump, P.; Golob, T. Determination of the geographical origin of Slovenian black locust, lime and chestnut honey. *Food Chem.* **2010**, *121*, 839–846. [CrossRef]
21. Etzold, E.; Lichtenberg-Kraag, B. Determination of the botanical origin of honey by Fourier-transformed infrared spectroscopy: An approach for routine analysis. *Eur. Food Res. Technol.* **2007**, *227*, 579–586. [CrossRef]
22. Lenhardt, L.; Zekovic, I.; Dramicanin, T.; Dramicanin, M.D.; Bro, R. Determination of the botanical origin of honey by front-face synchronous fluorescence spectroscopy. *Appl. Spectroscopy.* **2014**, *68*, 557–563. [CrossRef] [PubMed]
23. Ruoff, K. Authentication of the Botanical Origin of Honey. Doctoral Thesis, ETH Zurich, Zurich, Switzerland, 2006. [CrossRef]
24. Karoui, R.; Dufour, E.; Bosset, J.-O.; De Baerdemaeker, J. The use of front face fluorescence spectroscopy to classify the botanical origin of honey samples produced in Switzerland. *Food Chem.* **2007**, *101*, 314–323. [CrossRef]
25. Lenhardt, L.; Bro, R.; Zekovic, I.; Dramicanin, T.; Dramicanin, M.D. Fluorescence spectroscopy coupled with PARAFAC and PLS DA for characterization and classification of honey. *Food Chem.* **2015**, *175*, 284–291. [CrossRef] [PubMed]
26. Ruoff, K.; Karoui, R.; Dufour, E.; Luginbuhl, W.; Bosset, J.O.; Bogdanov, S.; Amadò, R. Authentication of the botanical origin of honey by front-face fluorescence spectroscopy, a preliminary study. *J. Agric. Food Chem.* **2005**, *53*, 1343–1347. [CrossRef]
27. Ruoff, K.; Luginbühl, W.; Künzli, R.; Bogdanov, S.; Bosset, J.O.; von der Ohe, K.; von der Ohe, W.; Amadò, R. Authentication of the botanical and geographical origin of honey by front-face fluorescence spectroscopy. *J. Agric. Food Chem.* **2006**, *54*, 6858–6866. [CrossRef]
28. Drami´canin, T.; Lenhardt Ackovic, L.; Zekovic, I.; Dramicanin, M.D. Detection of Adulterated Honey by Fluorescence Excitation-Emission Matrices. *J. Spectrosc.* **2018**, 8395212. [CrossRef]
29. Gebala, S.; Przybylowski, P.; Borawska, M.H.; Piekut, J. Klasyfikacja naturalnych miodów pszczelich na podstawie analizy kształtu widm fluorescencyjnych (Classification of natural honeys based on the analysis of surface spectrofluorimetry shapes). *Brom. Chem. Toksykol.* **2005**, 627–631.
30. Gębala, S. Measurements of solution fluorescence–a new concept. *Opt. Appl.* **2009**, *39*, 391–399.
31. Gębala, S.; Przybyłowski, P. Sposób Identyfikacji Odmian Miodu. Polska Patent nr 214784, 15 March 2010.

Sample Availability: Samples of the honeys are available from the authors.

© 2020 by the authors. Licensee MDPI, Basel, Switzerland. This article is an open access article distributed under the terms and conditions of the Creative Commons Attribution (CC BY) license (http://creativecommons.org/licenses/by/4.0/).

Classification of Unifloral Honeys from SARDINIA (Italy) by ATR-FTIR Spectroscopy and Random Forest

Marco Ciulu [1,*], Elisa Oertel [1], Rosanna Serra [2], Roberta Farre [2], Nadia Spano [2], Marco Caredda [3], Luca Malfatti [2] and Gavino Sanna [2]

1. Department of Animal Sciences, University of Göttingen, Kellnerweg 6, 37077 Göttingen, Germany; elisa.oertel@uni-goettingen.de
2. Dipartimento di Chimica e Farmacia, Università degli studi di Sassari, Via Vienna 2, 07100 Sassari, Italy; rosanna1981@live.it (R.S.); roberta.farre@tiscali.it (R.F.); nspano@uniss.it (N.S.); lucamalfatti@uniss.it (L.M.); sanna@uniss.it (G.S.)
3. AGRIS Sardegna, Loc. Bonassai S.S. 291 Km 18.6, 07100 Sassari, Italy; caredda.m@gmail.com
* Correspondence: marco.ciulu@uni-goettingen.de; Tel.: +49-05513926085

Abstract: Nowadays, the mislabeling of honey floral origin is a very common fraudulent practice. The scientific community is intensifying its efforts to provide the bodies responsible for controlling the authenticity of honey with fast and reliable analytical protocols. In this study, the classification of various monofloral honeys from Sardinia, Italy, was attempted by means of ATR-FTIR spectroscopy and random forest. Four different floral origins were considered: strawberry-tree (*Arbutus Unedo* L.), asphodel (*Asphodelus microcarpus*), thistle (*Galactites tormentosa*), and eucalyptus (*Eucalyptus calmadulensis*). Training a random forest on the infrared spectra allowed achieving an average accuracy of 87% in a cross-validation setting. The identification of the significant wavenumbers revealed the important role played by the region 1540–1175 cm^{-1} and, to a lesser extent, the region 1700–1600 cm^{-1}. The contribution of the phenolic fraction was identified as the main responsible for this observation.

Keywords: honey discrimination; strawberry-tree; thistle; eucalyptus; asphodel; attenuated total reflectance; Fourier transform infrared spectroscopy

1. Introduction

Since prehistoric times, honey has been one of the most popular foods for humans. In addition to its role as a food sweetener, several studies have classified honey as functional food because of its several biological and nutraceutical properties such as antioxidant, anti-ulcer, antibacterial and also anti-tumor activities [1]. The distinct organoleptic properties combined with its nutritional characteristics constitute the basis of a continuously growing demand for honey. In recent years, honey imports into the EU have increased at a rate of more than 10,000 tons per year [2]. The occurrence of fraudulent activities aimed at placing on the global market honey mislabeled with regard to its floral origin or adulterated with exogenous sugars have prompted the European Union to implement control measures. However, the official analytical procedures provided by current legislation present several limitations in the identification of the botanical and/or geographical origin of honey [3]. In addition, the traditional melissopalinological analysis is not effective in the authentication of filtered honeys and those whose pollen is underrepresented. For these reasons, for several years the scientific community has been developing instrumental analytical protocols aimed at relating specific markers or classes of compounds to the floral origin of honey, most of them summarized in a number of authoritative reviews [4–6]. In this context, chromatographic techniques have for years been a key tool in the search for specific markers. For instance, the use of HPLC-based protocols has in the past made possible to identify the origin of honey through the profile of carbohydrates [7–9], phenolic compounds [7,10,11], and macro- and micro-nutrients [12]. In addition, the use of gas-chromatography (especially when coupled to mass spectrometry) has proved to be fundamental for the identification of volatile compounds related to honey source [13–16]. As regards the use of the volatile fraction as a tool for botanical origin attribution, non-chromatographic techniques have been also used, such as MS-based electronic nose [17].

Some studies have demonstrated the effectiveness of spectroscopic approaches directed rather than the quantification of specific compounds, to the acquisition of finger prints containing the information necessary for the identification of honeys. Among these Raman spectroscopy [18,19], NMR [20,21], VIS/NIR spectroscopy [22], and mass spectrometry [23] are worthy of citation.

Attenuated total reflectance-Fourier transform infrared spectroscopy (ATR-FTIR) has proven to be particularly useful for the detection of honey adulteration [24–28]. Besides scientific literature offers various examples related to the employment of this technique in the classification of botanical origin of honey. For instance, the differentiation of honeys of various floral origins from India [29], Turkey [30], Poland [31], and Croatia [32] was performed by means of ATR-FTIR combined with chemometrics.

Strawberry-tree honey (*Arbutus unedo* L.) represents one of the most typical beekeeping products of Sardinia, Italy, and, more generally, of the Mediterranean region [33,34]. The refined bitter taste along with the scarce production make this honey one of the most expensive of Southern Europe [35]. In addition, the presence in high amounts of various bioactive compounds (e.g., phenolic acids, flavonoids, terpenes etc.) gives strawberry tree honey distinct nutraceutical and functional properties [36,37] comparable in some cases to those of the more famous Manuka honey [34]. The authentication of this honey has so far mainly been based on the qualitative and quantitative determination of its chemical marker, the homogentisic acid [10,38–40]. Other typical honeys of the area include asphodel (*Asphodelus microcarpus*), thistle (*Galactites tormentosa*), and eucalyptus (*Eucalyptus calmadulensis*) honeys [41,42]. Also for these products, the authentication of the botanical origin has been until now performed by the application of instrumental protocols aimed at identifying and/or quantifying specific chemical markers [33,43,44].

Multivariate data analysis and machine learning techniques have proved to be excellent strategies for honey discrimination [45]. Random forest is a classification algorithm based on the construction on several decision trees. A subset of independent features is used for the training phase in order to construct each tree. The deriving set of trees is then used to assign one class to an object (sample) on the basis of the most frequent classification among them [22,45–47]. The prediction accuracy obtained by a multitude of decision tree is generally higher than the one obtained with a single tree [46]. Despite the potential of this classification algorithm, only a few examples of the application of random forest for honey classification can be found in the literature [22,46,48].

To the best of our knowledge, a classification of the strawberry-tree honey using FTIR methods has been only once attempted but the low number of samples analyzed prevented from any reliable conclusion [49]. In this work, we report a classification of strawberry-tree honey along with three other typical floral origins from Sardinia (Italy) by means of ATR-FTIR spectroscopy combined with random forest.

2. Results

2.1. ATR-FTIR Spectra

Figure 1 shows raw representative ATR-FTIR spectra of the four selected honey types in the region 4000–400 cm^{-1}.

The visual analysis of the spectra allowed identifying the characteristic absorption bands of honey, based on the information provided by scientific literature [31]. More specifically, the region between 3000 and 2800 cm^{-1} includes the signals deriving from C-H stretching of carbohydrates [50], O-H stretching of carboxylic acids [51] and NH$_3$ stretching of free amino acids [50,52]. Bands in the region 1700–1600 cm^{-1} are instead attributable to the O-H stretching and bending of water [53], the stretching of carbonyls mainly belonging to carbohydrates [50] and the N-H bending of primary amides of proteins [54]. In the spectral region 1540–1175 cm^{-1} it is possible to observe the absorption bands related to the stretching and bending of not water-related hydroxyl groups [50,55], C-O and C-H stretching of carbohydrates [56], and the carbonyl stretching of ketones [55]. Ring vibrations (mainly attributable to carbohydrates) [50,55] along with the signals related to C-O and C-C stretching are visible in the region between 1175 and 950 cm^{-1} [56,57]. Finally, between 940 and 700 cm^{-1}, there is the anomeric region of carbohydrates [57,58] where the C-H bending [50,55,59] and ring vibrations produce signals [55].

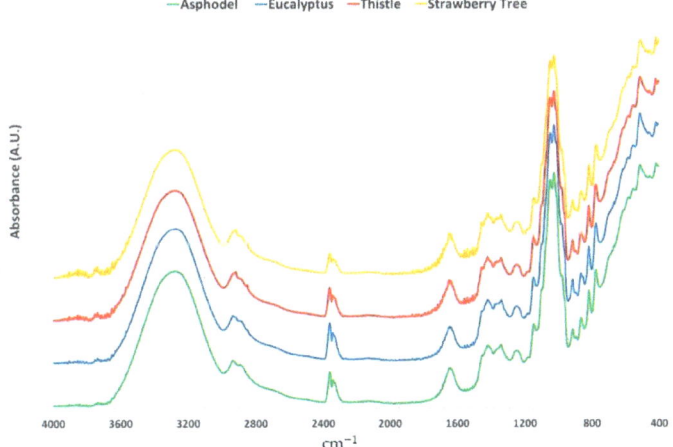

Figure 1. Average raw ATR-FTIR spectra of the selected floral origins.

2.2. Random Forest Classification

Across 100 runs, the random forest achieved a mean accuracy of 87% with a standard variation of 7%. The analysis was repeated with permuted labels. In this case, mean accuracy only reached 43%, indicating the former average accuracy was due to a real signal and no statistical artifact. Specificity and sensitivity values were 94.3% and 72.6% for asphodel, 93.9% and 87.3% for eucalyptus, 96.4% and 90.5% for thistle, and 99.9% and 91.6% for strawberry-tree, respectively.

For each run, the ten most important wavenumbers were identified. Table 1 lists preselected ranges and the frequency at which the most important wavelengths fell into them.

Table 1. Distribution of important wavelengths across preselected ranges.

Wavenumber Range (cm^{-1})	Frequency (%)
3000–2800	1.2
1700–1600	10.7
1600–1540	1.1
1540–1175	87
1175–940	0
940–700	0

3. Discussion

One of the greatest advantages of the adopted approach is given by the total absence of any sample pre-treatment along with the possibility to obtain, for each one of the samples, all the information required to build the classification model in a few seconds. To the best of our knowledge, this study provides the first example of the building of a machine learning model aimed at predicting the selected honey botanical sources without the support of any sample extraction and/or clean-up. The combination of ATR-FTIR with random forest algorithm proved to be successful for the classification of the selected floral origins. Although 87% prediction accuracy could be considered, in some data science scenarios, relatively low when k-fold or leave-one-out cross validation is performed, it is important to bear in mind that the above mentioned cross validation approaches tend to overestimate the general model performances when applied on small datasets [60]. The lack of previous studies aimed at classifying the selected floral origins by means of IR spectroscopy and random forest prevents us to make reasonable comparisons regarding the model accuracy. As regards the application of random forest in combination with other analytical techniques for honey classification, to the best of our knowledge only two contributions have been until now published. In the first study, electrophoresis was used in order to discriminate between two different honey types [48]. In the most recent contribution, the classification of honeys belonging to the different floral sources was successfully achieved (prediction accuracy of 98.2%) by means of laser induced breakdown spectroscopy and random forest [61]. However, in this study only ten samples were considered, and the specific attribution of the various floral sources was somehow missing, being some of them simply indicated as "flower honey" or "forest honey".

Our results further support the importance of the so-called fingerprint region in the definition of the floral origin of honey. In fact, also in previous studies [22,30], the interval 1800–750 cm^{-1} played a predominant role in the differentiation of honeys, although in those cases no classification technique was applied (i.e., only principal component and/or hierarchical cluster analysis). In our case, the major contribution of the various wavenumbers can be traced back to an even narrower range (1540–1175 cm^{-1}, 87%). As already explained above, the absorptions in this spectral region are mainly due to the stretching and bending of not water-related hydroxyl groups, C-O and C-H stretching of carbohydrates and the carbonyl stretching of ketones [30]. Flavanols and phenols contribute to this spectral region [62]. This observation is somehow supported by the conclusions obtained in a previous study, where total content of polyphenols was considered, among various chemical and physical parameters, one of the main discriminant factors between strawberry-tree, asphodel, thistle and eucalyptus honeys from Sardinia [42]. Although to a much lesser extent, also the 1700–1600 cm^{-1} range contributed to the classification of the four types of honey (10.7% of the wavenumbers). The presence of phenolic compounds has been in the past related to the bands in this region, supporting the hypothesis that the profile of polyphenols could underlie this honey differentiation. Average spectra in these regions are shown in Figure 2.

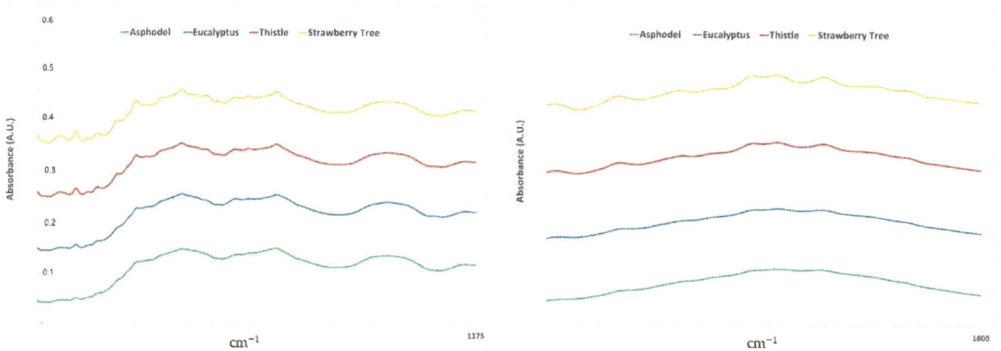

Figure 2. Average ATR-FTIR spectra in the 1540–1175 cm^{-1} (left) and 1700–1600 cm^{-1} (right) regions.

As a mere visualization aid, an unsupervised (i.e., with no classification and/or regression aims) random forest was performed and plotted after multidimensional scaling (MDS, Figure 3).

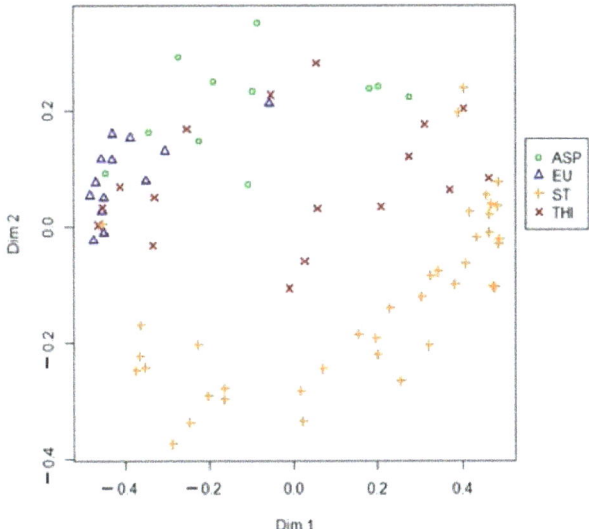

Figure 3. 2D visualization of unsupervised random forest.

The unsupervised proximity information plotted in 2D shows how strawberry-tree honeys are clearly identifiable, being the corresponding cluster visibly distinguishable from the other three. This observation could be the evidence of the influence of the distinct chemical characteristics of strawberry-tree honey on the IR spectrum. In fact, this honey stands out from the other local honeys due to its extremely high polyphenol content, which constitutes the main reason for its nutraceutical properties. This observation is also supported by the high sensitivity value recorded for strawberry-tree honeys, very close to 100%. On the other hand, asphodel honeys showed the lowest sensitivity resulting as the most misclassified among the four botanical origins, over the 100 iterations. These findings could be explained by the wide variability in the chemical composition of asphodel honey [33]. For example, in a previous study concerning the classification of the same floral sources by physicochemical determinations, this honey showed, compared to the others, a wider range in the total phenolic content, FRAP antioxidant activity and the DPPH radical scavenging activity [42]. Also, eucalyptus honeys showed a sensitivity lower than 90%. However, this case should be otherwise considered in comparison to the one just mentioned, since this lower sensitivity value can be attributable to one single eucalyptus honey sample which was repeatedly misclassified as asphodel honey over several cross-validation iterations, revealing a likely initial mislabeling.

4. Materials and Methods

4.1. Honey Samples

A total of 80 honey samples was collected from beekeepers from Sardinia (Italy) in the corresponding harvesting season (early spring for asphodel honey, late spring for thistle honey, summer for eucalyptus honey and autumn for strawberry tree honey). All samples were stored in the dark at 4 °C until analysis. The assignment of floral origin was accomplished on the basis of the information provided by local producers and melissopalinological analysis that provided, for each sample, data within the range measured by Floris et al. [41].

4.2. ATR-FTIR Spectroscopy

ATR-FTIR spectra were acquired by means of a Vertex 70 spectrophotometer equipped with a platinum ATR-QL diamond accessory (Bruker Optics, Ettlingen, Germany). Spectra were acquired in the region 4000–400 cm^{-1} by averaging 256 scans with a resolution of 4 cm^{-1}, including background subtraction of the diamond window. The diamond was cleaned between samples by using ethanol and ultrapure water.

4.3. Random Forest

A random forest was used in a cross-validation setting, where 70% (n = 56) of the samples were randomly chosen as training set, the remaining 30% (n = 24) were used for internal validation. More specifically, a shuffle-split cross-validation over 100 iterations was performed. The typical honey wavenumber intervals already described in Section 2.1 as features were used. Thus, from the original 3800 features/wavelength, a subset of 1182 remained. Selection of the most significant features/wavenumbers was achieved based on the highest mean decrease in accuracy per cross validation iteration. All statistical analysis was conducted in R with the package "randomForest" [63].

5. Conclusions

Given the paucity of contributions aimed to evaluate the potential of the combination of ATR-FTIR spectroscopy and random forest to classify honeys, this approach has been used on a representative sampling of the most renowned four unifloral honeys from Sardinia, Italy. The results demonstrated for a good level of prediction accuracy, obtaining for each sample the required analytical information in a short time. Aspects like a difficult classification on a palynological basis and the wide variability in the chemical composition of the asphodel honey should be taken into account when this botanical origin is included in the classification model. On the basis of the results reported here, further studies are required to assess the prediction accuracy of this approach for a larger number of botanical origins. As a final remark, since the phenolic fraction has been a key parameter for discriminating the selected honeys on the basis of the infrared spectra, attention will be paid in the future to the contribution of polyphenols towards the IR absorption spectra of unifloral honeys.

Author Contributions: Conceptualization, M.C. (Marco Ciulu), N.S. and G.S.; methodology, M.C (Marco Ciulu) and E.O.; software E.O.; validation M.C. (Marco Ciulu) and E.O.; formal analysis, M.C (Marco Ciulu), R.S. and E.O.; investigation, M.C. (Marco Ciulu), R.S., R.F. and E.O.; resources, G.S and L.M.; data curation, M.C. (Marco Ciulu) and E.O.; writing—original draft preparation, M.C (Marco Ciulu) and E.O.; writing—review and editing, M.C. (Marco Ciulu), E.O., R.S., R.F., M.C (Marco Caredda), N.S., L.M., and G.S.; visualization, M.C. (Marco Ciulu), R.S., and E.O.; supervision M.C. (Marco Ciulu) and G.S.; project administration, G.S.; funding acquisition, G.S. All authors have read and agreed to the published version of the manuscript.

Funding: This research received no external funding.

Data Availability Statement: The data used to support the findings of this study are available from the corresponding author upon request.

Acknowledgments: The authors gratefully thank Ignazio Floris of the University of Sassari (Italy) for providing the samples.

Conflicts of Interest: The authors declare no conflict of interest.

Sample Availability: Honey samples used for this study are not available from the authors.

References

1. Afrin, S.; Haneefa, S.M.; Fernandez-Cabezudo, M.J.; Giampieri, F.; al-Ramadi, B.K.; Battino, M. Therapeutic and preventive properties of honey and its bioactive compounds in cancer: An evidence-based review. *Nutr. Res. Rev.* **2020**, *33*, 50–76. [CrossRef] [PubMed]
2. García, N.L. The Current Situation on the International Honey Market. *Bee World* **2018**, *95*, 89–94. [CrossRef]

1. Aries, E.; Burton, J.; Carrasco, L.; De Rudder, L.; Maquet, A. *Scientific Support to the Implementation of a Coordinated Control Plan with a View to Establishing the Prevalence of Fraudulent Practices in the Marketing of Honey. Results of Honey Authenticity Testing by Liquid Chromatography-Isotope Ratio Mass Spectrometry*; European Commission: Brussels, Belgium, 2016; Available online: https://ec.europa.eu/food/sites/food/files/safety/docs/oc_control-progs_honey_jrc-tech-report_2016.pdf (accessed on 27 December 2020).
2. Anklam, E. A review of the analytical methods to determine the geographical and botanical origin of honey. *Food Chem.* **1998**, *63*, 549–562. [CrossRef]
3. Machado, A.M.; Miguel, M.G.; Vilas-Boas, M.; Figueiredo, A.C. Honey Volatiles as a Fingerprint for Botanical Origin—A Review on their Occurrence on Monofloral Honeys. *Molecules* **2020**, *25*, 374. [CrossRef]
4. Mădaș, M.N.; Mărghitaș, L.A.; Dezmirean, D.S.; Bobiș, O.; Abbas, O.; Danthine, S.; Francis, F.; Haubruge, E.; Nguyen, B.K. Labeling Regulations and Quality Control of Honey Origin: A Review. *Food Rev. Int.* **2020**, *36*, 215–240. [CrossRef]
5. De Sousa, J.M.B.; de Souza, E.L.; Marques, G.; de Toledo Benassi, M.; Gullón, B.; Pintado, M.M.; Magnani, M. Sugar profile, physicochemical and sensory aspects of monofloral honeys produced by different stingless bee species in Brazilian semi-arid region. *LWT Food Sci. Technol.* **2016**, *65*, 645–651. [CrossRef]
6. Nozal Nalda, M.J.; Bernal Yagüe, J.L.; Diego Calva, J.C.; Martín Gómez, M.T. Classifying honeys from the Soria Province of Spain via multivariate analysis. *Anal. Bioanal. Chem.* **2005**, *382*, 311–319. [CrossRef]
7. Escuredo, O.; Dobre, I.; Fernández-González, M.; Seijo, M.C. Contribution of botanical origin and sugar composition of honeys on the crystallization phenomenon. *Food Chem.* **2014**, *149*, 84–90. [CrossRef]
8. Tuberoso, C.I.G.; Bifulco, E.; Caboni, P.; Cottiglia, F.; Cabras, P.; Floris, I. Floral Markers of Strawberry Tree (Arbutus unedo L.) Honey. *J. Agric. Food Chem.* **2010**, *58*, 384–389. [CrossRef]
9. Pascual-Maté, A.; Osés, S.M.; Fernández-Muiño, M.A.; Sancho, M.T. Analysis of Polyphenols in Honey: Extraction, Separation and Quantification Procedures. *Sep. Purif. Rev.* **2018**, *47*, 142–158. [CrossRef]
10. Pohl, P. Determination of metal content in honey by atomic absorption and emission spectrometries. *TrAC Trends Anal. Chem.* **2009**, *28*, 117–128. [CrossRef]
11. Alissandrakis, E.; Tarantilis, P.A.; Pappas, C.; Harizanis, P.C.; Polissiou, M. Ultrasound-assisted extraction gas chromatography–mass spectrometry analysis of volatile compounds in unifloral thyme honey from Greece. *Eur. Food Res. Technol.* **2009**, *229*, 365–373. [CrossRef]
12. Aliferis, K.A.; Tarantilis, P.A.; Harizanis, P.C.; Alissandrakis, E. Botanical discrimination and classification of honey samples applying gas chromatography/mass spectrometry fingerprinting of headspace volatile compounds. *Food Chem.* **2010**, *121*, 856–862. [CrossRef]
13. Radovic, B.S.; Careri, M.; Mangia, A.; Musci, M.; Gerboles, M.; Anklam, E. Contribution of dynamic headspace GC–MS analysis of aroma compounds to authenticity testing of honey. *Food Chem.* **2001**, *72*, 511–520. [CrossRef]
14. Soria, A.C.; Martínez-Castro, I.; Sanz, J. Some aspects of dynamic headspace analysis of volatile components in honey. *Food Res. Int.* **2008**, *41*, 838–848. [CrossRef]
15. Ampuero, S.; Bogdanov, S.; Bosset, J.-O. Classification of unifloral honeys with an MS-based electronic nose using different sampling modes: SHS, SPME and INDEX. *Eur. Food Res. Technol.* **2004**, *218*, 198–207. [CrossRef]
16. Pierna, J.A.F.; Abbas, O.; Dardenne, P.; Baeten, V. Discrimination of Corsican honey by FT-Raman spectroscopy and chemometrics. *Biotechnol. Agron. Soc. Environ.* **2011**, *10*, 75–84.
17. Grazia Mignani, A.; Ciaccheri, L.; Mencaglia, A.A.; Di Sanzo, R.; Carabetta, S.; Russo, M. Dispersive Raman Spectroscopy for the Nondestructive and Rapid Assessment of the Quality of Southern Italian Honey Types. *J. Lightwave Technol.* **2016**, *34*, 4479–4485. [CrossRef]
18. Schievano, E.; Stocchero, M.; Morelato, E.; Facchin, C.; Mammi, S. An NMR-based metabolomic approach to identify the botanical origin of honey. *Metabolomics* **2012**, *8*, 679–690. [CrossRef]
19. Ren, H.; Yue, J.; Wang, D.; Fan, J.; An, L. HPLC and 1H-NMR combined with chemometrics analysis for rapid discrimination of floral origin of honey. *J. Food Meas. Charact.* **2019**, *13*, 1195–1204. [CrossRef]
20. Minaei, S.; Shafiee, S.; Polder, G.; Moghadam-Charkari, N.; van Ruth, S.; Barzegar, M.; Zahiri, J.; Alewijn, M.; Kuś, P.M. VIS/NIR imaging application for honey floral origin determination. *Infrared Phys. Technol.* **2017**, *86*, 218–225. [CrossRef]
21. Dinca, O.-R.; Ionete, R.E.; Popescu, R.; Costinel, D.; Radu, G.-L. Geographical and Botanical Origin Discrimination of Romanian Honey Using Complex Stable Isotope Data and Chemometrics. *Food Anal. Methods* **2015**, *8*, 401–412. [CrossRef]
22. Huang, F.; Song, H.; Guo, L.; Guang, P.; Yang, X.; Li, L.; Zhao, H.; Yang, M. Detection of adulteration in Chinese honey using NIR and ATR-FTIR spectral data fusion. *Spectrochim. Acta Part A Mol. Biomol. Spectrosc.* **2020**, *235*, 118297. [CrossRef] [PubMed]
23. Rios-Corripio, M.A.; Rojas-López, M.; Delgado-Macuil, R. Analysis of adulteration in honey with standard sugar solutions and syrups using attenuated total reflectance-Fourier transform infrared spectroscopy and multivariate methods. *CYTA J. Food* **2012**, *10*, 119–122. [CrossRef]
24. Riswahyuli, Y.; Rohman, A.; Setyabudi, F.M.C.S.; Raharjo, S. Indonesian wild honey authenticity analysis using attenuated total reflectance-fourier transform infrared (ATR-FTIR) spectroscopy combined with multivariate statistical techniques. *Heliyon* **2020**, *6*, e03662. [CrossRef] [PubMed]

27. Sahlan, M.; Karwita, S.; Gozan, M.; Hermansyah, H.; Yohda, M.; Yoo, Y.J.; Pratami, D.K. Identification and classification of honey's authenticity by attenuated total reflectance Fourier-transform infrared spectroscopy and chemometric method. *Vet. World* **2019**, *12*, 1304–1310. [CrossRef]
28. Li, Q.; Zeng, J.; Lin, L.; Zhang, J.; Zhu, J.; Yao, L.; Wang, S.; Yao, Z.; Wu, Z. Low risk of category misdiagnosis of rice syrup adulteration in three botanical origin honey by ATR-FTIR and general model. *Food Chem.* **2020**, *332*, 127356. [CrossRef]
29. Devi, A.; Jangir, J.; Anu-Appaiah, K.A. Chemical characterization complemented with chemometrics for the botanical origin identification of unifloral and multifloral honeys from India. *Food Res. Int.* **2018**, *107*, 216–226. [CrossRef]
30. Gok, S.; Severcan, M.; Goormaghtigh, E.; Kandemir, I.; Severcan, F. Differentiation of Anatolian honey samples from different botanical origins by ATR-FTIR spectroscopy using multivariate analysis. *Food Chem.* **2015**, *170*, 234–240. [CrossRef]
31. Kasprzyk, I.; Depciuch, J.; Grabek-Lejko, D.; Parlinska-Wojtan, M. FTIR-ATR spectroscopy of pollen and honey as a tool for unifloral honey authentication. The case study of rape honey. *Food Control* **2018**, *84*, 33–40. [CrossRef]
32. Svečnjak, L.; Bubalo, D.; Boranović, G.; Novosel, H. Optimization of FTIR-ATR spectroscopy for botanical authentication of unifloral honey types and melissopalinological data prediction. *Eur. Food Res. Technol.* **2015**, *240*, 1101–1115. [CrossRef]
33. Tuberoso, C.I.G.; Bifulco, E.; Jerković, I.; Caboni, P.; Cabras, P.; Floris, I. Methyl Syringate: A Chemical Marker of Asphodel (*Asphodelus microcarpus* Salzm. et Viv.) Monofloral Honey. *J. Agric. Food Chem.* **2009**, *57*, 3895–3900. [CrossRef] [PubMed]
34. Afrin, S.; Forbes-Hernandez, T.; Gasparrini, M.; Bompadre, S.; Quiles, J.; Sanna, G.; Spano, N.; Giampieri, F.; Battino, M. Strawberry-Tree Honey Induces Growth Inhibition of Human Colon Cancer Cells and Increases ROS Generation: A Comparison with Manuka Honey. *Int. J. Mol. Sci.* **2017**, *18*, 613. [CrossRef] [PubMed]
35. Tariba Lovaković, B.; Lazarus, M.; Brčić Karačonji, I.; Jurica, K.; Živković Semren, T.; Lušić, D.; Brajenović, N.; Pelaić, Z.; Pizent, A. Multi-elemental composition and antioxidant properties of strawberry tree (*Arbutus unedo* L.) honey from the coastal region of Croatia: Risk-benefit analysis. *J. Trace Elem. Med. Biol.* **2018**, *45*, 85–92. [CrossRef] [PubMed]
36. Rosa, A.; Tuberoso, C.I.G.; Atzeri, A.; Melis, M.P.; Bifulco, E.; Dessì, M.A. Antioxidant profile of strawberry tree honey and its marker homogentisic acid in several models of oxidative stress. *Food Chem.* **2011**, *129*, 1045–1053. [CrossRef] [PubMed]
37. Afrin, S.; Forbes-Hernández, T.Y.; Cianciosi, D.; Pistollato, F.; Zhang, J.; Pacetti, M.; Amici, A.; Reboredo-Rodríguez, P.; Simal-Gandara, J.; Bompadre, S.; et al. Strawberry tree honey as a new potential functional food. Part 2: Strawberry tree honey increases ROS generation by suppressing Nrf2-ARE and NF-B signaling pathways and decreases metabolic phenotypes and metastatic activity in colon cancer cells. *J. Funct. Foods* **2019**, *57*, 477–487. [CrossRef]
38. Cabras, P.; Angioni, A.; Tuberoso, C.; Floris, I.; Reniero, F.; Guillou, C.; Ghelli, S. Homogentisic Acid: A Phenolic Acid as a Marker of Strawberry-Tree (*Arbutus unedo*) Honey. *J. Agric. Food Chem.* **1999**, *47*, 4064–4067. [CrossRef]
39. Scanu, R.; Spano, N.; Panzanelli, A.; Pilo, M.I.; Piu, P.C.; Sanna, G.; Tapparo, A. Direct chromatographic methods for the rapid determination of homogentisic acid in strawberry tree (*Arbutus unedo* L.) honey. *J. Chromatogr. A* **2005**, *1090*, 76–80. [CrossRef]
40. Spano, N.; Casula, L.; Panzanelli, A.; Pilo, M.; Piu, P.; Scanu, R.; Tapparo, A.; Sanna, G. An RP-HPLC determination of 5-hydroxymethylfurfural in honeyThe case of strawberry tree honey. *Talanta* **2006**, *68*, 1390–1395. [CrossRef]
41. Floris, I. Honeys of Sardinia (Italy). *J. Apic. Res.* **2007**, 198–209. [CrossRef]
42. Ciulu, M.; Serra, R.; Caredda, M.; Salis, S.; Floris, I.; Pilo, M.I.; Spano, N.; Panzanelli, A.; Sanna, G. Chemometric treatment of simple physical and chemical data for the discrimination of unifloral honeys. *Talanta* **2018**, *190*, 382–390. [CrossRef] [PubMed]
43. Martos, I.; Ferreres, F.; Tomás-Barberán, F.A. Identification of Flavonoid Markers for the Botanical Origin of *Eucalyptus* Honey. *J. Agric. Food Chem.* **2000**, *48*, 1498–1502. [CrossRef] [PubMed]
44. Tuberoso, C.I.G.; Bifulco, E.; Caboni, P.; Sarais, G.; Cottiglia, F.; Floris, I. Lumichrome and Phenyllactic Acid as Chemical Markers of Thistle (*Galactites tomentosa* Moench) Honey. *J. Agric. Food Chem.* **2011**, *59*, 364–369. [CrossRef] [PubMed]
45. Maione, C.; Barbosa, F.; Barbosa, R.M. Predicting the botanical and geographical origin of honey with multivariate data analysis and machine learning techniques: A review. *Comput. Electron. Agric.* **2019**, *157*, 436–446. [CrossRef]
46. Batista, B.L.; da Silva, L.R.S.; Rocha, B.A.; Rodrigues, J.L.; Berretta-Silva, A.A.; Bonates, T.O.; Gomes, V.S.D.; Barbosa, R.M.; Barbosa, F. Multi-element determination in Brazilian honey samples by inductively coupled plasma mass spectrometry and estimation of geographic origin with data mining techniques. *Food Res. Int.* **2012**, *49*, 209–215. [CrossRef]
47. Breiman, L. Random Forests. *Mach. Learn.* **2001**, *45*, 5–32. [CrossRef]
48. Martinez-Castillo, C.; Astray, G.; Mejuto, J.C.; Simal-Gandara, J. Random Forest, Artificial Neural Network, and Support Vector Machine Models for Honey Classification. *eFood* **2019**, *1*, 69. [CrossRef]
49. Svečniak, L.; Biliškov, N.; Bubalo, D.; Barišic, D. Application of Infrared spectroscopy in Honey analysis. *Agric. Conspec. Sci.* **2011**, *76*, 191–195.
50. Gallardo-Velázquez, T.; Osorio-Revilla, G.; Loa, M.Z.; Rivera-Espinoza, Y. Application of FTIR-HATR spectroscopy and multivariate analysis to the quantification of adulterants in Mexican honeys. *Food Res. Int.* **2009**, *42*, 313–318. [CrossRef]
51. Movasaghi, Z.; Rehman, S.; ur Rehman, D.I. Fourier Transform Infrared (FTIR) Spectroscopy of Biological Tissues. *Appl. Spectrosc Rev.* **2008**, *43*, 134–179. [CrossRef]
52. Sivakesava, S.; Irudayaraj, J. Prediction of Inverted Cane Sugar Adulteration of Honey by Fourier Transform Infrared Spectroscopy. *J. Food Sci.* **2001**, *66*, 972–978. [CrossRef]
53. Cai, S.; Singh, B.R. A Distinct Utility of the Amide III Infrared Band for Secondary Structure Estimation of Aqueous Protein Solutions Using Partial Least Squares Methods. *Biochemistry* **2004**, *43*, 2541–2549. [CrossRef] [PubMed]

4. Philip, D. Honey mediated green synthesis of gold nanoparticles. *Spectrochim. Acta Part A Mol. Biomol. Spectrosc.* **2009**, *73*, 650–653. [CrossRef] [PubMed]
5. Tewari, J.; Irudayaraj, J. Quantification of Saccharides in Multiple Floral Honeys Using Fourier Transform Infrared Microattenuated Total Reflectance Spectroscopy. *J. Agric. Food Chem.* **2004**, *52*, 3237–3243. [CrossRef]
6. Tewari, J.C.; Irudayaraj, J.M.K. Floral Classification of Honey Using Mid-Infrared Spectroscopy and Surface Acoustic Wave Based z-Nose Sensor. *J. Agric. Food Chem.* **2005**, *53*, 6955–6966. [CrossRef]
7. Subari, N.; Mohamad Saleh, J.; Md Shakaff, A.; Zakaria, A. A Hybrid Sensing Approach for Pure and Adulterated Honey Classification. *Sensors* **2012**, *12*, 14022–14040. [CrossRef]
8. Mathlouthi, M.; Koenig, J.L. Vibrational Spectra of Carbohydrates. In *Advances in Carbohydrate Chemistry and Biochemistry*; Academic Press: London, UK, 1987; Volume 44, pp. 7–89. ISBN 978-0-12-007244-6.
9. Kelly, J.F.D.; Downey, G.; Fouratier, V. Initial Study of Honey Adulteration by Sugar Solutions Using Midinfrared (MIR) Spectroscopy and Chemometrics. *J. Agric. Food Chem.* **2004**, *52*, 33–39. [CrossRef]
10. Varma, S.; Simon, R. Bias in error estimation when using cross-validation for model selection. *BMC Bioinform.* **2006**, *7*, 91. [CrossRef]
11. Stefas, D.; Gyftokostas, N.; Couris, S. Laser induced breakdown spectroscopy for elemental analysis and discrimination of honey samples. *Spectrochim. Acta Part B At. Spectrosc.* **2020**, *172*, 105969. [CrossRef]
12. Tahir, H.E.; Xiaobo, Z.; Zhihua, L.; Jiyong, S.; Zhai, X.; Wang, S.; Mariod, A.A. Rapid prediction of phenolic compounds and antioxidant activity of Sudanese honey using Raman and Fourier transform infrared (FT-IR) spectroscopy. *Food Chem.* **2017**, *226*, 202–211. [CrossRef]
13. Liaw, A.; Wiener, M. Classification and Regression by randomForest. *R News* **2002**, *2*, 18–22.

Article

The Comparison of Physicochemical Parameters, Antioxidant Activity and Proteins for the Raw Local Polish Honeys and Imported Honey Blends

Michał Miłek [1,*], Aleksandra Bocian [2], Ewelina Kleczyńska [1], Patrycja Sowa [3] and Małgorzata Dżugan [1]

1. Department of Chemistry and Food Toxicology, Institute of Food Technology and Nutrition, University of Rzeszów, Ćwiklińskiej 1a, 35-601 Rzeszów, Poland; ewelina1213@poczta.onet.pl (E.K.); mdzugan@ur.edu.pl (M.D.)
2. Department of Biotechnology and Bioinformatics, Faculty of Chemistry, Rzeszów University of Technology, Powstańców Warszawy 6, 35-959 Rzeszów, Poland; bocian@prz.edu.pl
3. Department of Bioenergetics Food Analysis and Microbiology, Institute of Food Technology and Nutrition, University of Rzeszów, Ćwiklinskiej 2D, 35-601 Rzeszów, Poland; psowa@ur.edu.pl
* Correspondence: mmilek@ur.edu.pl; Tel.: +48-17-872-1730

Abstract: Many imported honeys distributed on the Polish market compete with local products mainly by lower price, which can correspond to lower quality and widespread adulteration. The aim of the study was to compare honey samples (11 imported honey blends and 5 local honeys) based on their antioxidant activity (measured by DPPH, FRAP, and total phenolic content), protein profile obtained by native PAGE, soluble protein content, diastase, and acid phosphatase activities identified by zymography. These indicators were correlated with standard quality parameters (water, HMF, pH, free acidity, and electrical conductivity). It was found that raw local Polish honeys show higher antioxidant and enzymatic activity, as well as being more abundant in soluble protein. With the use of principal component analysis (PCA) and stepwise linear discriminant analysis (LDA) protein content and diastase number were found to be significant ($p < 0.05$) among all tested parameters to differentiate imported honey from raw local honeys.

Keywords: honey; quality standards; protein; amylase; acid phosphatase; native PAGE

1. Introduction

Honey is a product with a diverse chemical composition, which depends mainly on the type and species of plant from which it originates. Poland is distinguished by particularly large beekeeping traditions and Polish honeys invariably have a good reputation in foreign markets. Lately, a lot of low-price imported honeys available on the Polish and EU market have been competing with local products [1]. When the honey is a blend of honeys harvested from more than one country, placing exact information concerning the country of origin on the label is not required [2]. Imported honey is usually labeled as: "blend of EU honeys", "blend of non-EU honeys", or "blend of EU and non-EU honeys". Such honeys can be of poor quality due to processing to extend their shelf life. Research conducted by Dżugan et al. (2018) showed that imported honeys had an increased content of HMF (5-hydroxymethylfurfural) and reduced diastase number, electrical conductivity, and total acidity as compared to raw local honeys [3]. Imported honeys are frequently thermally processed to kill certain types of bacteria or yeast responsible for fermentation and prevent crystallization during storage. However, such processing also removes the natural flavors and reduces antibacterial properties, nutrients, and antioxidants content. Contrarily, local raw honey is seen as a high-quality and less allergenic product due to its pollen origin from the immediate locations. Moreover, local raw honey containing pollen from the surrounding area is known to immunize allergy sufferers, especially children [4,5].

Until now the comparison between local and imported blends has been rarely performed. Simultaneously, in numerous studies multivariate statistical analysis has been applied for differentiating honey samples based on physicochemical properties and the content of biologically active compounds [6–8]. Such classification is performed mainly based on the biological origin of the samples, less often the geographical one. These analyses can be also used to detect adulteration in honey. Despite a large spectrum of indicators used in the honey analysis, starting with simple physicochemical parameters evaluation, pollen analysis, sugar profiles by HPLC analysis, up to stable carbon isotope ratio analysis (SCIRA) based on the calculation of the 13C/12C isotopes ratio [9–11], there is still an urgent need to implement an effective method to differentiate the quality of honey samples. Thus, the big challenge is applying chemometric analyses in the area of the differentiation of raw local honeys and available on the market blends.

Raw honeys are generally more abundant in proteins which, due to their thermolability, seem to be a sensitive but not frequently used marker of honey quality. Honey proteins are mainly of bee origin, and only part of them come from nectar [12]. The protein content is variety-dependent (0.2–0.4 mg/100 g for blossom and 0.4–0.7 mg/100 g for honeydew honey) [13] and thermal processing affects protein level negatively. These molecules are found in honey in small amounts, but they are partly responsible for the healing properties of honey. Although natural honeys contain a small amount of enzymes, including diastase, invertase, glucose oxidase, acid phosphatase, catalase, and β-glucosidase, they are very important in creating honey bioactivity [14]. Only diastase activity is included in honey standards [15]. By using electrophoresis SDS-PAGE, protein fractions can be obtained, and the number of proteins and polypeptides, as well as their molecular weight, can be determined. However, electrophoretic techniques are rarely used in honey analysis, although they can be a good tool for assessing the protein profile and even zymographic detection of individual enzymes [16]. Confirmation of the suitability of this technique would provide a promising tool to evaluate the quality of honey.

The aim of this study was to compare the quality of raw local honeys and imported honey blends available on the Polish market based on antioxidant potential, amylase and phosphatase activities, and protein profiling by native PAGE. Multivariate analysis (PCA and LDA) applied to the obtained results allowed us to verify tested samples according to their origin.

2. Results

2.1. Standard Quality Parameters

In order to assess the quality of tested imported honey in accordance with legal regulations, their physicochemical parameters were determined. These data were compared to reference honey samples originating from an ecological apiary of the Podkarpackie region (Table 1).

Tested honeys predominantly fulfilled applicable legal standards regarding their physicochemical properties. The water content in most of the tested honeys was within the legal limits which was set to be below 20%, except for heather honey (maximum of 23%) [2], but 25% of samples (including controls) slightly exceeded the set limit. However, the water content was variety-dependent; the lowest was determined in acacia honey (17.71% on average) as well as two samples of honeydew honey (17.55%), and the highest in buckwheat and linden (average values 20 and 19.94%, respectively). Increased water content may be caused by adverse weather conditions prevailing when honey was produced by bees or immaturity resulting from early acquisition from the hive [13]. The results of the analysis of free acids contained in the tested honeys prove that these honeys were mostly within the norm and were comparable to our earlier findings [3,17,18]. The acidity of honey depends on the type of raw material, the season in which it was obtained, and the degree of its maturity [13]. Organic acids contained in honey lower its pH, which prevents the growth of microorganisms and extends the product's shelf life. As honey conductivity should be within the range of 0.2 to 0.8 mS/cm for nectar and above 0.8 mS/cm for honeydew honey [2], all tested honeys fell within these parameters. This parameter allows the distinguishing between nectar and honeydew honeys easily. The analysis by Tomczyk et al. [18] of the physicochemical properties of selected nectar honey varieties from the Podkarpackie region showed the conductivity of nectar honey ranged from 0.23 (for rape) to 0.82 mS/cm (for forest honey) which is comparable to the values obtained in the present study.

Table 1. The physicochemical parameters of imported honey blends and raw local honey compared to applicable EU regulations.

	Honey Sample	Moisture Content [%]	pH	Free Acidity [mval/kg]	Electrical Conductivity [mS/cm]	HMF Content [mg/kg]
Nectar Honey	A1	17.20 ± 0.00	3.99 ± 0.02 *	16.00 ± 1.40 *	0.40 ± 0.00 *	5.82 ± 0.00 *
	A2	18.75 ± 0.05 *	4.23 ± 0.01 *	10.50 ± 0.70 *	0.31 ± 0.02	35.20 ± 0.00 *
	AC	**17.17 ± 0.30**	**3.77 ± 0.03**	**20.85 ± 6.50**	**0.31 ± 0.18**	**22.66 ± 0.00**
	B1	19.80 ± 0.00	4.01 ± 0.02 *	21.50 ± 0.70 *	0.54 ± 0.00 *	6.40 ± 0.60 *
	B2	19.70 ± 0.00	3.83 ± 0.00	23.50 ± 0.70 *	0.54 ± 0.01 *	34.18 ± 0.00 *
	BC	**20.50 ± 0.00**	**4.04 ± 0.00 ***	**26.50 ± 0.00**	**0.43 ± 0.01**	**20.02 ± 0.00**
	L1	20.50 ± 0.10	4.32 ± 0.00	14.50 ± 0.70 *	0.25 ± 0.02 *	19.33 ± 0.00 *
	L2	19.05 ± 0.15 *	4.42 ± 0.02	14.50 ± 0.70 *	0.23 ± 0.02 *	23.23 ± 0.00 *
	LC	**20.30 ± 0.10**	**4.13 ± 0.00**	**25.50 ± 0.00**	**0.64 ± 0.00**	**13.20 ± 0.00**
	M1	18.85 ± 0.05 *	4.30 ± 0.02	10.50 ± 2.10 *	0.79 ± 0.02 *	73.92 ± 0.00 *
	M2	18.35 ± 0.05 *	3.75 ± 0.02 *	16.50 ± 0.70 *	0.75 ± 0.01 *	47.55 ± 0.48 *
	MC	**20.20 ± 0.00**	**4.11 ± 0.00**	**29.50 ± 0.00**	**0.67 ± 0.00**	**12.50 ± 0.00**
	Applicable limits [2]	20%		<50	<0.8	<40
Honeydew Honey	H1	17.55 ± 0.05 *	4.54 ± 0.02 *	29.50 ± 0.70 *	1.70 ± 0.01 *	11.78 ± 0.12 *
	H2	17.55 ± 0.05 *	4.44 ± 0.01 *	38.00 ± 1.40 *	1.95 ± 0.06 *	35.38 ± 0.36 *
	H3	18.75 ± 0.15 *	4.68 ± 0.00 *	29.50 ± 2.10 *	1.99 ± 0.07 *	15.04 ± 0.14
	HC	**19.90 ± 0.00**	**3.64 ± 0.00**	**30.50 ± 0.00**	**1.16 ± 0.00**	**19.80 ± 0.00**
	Applicable limits [2]	20%		<50	>0.8	<40

* Means marked with the symbol differ statistically significantly from a suitable high-quality local control sample (marked with bold): AC for acacia, BC for buckwheat, LC for linden, MC for multifloral, and HC for honeydew honey (t-test, $p < 0.05$).

The HMF (5-hydroxymethylfurfural) content in honey must not exceed 40 mg/kg [2]. Among the samples tested, only two imported multifloral honeys showed HMF content higher than permissible standards. However, in most cases, raw honeys from the Podkarpackie region contained less HMF than their imported counterparts, except for the lower values for single samples of acacia, buckwheat, and honeydew honey. As an HMF increase can result from long storage in inappropriate conditions, adulteration with corn syrup, or prolonged heating, it is an important parameter used to control honey overheating. Such processing may cause a decrease in the nutritional value by degradation of thermolabile vitamins and bio-nutrients, and also contribute to a decrease in diastase activity [19]. It is in agreement with the findings of Sanz et al. for honeys obtained directly from beekeepers, which contained approximately five times lower HMF than honeys purchased in a supermarket [20]. The increased level of HMF content in the case of some imported honeys may result from their long storage or from the use of heating the honey in the production of blends.

2.2. Antioxidant Properties

Antioxidant properties of honey are not specified in legal regulations; however, they have been proposed as a useful indicator in the authentication of honey botanical origin [7]. On the other hand, antioxidant activity can serve simply as an indicator of the biological activity of honey measured in vitro. A higher antioxidant capacity determines better antimicrobial and anti-inflammatory activity of honey [21,22]. The data regarding the antioxidant properties of all analyzed honeys are shown in Table 2. The total phenolic content was significantly correlated with the results of FRAP and DPPH analysis (Pearson coefficient 0.952 and 0.558, respectively). The different strength of the correlation results from the different mechanisms of the two methods used, which differ in sensitivity against the hydrophilic and hydrophobic antioxidants fraction. The study showed a diverse content of phenolic compounds depending on the honey type. Dark honeys (buckwheat and honeydew) showed a higher content of these compounds. This is a well-known feature of honey: the darker the honey, the richer in polyphenols, which was previously presented by several authors [22–24]. Comparing ecological Podkarpackie honeys with imported honey blends within the same variety, even the several times higher polyphenol content and antioxidant activity of Polish nectar honeys ($p < 0.05$) measured by FRAP method can be noticed. For honeydew honeys, smaller differences were observed. Smaller differences for DPPH assay results were observed, which follows from the manner of the results' expression, as a percent of radical inhibition for direct comparison of the same honey dilution, without calculating it to the honey mass unit. Based on the obtained results of antioxidant potential, health benefits can be expected from consuming local honeys of high quality. Special pro-health properties of honey with a high content of antioxidant compounds were previously proved in the example of buckwheat honey which showed a protective effect against oxidative stress during an in vitro study using a yeast biological model [22].

Table 2. The content of soluble protein, enzyme activity, and antioxidant activity of imported honey blends and raw local honey.

Honey Sample	Protein Content [mg/100 g]	Enzymatic Activity		Antioxidant Activity		
		Diastase Activity (as Diastase Number DN)	Acid Phosphatase Activity (mmol/g/min)	Total Phenolic Content (mg GAE/kg)	FRAP (µmol TE/kg)	DPPH (% Radical Inhibition)
A1	27.27 ± 2.57 *	8.3 ± 2.3 *	5.0 ± 1.9	58.56 ± 0.56 *	203.85 ± 2.04 *	10.82 ± 0.22 *
A2	19.09 ± 2.57 *	2.7 ± 0.6 *	7.5 ± 2.6	98.20 ± 2.95 *	317.31 ± 6.35 *	11.51 ± 0.12 *
AC	37.73 ± 0.64	5.63 ± 1.8	7.7 ± 0.5	140.81 ± 11.41	531.96 ± 18.50	19.64 ± 6.13
B1	224.55 ± 6.43 *	13.9 ± 3.8 *	20.7 ± 0.4 *	516.22 ± 5.02 *	1113.46 ± 33.40 *	33.99 ± 1.02 *
B2	175.46 ± 14.14 *	10.3 ± 2.2 *	19.5 ± 1.7 *	653.15 ± 6.53 *	623.08 ± 11.98 *	39.55 ± 4.00 *
BC	475.456 ± 3.86	66.43 ± 5.7	31.8 ± 2.6	2075.70 ± 19.90	4973.10 ± 46.73	59.69 ± 2.51
L1	23.18 ± 0.64 *	4.3 ± 1.1 *	6.7 ± 0.9 *	125.23 ± 1.50 *	615.38 ± 12.31 *	17.76 ± 1.60 *
L2	23.634 ± 3.856 *	8.0 ± 1.2 *	5.2 ± 1.2 *	102.70 ± 1.13 *	607.69 ± 18.23 *	17.92 ± 0.36 *
LC	89.55 ± 1.93	28.94 ± 3.8	17.8 ± 2.6	436.90 ± 4.25	1555.80 ± 14.82	23.30 ± 0.76
M1	10.91 ± 0.00 *	2.7 ± 0.7 *	9.8 ± 1.7 *	161.26 ± 2.01 *	425.00 ± 8.50 *	13.73 ± 0.14 *
M2	53.18 ± 3.21 *	3.1 ± 0.9 *	10.3 ± 1.4	117.20 ± 1.18 *	334.62 ± 6.69 *	13.41 ± 0.11 *
MC	133.18 ± 4.50	29.51 ± 4.0	11.5 ± 0.2	496.80 ± 5.00	1470.20 ± 14.93	19.52 ± 0.26
H1	89.09 ± 6.43 *	16.3 ± 2.2 *	16.7 ± 0.5	310.81 ± 2.89 *	1390.38 ± 41.71 *	79.37 ± 4.76
H2	105.456 ± 21.86 *	12.5 ± 1.9 *	17.3 ± 2.8 *	597.30 ± 11.95	1934.62 ± 77.38 *	72.32 ± 0,71
H3	70.46 ± 8.36	14.6 ± 1.0	16.3 ± 1.9	568.47 ± 6.25 *	1488.46 ± 29.77 *	12.73 ± 0.10 *
HC	70.91 ± 2.57	17.45 ± 2.9	14.0 ± 1.9	646.80 ± 5.87	1526.00 ± 16.97	69.58 ± 0.35

* Means marked with the symbol differ statistically significantly from a high quality local control sample (marked with bold): AC for acacia, BC for buckwheat, LC for linden, MC for multifloral, and HC for honeydew honey (t-test, $p < 0.05$).

2.3. Protein Content and Enzyme Activities

Table 2 summarizes also the total protein content and activity of selected enzymes (amylase and acid phosphatase) in tested honeys.

Based on the obtained results, it was found that the protein content strongly depends on the honey variety. The largest amounts of protein were determined in buckwheat and honeydew honeys, which belong to the dark honeys. Acacia honey contained the lowest amount of protein. Comparing ecological honeys with imported blends regardless of the variety, a lower protein content was determined, excluding honeydew honey. Statistically significant differences occurred for honeydew, multi-flower, buckwheat, linden, and acacia honeys. However, the largest difference in protein content between the control and other samples was found in buckwheat honey. Based on the obtained data, it can be assumed that the amount of protein in honey strongly depends on its botanical origin. Cimpoiu et al. presented a similar opinion; analyzing numerous samples of honey they noticed a significant relationship between the amount of protein and the variety of honey [25]. The total protein content was also determined using the Bradford method by Flanjak et al., who demonstrated that the protein content in honey was in the ranges: 21–43 mg/100 g of honey (acacia) and 30–95 mg/100 g of honey (honeydew) [26] which show a similar order of magnitude to the data presented in this study. Especially in the case of buckwheat honey, enhanced protein content was observed. A recognized indicator of honey quality, included in Polish and international legal requirements for honey, is diastase (α-amylase), the enzyme responsible for the hydrolysis of complex sugars. Natural honey does not contain complex sugars, and the function of this enzyme in honey is not fully known [27]. However, the strong presence of amylase was confirmed in tested raw Polish honeys regardless of the variety [17]. The found values of diastase numbers were very diverse, ranging from 2.7 for acacia and multifloral honey up to 66 for buckwheat. Similarly, in the presented study, the highest enzymatic activity of diastase was observed for all tested buckwheat honeys, control honeydew honey, multifloral, and linden honey. Comparing raw local honeys to imported blends, a much lower diastase number was determined in imported honey samples, excluding acacia honey. This phenomenon could be a result of honey thermal processing. Flanjak et al. investigated the enzymatic activity of amylase, comparing the catalytic capacity of amylase in various types of honey [26], and found the high amylolytic activity of honeydew honeys (DN from approx. 12–37) and low activity of acacia honeys (DN 7.5–14). In turn, Bonta et al. found for acacia honey that the values of the diastase number were below the limit specified in the regulations for 60% of tested samples (from 2.6 to 6) [28].

Honey acid phosphatase activity is related to the fermentation processes of honey. This enzyme originates mainly from nectar and pollen and can be used as a parameter for honey characterization [29]. Buckwheat and honeydew honeys possessed the highest enzymatic activity of acid phosphatase, while the smallest activity was found in linden and acacia honeys. It is worth noting that for most honeys obtained from the ecological apiary, a higher value of enzyme activity was determined than for their imported counterparts. Comparable results were obtained by Flanjak et al. for honeys of different varieties [26]. The authors indicate a wide discrepancy in the results of enzyme activity, but also draw attention to the clearly greater enzymatic activity of acid phosphatase in honeydew honeys than in acacia honey, which is in agreement with the results. Similar conclusions were obtained in the further studies of Dżugan and Wesołowska, where the highest acid phosphatase activity was showed in buckwheat, following by honeydew, linden, and multiflorous honey [30].

2.4. Protein Profile by PAGE

Native electrophoresis gels were stained for protein profile using a colloidal Coomassie Brilliant Blue dye (Figure 1a). It was found that the tested honeys strongly differed in protein profiles, which were manifested in the number of bands and/or their intensity. It was especially visible for buckwheat (especially local BC) and honeydew honeys. The lowest protein content was observed in linden and acacia honeys. It was also noted that organic honey had definitely more protein than imported honey samples of the same variety. The electrophoregrams clearly show that the total number of bands for individual samples strongly varied. The smallest number of bands (3–4) was observed for multiflower honeys and acacia honeys, and the greatest (approx. 6–7) for buckwheat and honeydew honeys. Moreover, organic honey exhibited a richer protein profile than imported ones, excluding honeydew honey. In all samples of Polish honey, the bands visible on the gel are clearer and stronger than in imported samples (Figure 1a). The results obtained indicate that despite not using a marker during native electrophoresis, which is because many different factors affect the speed of protein migration (mass, shape, and charge), it is still an extremely useful technique for screening. Based on electrophoretic separation, it is possible to observe differences in both the amount of protein in samples and the protein profiles of individual honey.

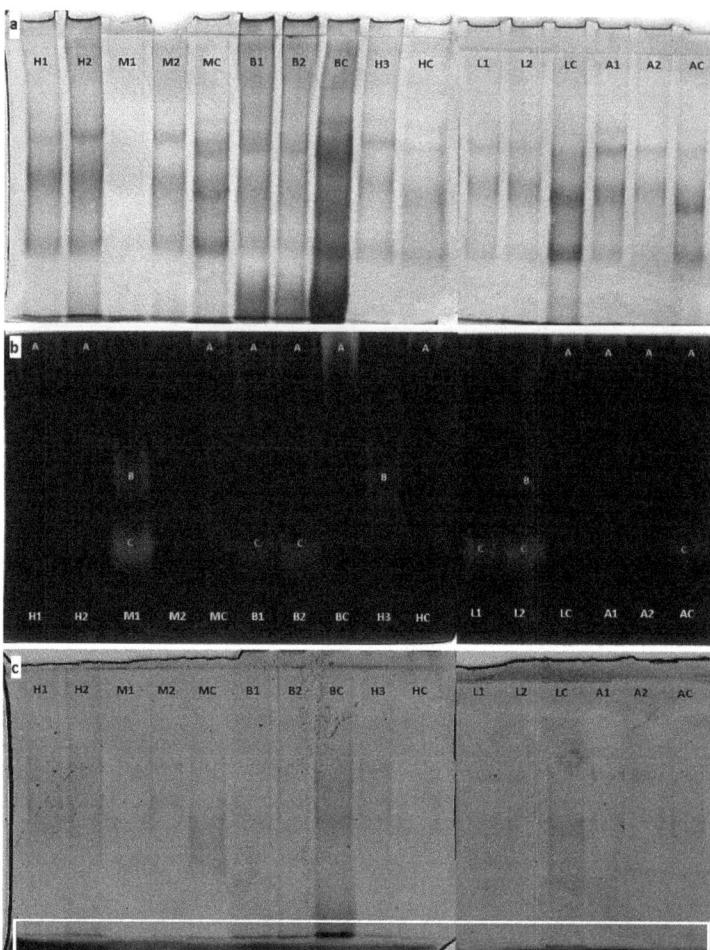

Figure 1. Gels from native PAGE electrophoresis (**a**) for total protein stained with Coomassie Brilliant Blue, (**b**) for amylase activity, and (**c**) for acid phosphatase activity. A,B,C—amylase fraction, the white box marks the location of the acid phosphatase band.

2.5. Native Enzymes Detection by PAGE

Amylase activity was detected by native electrophoresis on gels with the addition of starch which were stained with Lugol's solution (Figure 1b). After staining the gels, bright spots on the gel formed in places where the starch present in the polyacrylamide gel had been digested by the active amylase. This type of electrophoresis is called zymography. Based on the image, it can be assumed that three amylase isoforms (A, B, and C) which differed in electrophoretic migration rates and molecular weights resulted in different positions on the gel. For local honeys, isoform A was specific, excluding acacia honey where isoform C also occurred. Meanwhile, for imported honeys, the isoforms B and C were detected for three and five samples, respectively. Based on the intensity of the bright band, which depends on the activity of enzyme protein, it can be confirmed that Polish buckwheat honey contained the highest amount of active amylase protein, followed by honeydew and linden honeys. The different forms of amylase detected in some imported honeys indicated different origins of the enzyme. The amylase in honey is considered to be mainly of bee origin and is secreted by the salivary and hypopharyngeal glands [31]. However, the presence of proteins with amylolytic activity derived from plants or microorganisms cannot be ruled out [32]. Gels stained for acid phosphatase activity are shown in Figure 1c. In all honey samples, the band corresponding to acid phosphatase was detected in the same place at the bottom of the gel. It may indicate the common source of an acid phosphatase present in honey, and confirmed the idea that this enzyme originated from the honey bee digestive tract [33]. Depending on the honey variety, the color of the bands was more or less intense, which was related to the content of the active enzyme in the samples. Buckwheat honeys were characterized by the highest enzyme activity, whereas lime and acacia honeys were less abundant in acid phosphatase. The ecological Polish honeys (marked with the symbol C) exhibited higher enzyme activity ($p < 0.05$) compared to imported samples of the same botanical origin.

The native electrophoresis is rarely used in the study of honey but allows under non-denaturing conditions to separate isoforms of native proteins with preserved enzymatic activity. Borutinskaite et al. [34] analyzed the enzymatic activity of catalase and glucose oxidase in buckwheat honey using the native PAGE technique and performed an electrophoretic separation of proteins found in buckwheat honey. They showed that buckwheat honey is rich in proteins, as evidenced by a large number of intensely colored protein fractions, and also detected the activity of selected enzymes (catalase and glucose oxidase) on the gel. The use of electrophoresis in denaturing conditions (SDS-PAGE) is more frequently used to verify the honey variety as well as its geographical origin [12,35–37]. Some attempts to analyze honey proteins by 2D electrophoresis techniques are also known [16,38]. The authors confirmed the differences in the honeydew and nectar honey proteomes, and they also selected a set of proteins useful for differentiating honey varieties. Based on the PAGE gels presented in this study, it was indicated that native electrophoresis can be a useful tool for the differentiation of local organic honeys and imported blends.

2.6. Multivariate Statistical Analysis of Obtained Results

Based on the determined parameters the chemometrics analysis was used to separate imported honey blends from raw honey produced by local beekeepers. MANOVA analysis was applied to determine which variables were statistically dependent ($p < 0.05$) in terms of the botanical or geographical origin of honey samples. The moisture and HMF content, as well as DPPH, were significant only in terms of botanical origin. Other variables were statistically significant in both cases, excluding pH value. None of the variables were statistically significant only due to the geographical origin. For the multivariate analysis, only the seven significant variables (free acidity, electrical conductivity, protein content, diastase number, acid phosphatase activity, total phenolic content, and antioxidant activity-FRAP) were used. Principal component analysis (PCA) was performed to analyze the similarities between the tested honey samples and the relationship between defined statistically significant variables. Using the Kaiser criterion, the two principal components (PCs) accounting for 91.87% of the total variance were chosen: PC1 (including protein content, diastase number, acid phosphatase activity, total phenolic content (TPC), and antioxidant activity (FRAP assay)) and PC2 (electrical conductivity and free acidity), explained 70.19 and 21.68% of the variance, respectively (Figure 2a).

The honey samples were divided into four separate groups (Figure 2b). Samples with the low values of tested parameters are located on the upper-right part of the graph. These were mainly imported samples (linden, multifloral, and acacia) which exhibited lower values of studied parameters compared to samples from ecological apiaries, especially the parameters that are responsible for the health-promoting properties of honey (such as enzymatic and antioxidant activity). A particularly significant difference was observed in the case of buckwheat honey (BC vs. B1 and B2 location on the plot) but linden and multifloral honey are also located in two different sides of the plot (MC and LC left side of PC1, M1, M2, L1, and L2 right side of PC1). Buckwheat honey from ecological apiary was characterized by the highest amount of protein and total phenolic content and diastase number, as well as highest antioxidant and acid phosphatase activity. These variables were highly positively correlated, with a Pearson correlation coefficient *r* above 0.86. Honeydew honey samples are located in the bottom part of the graph, which is linked to PC2. These honey samples had the highest electrical conductivity. This parameter was only correlated with free acidity ($r = 0.71$). This means that this variable depended more on botanical origin than the place where the samples were bought.

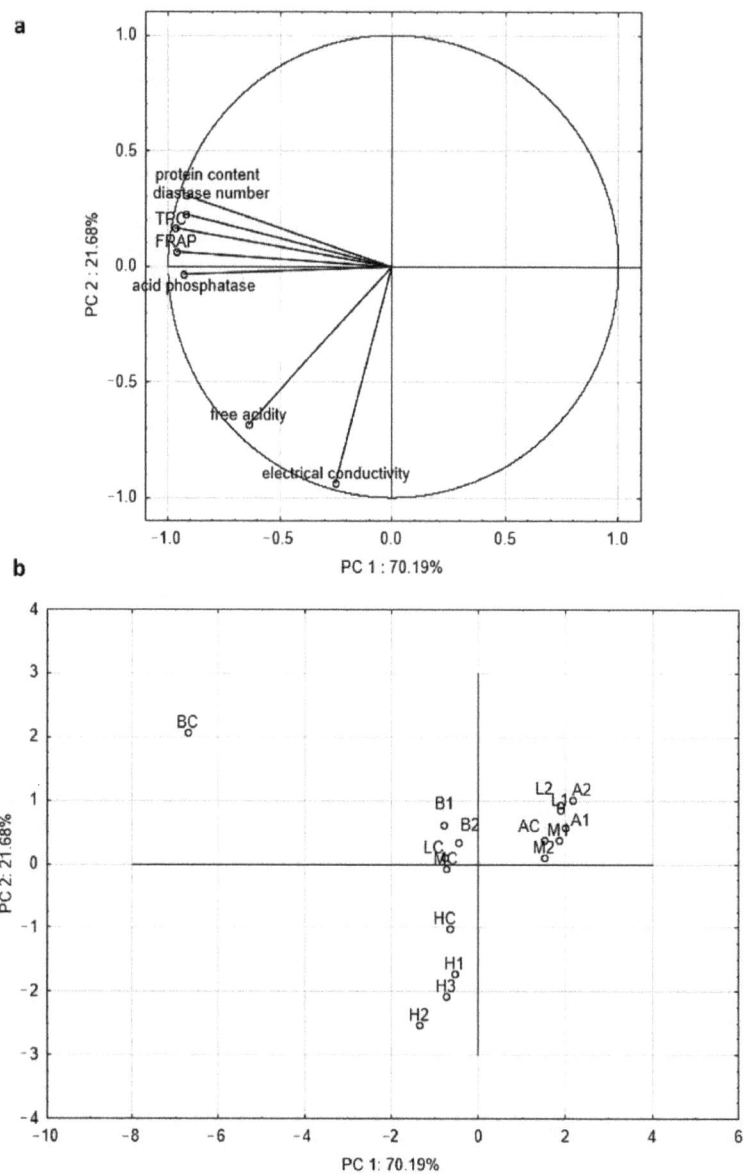

Figure 2. Statistical analysis results: (**a**) projection of chosen variables as a function of the PC1 vs. PC2, (**b**) PCA score plot of imported honey samples available on the Polish market and their ecological local counterparts.

Clear separation of imported samples from those bought in ecological apiaries was not observed, because many variables were varietal-dependent. However, if we consider the analysis of individual honey varieties, variables linked to PC1 can be considered as parameters used for the differentiation of samples of low quality.

The stepwise linear discriminant analysis (LDA) was applied to determine which variables could be used to distinguish imported honey samples available on the Polish market from their ecological local counterparts. The best discriminant variables were selected depending on their influence on the classification of samples based on the Wilks' lambda criterion. Seven significant parameters were used as independent variables, and the geographical origin of the sample was chosen as a dependent variable. The results show that only two variables, protein content and diastase number, were found to be significant ($p < 0.05$) for the discrimination of tested honey samples. One discriminant function was formed: Wilk's lambda = 0.495, χ^2= 9.833, df = 2, and p = 0.07. The discriminant function was used for the classification of honey samples according to the place of origin because it explained 100% of the total variance, providing an eigenvalue higher than 1.

According to the classification matrix, all imported samples were classified correctly (100% correct classification rate), while for local honeys the classification rate was 80%. One sample (acacia honey) was incorrectly classified (Table 3). The tested acacia honey (AC) was characterized by low enzymatic activity and low protein content, at a very similar level as that found in imported honey. Moreover, in PCA this sample was grouped with imported honey. Based on the obtained results, it can be stated that protein content and diastase number could be considered as markers for the identification of poor-quality samples within the specific honey variety. Furthermore, stepwise linear discriminant analysis proved to be an effective tool in distinguishing imported blends and local organic honeys (100% correct classification). Similarly, stepwise LDA was successfully used by Manzanares et al. [6], who differentiated honeydew from blossom honey based on physicochemical parameters.

Table 3. Results of stepwise discriminant analysis (LDA) for all the samples considered of different origin, and classification matrix for individual honey samples.

Original Group	Predicted Classification (Number of Samples)		Correct Classification (%)
	Imported	Local	
imported	11	0	100
local	1	4	80
Total	12	4	93.75

Honey Samples	Original Group	Predicted Classification
A1	imported	imported
A2	imported	imported
AC *	local	imported
B1	imported	imported
B2	imported	imported
BC	local	local
L1	imported	imported
L2	imported	imported
LC	local	local
M1	imported	imported
M2	imported	imported
MC	local	local
H1	imported	imported
H2	imported	imported
H3	imported	imported
HC	local	local

* Incorrect classifications are marked.

3. Materials and Methods

3.1. Honey

Eleven imported honeys available on the Podkarpackie market in 2018 labeled as mixtures of honeys originating in the EU and not originating in the EU were used. As control samples, five raw local honeys produced in organic apiaries in the Podkarpackie Voivodeship were used. Local honeys were selected as representative samples based on our earlier study. Honey was kept in a dark place at room temperature until analysis. The used markings, varieties, origin, and appearance of tested honeys samples are listed in Table 4.

Table 4. List of honeys used for research.

Symbol	Variety	Type	Origin	Color	Crystallization State
A1	acacia	blend, imported	from outside the EU	white	liquid
A2	acacia	blend, imported	EU and from outside the EU	white	liquid
AC	acacia	raw, local	Podkarpackie, Poland	white	liquid
B1	buckwheat	blend, imported	EU and from outside the EU	dark amber	partially crystallized
B2	buckwheat	blend, imported	EU and from outside the EU	dark amber	crystallized
BC	buckwheat	raw, local	Podkarpackie, Poland	dark amber	partially crystallized
L1	linden	blend, imported	EU and from outside the EU	white	half crystallized
L2	linden	blend, imported	EU and from outside the EU	extra light amber	half crystallized
LC	linden	raw, local	Podkarpackie, Poland	extra light amber	liquid
M1	multifloral	blend, imported	EU and from outside the EU	amber	crystallized
M2	multifloral	blend, imported	EU and from outside the EU	light amber	crystallized
MC	multifloral	raw, local	Podkarpackie, Poland	light amber	crystallized
H1	honeydew	blend, imported	EU and from outside the EU	dark amber	crystallized
H2	honeydew	blend imported	EU	dark amber	liquid
H3	honeydew	blend, imported	EU	dark amber	liquid
HC	honeydew	raw, local	Podkarpackie, Poland	dark amber	crystallized

3.2. Refractometric Determination of the Water Content in Honey

The determination of the water content was done by the refractometric method, using a RHN1-ATC refractometer (refraktometr.eu, Hradec Kralove, Czech Republic).

3.3. Active and Free Acidity

For the determination of acidity, 20% solutions of honey in distilled water were prepared. To determine active acidity, a pH measurement was used using a CP-401 pH meter (Elmetron, Zabrze, Poland). To determine the free acidity, 50 mL of 20% honey solution was titrated by 0.1 M NaOH to reach a pH of 8.3 measured by pH meter. The results were expressed in mval/kg.

3.4. Conductivity

To determine the specific electrical conductivity, 20% solutions of honey in distilled water were used. The conductivity of each honey solution was measured using a conductometer CP-401 (Elmetron, Zabrze, Poland) and the results (in mS/cm) were calculated using a conductivity constant ($K = 0.938$ cm^{-1}).

3.5. HMF Determination

The determination of 5-hydroxymethylfurfural (HMF) content was carried out by HPLC in accordance with the guidelines of the regulations of the Polish Ministry of Agriculture and Development of Rural Areas [39]. HPLC analysis was performed at the Plant Biotechnology Laboratory "Aeropolis" using a Gilson chromatograph (Gilson Inc., Middleton, WI, USA) equipped with a binary pump (Gilson 322), DAD detector (Gilson 172), column thermostat (Knauer, Berlin, Germany), and autosampler with a fraction collector (GX-271 Liquid). The separation was carried out using a Knauer Eurosphere II RP-18H 100-5 column (250 × 4.6 mm) with a pre-column (Gilson) at 35 °C, with mobile phase water: methanol (90:10, *v/v*), the isocratic flow was 1 mL/min, analysis time was 15 min, injection volume was 20 µL, and the detection was carried out at wavelength λ = 285 nm. The method was calibrated for the HMF standard (Sigma-Aldrich, Saint Louis, MO, USA) in the range of 0.25 to 6 µg (y = 5123.8x, R^2 = 0.9989).

3.6. Total Protein Content

Protein concentration in the tested honey samples was determined by the Bradford method [40], using 10% solutions of all tested honeys. To 20 µL of honey solution, 1 mL of Bradford's reagent (BioRad, Hercules, CA, USA) was added and mixed thoroughly. Then, after 5 min the absorbance at λ = 595 nm was measured against a blank using a Biomate 3 spectrophotometer (Thermo Scientific, Waltham, MA, USA). The protein content of the samples was calculated based on the calibration curve (y = 0.0011x, R^2 = 0.992) made for bovine albumin in the range of 62.5 to 1000 µg.

3.7. Diastase Number Determination

Diastase number was determined by a spectrophotometric method with the Phadebas Honey Diastase test (Magle AB, Lund, Sweden) according to the manufacturer's instructions. Five ml of a 1% honey solution in 0.1 M acetate buffer was heated for 5 min at 40 °C in a water bath. A Phadebas Honey Diastase test tablet was then added to each sample and after thorough mixing, incubated at 40 °C for 30 min. Then, 1 mL 0.5 M NaOH was added, mixed, and filtered into tubes and the absorbance of the filtrate was measured at wavelength λ = 620 nm against a blank (acetate buffer) using a Biomate 3 spectrophotometer (Thermo Scientific, Waltham, MA, USA). The values of the diastase number were calculated from the formulas: Equation (1) when the value of the diastase number did not exceed 8:

$$DN = 35.2 \times A - 0.46 \tag{1}$$

or Equation (2) when diastase number was above 8:

$$DN = 28.2 \times A + 2.64 \tag{2}$$

3.8. Acid Phosphatase Activity Assay

Acid phosphatase activity was determined using 4-nitrophenyl phosphate as a substrate, according to Alonso Torre et al. [29] with slight modification. The substrate solution of 5 mM in 0.2 M citrate buffer, pH = 4.5, was used. A total of 100 µL of the test sample (20% *w/v* honey in water) was mixed with 100 µL of a substrate and incubated for 10 min at 37 °C. After this time, 1 mL of 0.25 M NaOH was added to all samples and the absorbance at λ = 400 nm was measured using a Biomate 3 spectrophotometer (Thermo Scientific, Waltham, MA, USA). Acid phosphatase activity was expressed in µmol/g × min, using the molar extinction coefficient 18,000 dm^3/mol × cm.

3.9. Total Phenolic Content Determination

Total phenolic content was determined by the Folin–Ciocalteu method as per Singleton and Rossi [41]. In the test tube, 0.2 mL of 5% honey solution, 1 mL of Folin–Ciocalteu reagent (10%), and 0.8 mL of 7.5% Na_2CO_3 were mixed. After 2 h, the absorbance of the test samples against a blank was measured at 760 nm using a Biomate 3 spectrophotometer (Thermo Scientific, Waltham, MA, USA). The total content of polyphenols was expressed as gallic acid equivalents, using a calibration curve made in the concentration range of 0 to 100 mg/dm^3 (y = 0.0555x, R^2 = 0.9976).

3.10. Antioxidant Assays

For antioxidant potential determination, two methods (DPPH radical scavenging test and FRAP reducing power assay), frequently used in honey analyses, were selected.

3.10.1. DPPH Test

DPPH (2, 2-diphenyl-1-picrylhydrazyl) radical inhibition was measured according to the assay described by Dżugan et al. [7]. A solution of DPPH radical (1.8 mL) was added to the proper samples (0.2 mL of 5% honey solution), and after 30 min absorbance (A) was measured at a wavelength λ = 517 nm relative to the control (A0). The percentage inhibition of (% A) DPPH radical was calculated from the Equation (3):

$$[\% A] = (A0 - A)/A0 \cdot 100\% \tag{3}$$

3.10.2. FRAP Assay

The FRAP assay (ferric reducing ability of plasma) was carried out as per Bertoncelj et al. [42]. The FRAP reagent contained 2.5 mL of a 10 mM 2,4,6-tripyridyltriazine (TPTZ) solution in 40 mM HCl, 2.5 mL of 20 mM $FeCl_3$, and 25 mL of 0.3 M acetate buffer (pH 3.6). In the test tube, 0.2 mL of diluted honey (5% in distilled water) was mixed with 1.8 mL of FRAP reagent. After 10 min of incubation at 37 °C, absorbance at λ = 593 nm was measured against a blank with the use of a Biomate 3 spectrophotometer (Thermo Scientific, Waltham, MA, USA). The results were expressed as µmol of Trolox (TE) equivalents per kilogram of honey (µmol/kg), based on the calibration curve (y = 0.026x, R^2 = 0.998) prepared for 0.1 mM Trolox in the range of 15 to 200 nmol.

3.11. Electrophoretic Analyses

3.11.1. Native Protein Electrophoresis

All the honey samples were diluted with water in a ratio of 1 g of honey per 1 mL of deionized water and mixed thoroughly. A pinch of bromphenol blue was added to the samples as an electrophoretic indicator and 20 µL of the prepared sample was placed in each well. Electrophoresis was carried out on polyacrylamide native gels (10% separating gel and 5% stacking gel, both without SDS) using Tris-glycine running buffer in Mini-Protean II apparatus (Bio-Rad Laboratories, Hercules, CA, USA). The separation was carried out for 2.5 h at 100 V and after electrophoresis, the gels were incubated overnight in a colloidal solution of CBB G-250 [43] and then destained with deionized water for 24 h.

3.11.2. Amylase Electrophoretic Detection

Electrophoresis was performed as described above with one modification—separating gel containing 0.2% of starch was prepared for amylase zymography. Staining for amylases was performed according to Rafiei et al. [44] with slight modifications. When the separation was completed the gels were washed twice with 1% Triton X-100 solution and once with water (10 min each time). Next, gels were washed with 0.25 M acetate buffer, pH 5.5, and placed in a heater (37 °C) overnight. During this time the amylases hydrolyzed the starch present in the gel. The next day, the gels were covered with a solution of iodine in potassium iodide (Lugol's solution), which resulted in the dark coloration of the whole gel, except the places where the amylases hydrolyzed the starch.

3.11.3. Acid Phosphatase Electrophoretic Detection

Preparation of samples, gels, and electrophoresis was carried out analogously to that for native proteins. Staining for acid phosphatases was performed as per Kalinowski et al. [45]. After the separation, the gels were rinsed twice with 1% Triton X-100 solution for 10 min and with water, also for 10 min. Next, the gels were covered with a solution of 0.5% α-naphthyl phosphate, 0.01% Fast Blue RR and 0.5% polyvinylpyrrolidone, in 0.01 M acetate buffer, pH 5.5, and left overnight. The next day the gels were rinsed with distilled water. Dark bands were observed in places where the migration of proteins with acid phosphatase activity stopped.

3.12. Statistical Analysis

Each honey sample was analyzed in triplicate and the results were expressed as means ± standard deviations (SD). The significant differences were obtained by t-test ($p < 0.05$). MANOVA was applied to all investigated parameters, as a pre-treatment procedure, to point out the significant data ($p < 0.05$) according to sample origin. Principal component analysis (PCA) and linear discriminant analysis (LDA) using the stepwise method were carried out to differentiate imported honey from raw local honey. The correlation between studied parameters was calculated using Pearson's correlation test. Statistical analysis was perform using Statistica 13.1 software (StatSoft, Inc., Tulsa, OK, USA).

4. Conclusions

Comparing raw local honeys with their imported counterparts (blends) within the same variety, significant differences were found, mainly in antioxidant potential, enzymatic activities, and HMF content, in favor of local honeys. Moreover, Polish honeys were characterized by higher protein content regardless of the variety, excluding acacia honey, which was also confirmed using the native PAGE method. For the first time, the possibility of zymographic identification of native amylase and acid phosphatase isoenzymes in honey was proposed. The usefulness of in-depth protein analysis to assess the quality of honey and to distinguish local and imported honeys was confirmed by multivariate statistical analysis (PCA and LDA). Considering the antioxidant properties and enzymatic activity of honey, responsible for its pro-health value which can be reduced during thermal processing, it can be concluded that raw local honeys are more valuable products than imported honey blends.

Author Contributions: Conceptualization, M.M. and M.D.; methodology, M.M. and A.B.; software, M.M. and P.S.; validation, M.M. and A.B.; formal analysis, M.M. and P.S.; investigation, M.M., E.K., P.S. and A.B.; resources, M.M.; data curation, P.S.; writing—original draft preparation, M.M. and P.S.; writing—review and editing, M.D. and A.B.; visualization, M.M. and P.S.; supervision, M.D.; project administration, M.D.; funding acquisition, M.D. All authors have read and agreed to the published version of the manuscript.

Funding: This research was funded by the project financed under the program of the Minister of Science and Higher Education entitled "Regional Initiative of Excellence" in 2019–2022 (project no. 026/RID/2018/19). Amount of financing: 9 542 500 PLN.

Institutional Review Board Statement: Not applicable.

Informed Consent Statement: Not applicable.

Data Availability Statement: The data presented in this study are available in the article.

Conflicts of Interest: The authors declare no conflict of interest.

Sample Availability: Samples of honeys are available from the authors for a limited time.

References

1. Zábrodská, B.; Vorlová, L. Adulteration of honey and available methods for detection—A review. *Acta Vet. Brno.* **2014**, *83*, S85–S102. [CrossRef]
2. Directive 2014/63/EU of the European Parliament and of the Council amending Council Directive2001/110/EC relating to honey. *Off. J. Eur. Communities* **2014**, *57*, L164/1.
3. Dżugan, M.; Ruszel, A.; Tomczyk, M. Quality of imported honeys obtainable on the market in the podkarpacie region. *Żywn. Nauk. Technol. Jakosc/Food. Sci. Technol. Qual.* **2018**, *25*, 127–139. [CrossRef]
4. Asha'Ari, Z.A.; Ahmad, M.Z.; Wan Din, W.S.J.; Che Hussin, C.M.; Leman, I. Ingestion of honey improves the symptoms of allergic rhinitis: Evidence from a randomized placebo-controlled trial in the East Coast of Peninsular Malaysia. *Ann. Saudi Med.* **2013**, *33*, 469–475. [CrossRef]
5. Newman, T.J. Local, Unpasteurized Honey as a Treatment for Allergic Rhinitis: A Systematic Review Local, Unpasteurized Honey as a Treatment for Allergic Rhinitis. Master's Thesis, Pacific University Oregon, Forest Grove, OR, USA, 2014.

6. Majewska, E.; Drużyńska, B.; Wołosiak, R. Determination of the botanical origin of honeybee honeys based on the analysis of their selected physicochemical parameters coupled with chemometric assays. *Food Sci. Biotechnol.* **2019**, *28*, 1307–1314. [CrossRef] [PubMed]
7. Dżugan, M.; Tomczyk, M.; Sowa, P.; Grabek-Lejko, D. Antioxidant activity as biomarker of honey variety. *Molecules* **2018**, *23*, 2069. [CrossRef] [PubMed]
8. Kaczmarek, A.; Muzolf-Panek, M.; Tomaszewska-Gras, J.; Konieczny, P. Predicting the botanical origin of honeys with chemometric analysis according to their antioxidant and physicochemical properties. *Pol. J. Food Nutr. Sci.* **2019**, *69*, 191–201. [CrossRef]
9. Mai, Z.; Lai, B.; Sun, M.; Shao, J; Guo, L. Food adulteration and traceability tests using stable carbon isotope technologies. *Trop. J. Pharm. Res.* **2019**, *18*, 1771–1784. [CrossRef]
10. Wu, L.; Du, B.; Vander Heyden, Y.; Chen, L.; Zhao, L.; Wang, M.; Xue, X. Recent advancements in detecting sugar-based adulterants in honey—A challenge. *Trac. Trends Anal. Chem.* **2017**, *86*, 25–38. [CrossRef]
11. Guler, A.; Kocaokutgen, H.; Garipoglu, A.V.; Onder, H.; Ekinci, D.; Biyik, S. Detection of adulterated honey produced by honeybee (*Apis mellifera* L.) colonies fed with different levels of commercial industrial sugar (C3 and C4 plants) syrups by the carbon isotope ratio analysis. *Food Chem.* **2014**, *155*, 155–160. [CrossRef]
12. Baroni, M.V.; Chiabrando, G.A.; Costa, C.; Wunderlin, D.A. Assessment of the floral origin of honey by SDS-page immunoblot techniques. *J. Agric. Food Chem.* **2002**, *50*, 1362–1367. [CrossRef] [PubMed]
13. Bogdanov, S. Physical properties of honey: Honey composition. In *The Honey Book*; University of Belgrade: Belgrade, Serbia, 2011.
14. Machado De-Melo, A.A.; Almeida-Muradian, L.B.D.; Sancho, M.T.; Pascual-Maté, A. Composition and properties of *Apis mellifera* honey: A review. *J. Apic. Res.* **2018**, *57*, 5–37. [CrossRef]
15. Thrasyvoulou, A.; Tananaki, C.; Goras, G.; Karazafiris, E.; Dimou, M.; Liolios, V.; Kanelis, D.; Gounari, S. Legislation of honey criteria and standards. *J. Apic. Res.* **2018**, *57*, 88–96. [CrossRef]
16. Bocian, A.; Buczkowicz, J.; Jaromin, M.; Hus, K.K.; Legáth, J. An Effective Method of Isolating Honey Proteins. *Molecules* **2019**, *24*, 2399. [CrossRef]
17. Wesołowska, M.; Dżugan, M. Activity and thermal stability of diastase present in honey from Podkarpacie region. *Zywn. Nauk. Technol. Jakosc/Food. Sci. Technol. Qual.* **2017**, *24*, 103–112. [CrossRef]
18. Tomczyk, M.; Tarapatskyy, M.; Dżugan, M. The influence of geographical origin on honey composition studied by Polish and Slovak honeys. *Czech J. Food Sci.* **2019**, *37*, 232–238. [CrossRef]
19. Eshete, Y.; Eshete, T. A Review on the Effect of Processing Temperature and Time duration on Commercial Honey Quality. *Madr. J. Food Technol.* **2019**, *4*, 158–162. [CrossRef]
20. Sanz, M.L.; Del Castillo, M.D.; Corzo, N.; Olano, A. 2-Furoylmethyl Amino Acids and Hydroxymethylfurfural As Indicators of Honey Quality. *J. Agric. Food Chem.* **2003**, *51*, 4278–4283. [CrossRef] [PubMed]
21. Samarghandian, S.; Farkhondeh, T.; Samini, F. Honey and Health: A Review of Recent Clinical Research. *Pharmacogn. Res.* **2017**, *9*, 121–127.
22. Dżugan, M.; Grabek-Lejko, D.; Swacha, S.; Tomczyk, M.; Bednarska, S.; Kapusta, I. Physicochemical quality parameters, antibacterial properties and cellular antioxidant activity of Polish buckwheat honey. *Food Biosci.* **2020**, *34*, 100538. [CrossRef]
23. Pauliuc, D.; Dranca, F.; Oroian, M. Antioxidant activity, total phenolic content, individual phenolics and physicochemical parameters suitability for Romanian honey authentication. *Foods* **2020**, *9*, 306. [CrossRef]
24. Cabrera, M.; Perez, M.; Gallez, L.; Andrada, A.; Balbarrey, G. Colour, antioxidant capacity, phenolic and flavonoid content of honey from the Humid Chaco Region, Argentina. *Phyton* **2017**, *86*, 124–130. [CrossRef]
25. Cimpoiu, C.; Hosu, A.; Miclaus, V.; Puscas, A. Determination of the floral origin of some Romanian honeys on the basis of physical and biochemical properties. *Spectrochim. Acta Part A Mol. Biomol. Spectrosc.* **2013**, *100*, 149–154. [CrossRef] [PubMed]
26. Flanjak, I.; Strelec, I.; Kenjerić, D.; Primorac, L. Croatian produced unifloral honeys characterised according to the protein and proline content and enzyme activities. *J. Apic. Sci.* **2016**, *60*, 39–48. [CrossRef]
27. White, J.W. Quality evaluation of honey: Role of HMF and diastase assays. *Am. Bee J.* **1992**, *132*, 737–743. [CrossRef]
28. Bonta, V.; Dezmirean, D.S.; Marghitas, L.A.; Urcan, A.C.; Bobis, O. Sugar Spectrum, Hydroxymethylfurfural and Diastase Activity in Honey: A Validated Approach as Indicator of Possible Adulteration. *Bull. Univ. Agric. Sci. Vet. Med. Cluj-Napoca. Anim. Sci. Biotechnol.* **2020**, *77*, 44. [CrossRef]
29. Alonso-Torre, S.R.; Cavia, M.M.; Fernández-Muiño, M.A.; Moreno, G.; Huidobro, J.F.; Sancho, M.T. Evolution of acid phosphatase activity of honeys from different climates. *Food Chem.* **2006**, *97*, 750–755. [CrossRef]
30. Dżugan, M.; Wesołowska, M. *Jakość Miodów Produkowanych na Podkarpaciu (The Quality of Honeys Produced in Podkarpackie Region, in Polish)*; Uniwersytet Rzeszowski: Rzeszów, Poland, 2016; pp. 52–54.
31. Saxena, S.; Gautam, S. A Preliminary Cost Effective Qualitative Assay for Diastase Analysis in Honey. *Insights Enzym. Res.* **2018**, *01*, 1–4. [CrossRef]
32. Voldřich, M.; Rajchl, A.; Čižkova, H.; Cuhra, P. Detection of Foreign Enzyme Addition into the Adulterated Honey. *Czech J. Food Sci.* **2009**, *27*, S280–S282. [CrossRef]
33. Jimenez, D.R.; Gilliam, M. Ultrastructure of the ventriculus of the honey bee, *Apis mellifera* (L.): Cytochemical localization of acid phosphatase, alkaline phosphatase, and nonspecific esterase. *Cell Tissue Res.* **1990**, *261*, 431–443. [CrossRef]
34. Borutinskaite, V.; Treigyte, G.; Čeksteryte, V.; Kurtinaitiene, B.; Navakauskiene, R. Proteomic identification and enzymatic activity of buckwheat (*Fagopyrum esculentum*) honey based on different assays. *J. Food Nutr. Res.* **2018**, *57*, 57–69.

15. Nisbet, C.; Guler, A.; Ciftci, G.; Yarim, G.F. The Investigation of Protein Prophile of Different Botanic Origin Honey and Density Saccharose-Adulterated Honey by SDS-PAGE Method. *Kafkas Univ. Vet.Fak. Derg.* **2009**, *15*, 443–446.
16. Wang, J.; Kliks, M.M.; Qu, W.; Jun, S.; Shi, G.; Li, Q.X. Rapid determination of the geographical origin of honey based on protein fingerprinting and barcoding using MALDI TOF MS. *J. Agric. Food Chem.* **2009**, *57*, 10081–10088. [CrossRef]
17. Di Girolamo, F.; D'Amato, A.; Righetti, P.G. Assessment of the floral origin of honey via proteomic tools. *J. Proteom.* **2012**, *75*, 3688–3693. [CrossRef] [PubMed]
18. Azevedo, M.S.; Valentim-Neto, P.A.; Seraglio, S.K.T.; da Luz, C.F.P.; Arisi, A.C.M.; Costa, A.C.O. Proteome comparison for discrimination between honeydew and floral honeys from botanical species *Mimosa scabrella* Bentham by principal component analysis. *J. Sci. Food Agric.* **2017**, *97*, 4515–4519. [CrossRef] [PubMed]
19. Regulation 2009/17/94 of the Minister of Agriculture and Rural Development of Poland regarding methods analyzes related to honey assessment. *J. Laws Repub. Poland* **2009**, *17*, 94.
20. Bradford, M.M. A rapid and sensitive method for the quantitation of microgram quantities of protein utilizing the principle of protein—Dye binding. *Anal. Biochem.* **1979**, *10*, 248–254. [CrossRef]
21. Singleton, V.L.; Rossi, J.A. Colorimetry of Total Phenolics with Phosphomolybdic-Phosphotungstic Acid Reagents. *Am. J. Enol. Vitic.* **1965**, *16*, 144–158.
22. Bertoncelj, J.; Doberšek, U.; Jamnik, M.; Golob, T. Evaluation of the phenolic content, antioxidant activity and colour of Slovenian honey. *Food Chem.* **2007**, *105*, 822–828. [CrossRef]
23. Neuhoff, V.; Arold, N.; Taube, D.; Ehrhardt, W. Improved staining of proteins in polyacrylamide gels including isoelectric focusing gels with clear background at nanogram sensitivity using Coomassie Brilliant Blue G-250 and R-250. *Electrophoresis* **1988**, *9*, 255–262. [CrossRef]
24. Rafiei, B.; Ghadamyari, M.; Imani, S.; Hosseininaveh, V.; Ahadiyat, A. Purification and characterization of α-amylase in Moroccan locust, *Dociostaurus maroccanus* Thunberg (Orthoptera: Acrididae) and its inhibition by inhibitors from *Phaseolus vulgaris* L. *Toxin Rev.* **2016**, *35*, 90–97. [CrossRef]
25. Kalinowski, A.; Bocian, A.; Kosmala, A.; Winiarczyk, K. Two-dimensional patterns of soluble proteins including three hydrolytic enzymes of mature pollen of tristylous *Lythrum salicaria*. *Sex. Plant Reprod.* **2007**, *20*, 51–62. [CrossRef]

Article

LC-ESI/LTQ-Orbitrap-MS Based Metabolomics in Evaluation of Bitter Taste of *Arbutus unedo* Honey

Paola Montoro [1], Gilda D'Urso [1], Adam Kowalczyk [2] and Carlo Ignazio Giovanni Tuberoso [3,*]

[1] Department of Pharmacy, University of Salerno, via Giovanni Paolo II, 132, 84084 Fisciano, Italy; pmontoro@unisa.it (P.M.); gidurso@unisa.it (G.D.)
[2] Department of Pharmacognosy and Herbal Medicines, Wrocław Medical University, ul. Borowska, 211, 50-556 Wrocław, Poland; adam.kowalczyk@umed.wroc.pl
[3] Department of Life and Environmental Sciences, University of Cagliari, University Campus, S.P. Monserrato-Sestu km 0.700, 09042 Monserrato, Italy
* Correspondence: tuberoso@unica.it; Tel.: +39-070-675-8644

Abstract: Strawberry tree honey is a high-value honey from the Mediterranean area and it is characterised by a typical bitter taste. To possibly identify the secondary metabolites responsible for the bitter taste, the honey was fractionated on a C18 column and the individual fractions were subjected to sensory analysis and then analysed by liquid chromatography coupled with high-resolution tandem mass spectrometry in negative ion mode, using a mass spectrometer with an electrospray source coupled to a hybrid high resolution mass analyser (LC-ESI/LTQ-Orbitrap-MS). A chemometric model obtained by preliminary principal component analysis (PCA) of LC-ESI/LTQ-Orbitrap-MS data allowed the identification of the fractions that caused the perception of bitterness. Subsequently, a partial least squares (PLS) regression model was built. The studies carried out with multivariate analysis showed that unedone (2-(1,2-dihydroxypropyl)-4,4,8-trimethyl-1-oxaspiro [2.5] oct-7-en-6-one) can be considered responsible for the bitter taste of strawberry tree honey. Confirmation of the bitter taste of unedone was obtained by sensory evaluation of a pure standard, allowing it to be added to the list of natural compounds responsible for giving the sensation of bitterness to humans.

Keywords: unedone; bitter taste; strawberry tree honey; LC-ESI/LTQ-Orbitrap-MS; PCA; PLS

1. Introduction

Strawberry tree (*Arbutus unedo* L.) (ST) honey is a peculiar bitter tasting honey produced in the area of the Mediterranean basin. This highly valuable honey is well recognised for its antioxidant [1–3], anti-inflammatory, and antimicrobial activities [4], and shows anti-xanthine oxidase and antityrosinase activities [5]. Several studies have been performed to characterise the chemical composition of ST honey, which is characterised by phenolic compounds (hydroxy derivatives of benzoic and cinnamic acids, and flavonoids), isoprenoids, and free amino acids [6,7]. Homogentisic acid (2,5-dihydroxyphenylacetic, HGA), unedone (2-(1,2-dihydroxypropyl)-4,8,8-trimethyl-1-oxaspiro [2.5]oct-4-en-6-one), (±)-2-*cis*, 4-*trans*-abscisic acid (*c*,*t*-ABA), and (±)-2-*trans*, 4-*trans*-abscisic acid (*t*,*t*-ABA) have been considered as chemical markers of the botanical origin of ST honey [8,9].

Although ST honey has been investigated for its chemical composition, no studies have been reported as regards the possible compounds responsible for its bitter taste so far. Arbutin, the glucosylated form of hydroquinone, has been suggested as possibly responsible for the bitter taste of ST honey [4]. However, although the bitter taste perception of arbutin is very strong [10] and this compound can be found abundantly in *A. unedo* plant, its presence in ST honey is variable, often insignificant, and sometimes it is totally absent [4]. Thus, the involvement of arbutin in the bitter taste of ST honey is very unlikely. Literature data indicate many plant-derived bitter-tasting compounds [11,12]. They can be represented by alkaloids, terpenoids, phenols, amino acids, and peptides, and they activate the bitter taste receptors (T2R). Comparison of the known natural compounds responsible for the bitter taste with the molecules reported in the literature for ST honey did not make it possible to speculate which compound was responsible for the typical bitter taste of this honey.

Given the above, investigation of ST honey using the metabolomic approach could help in detecting the compounds responsible for the bitter taste in ST honey. Metabolomics, as an emerging discipline of omics science, is a valid tool for the characterization of complex biological samples as it allows the production of a molecular fingerprint for samples by using innovative analytical techniques, such as mass spectrometry (MS) and nuclear magnetic resonance (NMR) [13,14]. Particularly, liquid chromatography-high resolution mass spectrometry metabolic profiling has begun to be used to discover possible markers in foods, especially those most likely responsible for the biological activities [15–18].

The aim of this study was to develop an LC-ESI/LTQ-Orbitrap-MS based metabolomic approach in the evaluation of the compounds responsible for the bitter taste of *Arbutus unedo* honey. To this purpose, polar compounds from strawberry tree honey were separated on a column containing C18 resin and the different fractions obtained were submitted to sensory analysis and investigated by high resolution mass spectrometry ((HR) LC-ESI/LTQ-Orbitrap-MS, (HR) LC-ESI/LTQ-Orbitrap-MS/MS) and by HPLC-DAD. In addition, principal component analysis (PCA) of LC-ESI/LTQ-Orbitrap-MS data and a partial least squares (PLS) regression model were used to identify compounds responsible for the bitter taste. Finally, sensorial analysis on targeted pure compounds was performed to confirm the molecule responsible for the bitter taste in strawberry tree honey.

2. Results and Discussion

2.1. LC-ESI/LTQ-Orbitrap-MS

The 21 fractions obtained from the fractioning on C18 column were tasted to evaluate the impact of bitterness and were classified by comparison with quinine solutions (see Section 3.3) in five groups of bitter perception (from 0 to 4) as follows: no bitter taste (0, fractions A, B and C), barely detectable (1, fractions D, E, M, N, O, Q, R, S, T and U), weak perception (2, fractions F and V), moderate perception of bitterness (3, fractions G, H, I, L and P), and strong perception (4, fraction Z) (Table S1). Fraction E showed an LC-MS profile similar to fraction D, fractions M, N, O, Q, R, S, T, and U showed profiles similar to fraction P, while fractions G and I showed profiles similar to fraction F; therefore, only representative fractions with different profiling characteristics are shown in Figure 1.

The accurate mass measurement (ppm ≤ 5) and the MS/MS experiments in negative ionization mode, together with the comparison with the data present in the literature and databases, such as KNApSAcK [19], allowed to identify 29 chemical constituents reported in Table 1. Analysis of the samples was performed also in positive ion mode, but the better answer from the instrument was in negative ion mode. For this reason, the study was carried out using the negative polarity. An identification level was assigned to each sample, referring to the usual four levels of identification in metabolomic analyses [20].

Table 1. Chemical compounds identified in fractions of strawberry-tree honey by LC-ESI/LTQ-Orbitrap-MS and LC-ESI/LTQ-Orbitrap-MS/MS.

N°	Rt	[M − H]	Molecular Formula	ppm	Identification	MS/MS	Fraction	L.I.	Reference
1	3.97	329.0868	$C_{14}H_{18}O_9$	0.4	glucopiranosyl vanillic acid	167	A	2	[21]
2	6.38	167.0346	$C_8H_8O_4$	4.3	homogenistic acid	123	A	1	[8]
3	6.86	481.1310	$C_{22}H_{26}O_{12}$	4.9	arbutin peracetate	271	A	2	[22]
4	10.27	285.1333	$C_{14}H_{22}O_6$	0.1	methacrylic acid, diester with triethylene glycol	-	L	3	[23]
5	11.34	199.0972	$C_{10}H_{16}O_4$	3.7	camphoric acid	155	B, Z	2	[24]
6	13.76	301.1798	$C_{19}H_{26}O_3$	−0.03	allethrin	133	V	2	[25]
7	13.85	275.1280	$C_{16}H_{20}O_4$	0.9	propenoic acid, dimethoxyphenyl-methyl-butenyl ester	71	D	3	[26]
8	14.32	447.1277	$C_{22}H_{24}O_{10}$	−1.4	sakuranin	285	Z	2	[27]
9	14.62	303.1228	$C_{17}H_{20}O_5$	0.8	(±)-oleocanthal isomer	137/119	F	3	[28]
10	14.71	263.1278	$C_{15}H_{20}O_4$	0.9	(⊥)-2-cis, 4-trans-abscisicacid (c,t-ABA)	219/204/153	A/B/C	1	[9]
11	15.05	335.1126	$C_{17}H_{20}O_7$	0.3	tutin, 6-acetate	293	Z	3	[29]
12	15.40	153.0922	$C_9H_{14}O_2$	4.8	2-hydroxyisophorone	135	A/B/C/D	2	[30]
13	15.44	219.1385	$C_{14}H_{20}O_2$	2.7	di-*tert*-butyl-benzoquinone	107	L	2	[24]
14	15.78	263.1281	$C_{15}H_{20}O_4$	1	(±)-2-*trans*, 4-*trans*-abscisic acid (t,t-ABA)	219/204/153	A/B/C/D/E/F/H	1	[9]
15	15.83	239.091	$C_{13}H_{20}O_4$	1.2	unedone	151/107	Z	2	[9]
16	15.83	359.1489	$C_{20}H_{24}O_6$	0.29	triptolide	340/329/311	Z	2	[31]
17	16.18	287.1642	$C_{18}H_{24}O_3$	−0.2	estriol	171	V	2	[32]
18	16.39	415.2107	$C_{24}H_{32}O_6$	−1.6	desonide	397	L	2	[33]
19	16.39	201.1280	$C_{14}H_{18}O$	3.5	amylcinnamaldehyde	183	A	2	[34]
20	16.39	219.1386	$C_{14}H_{20}O_2$	2.9	di-*tert*-butyl-benzoquinone	107	A	2	[24]
21	16.99	241.1225	$C_{16}H_{18}O_2$	0.9	Bisphenol B	211	P	2	[35]
22	17.04	177.0917	$C_{11}H_{14}O_2$	3.6	4-*tert*-butylbenzoic acid	121	V	2	[36]
23	17.04	417.2269	$C_{24}H_{34}O_6$	−0.9	deoxyphorbol-isobutyrate	347	V	2	[37]
24	19.11	219.1385	$C_{14}H_{20}O_2$	2.4	di-*tert*-butyl-benzoquinone isomer	107	P	2	[24]
25	19.11	415.2110	$C_{24}H_{32}O_6$	−0.8	desonide	397/197	P	2	[33]
26	20.71	325.1438	$C_{20}H_{22}O_4$	0.8	hydroxy-methyl-butenyl-oxyphenyl-ethenyl-methoxyphenol	153	Z	2	[38]
27	21.61	287.2220	$C_{16}H_{32}O_4$	1.1	dihydroxypalmitic acid	147/121/109	Z/V	2	[39]
28	22.60	253.0497	$C_{15}H_{10}O_4$	0.8	chrysin	255/153	Z	2	[40]
29	22.60	437.1952	$C_{26}H_{30}O_6$	−0.7	kurarinone	301	Z	2	[41]

L.I.: Level of identification.

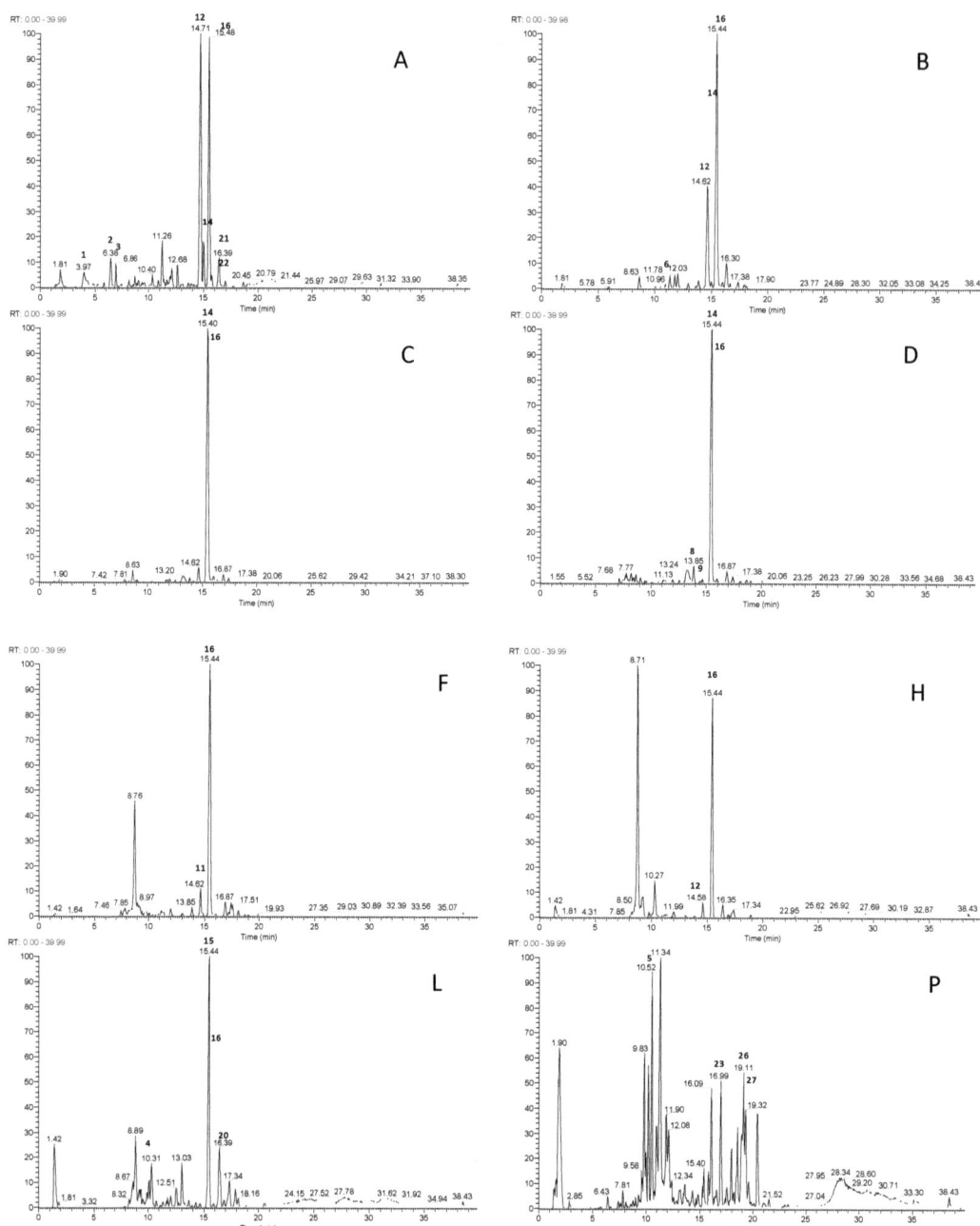

Figure 1. Cont.

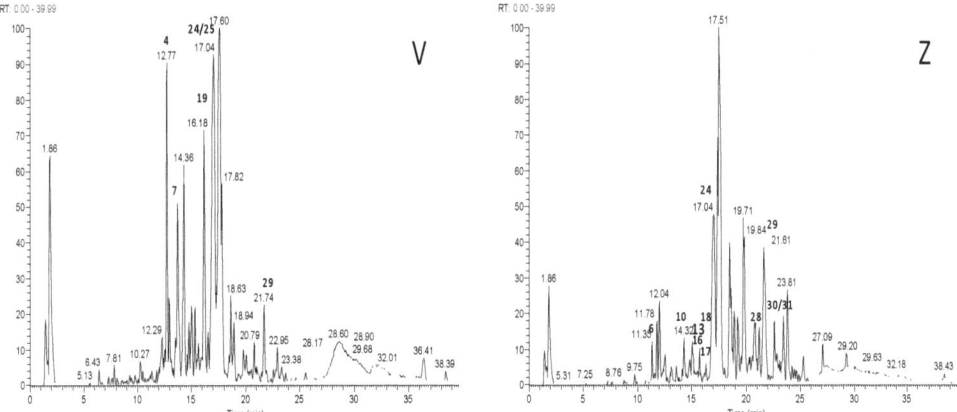

Figure 1. LC-ESI-LTQ-Orbitrap-MS base peak profiles of ten representative strawberry tree honey fractions (A, B, C, D, F, H, L, P, V, Z). Base peak intensity was fixed at NL 7.02E^6 for all the chromatograms.

Compound **2** showed an ion [M − H]$^-$ at m/z 167.0346 corresponding to the molecular formula C$_8$H$_8$O$_4$,. The compound was identified as a homogentisic acid, already reported in ST honey [8] and considered one of the marker compounds to evaluate the botanical origin of this honey along with *c*,*t*-ABA (**10**) and *t*,*t*-ABA (**14**) [9]. Compounds **2**, **10** and **14** were found to be present in fractions A, B, and C, the sweetest fractions. Compound **3** and compound **22** have already been reported in the literature [22]. In particular, compound **3** showed an ion [M − H]$^-$ at m/z 481.1310 corresponding to the molecular formula C$_{22}$H$_{26}$O$_{12}$ and was identified as arbutin peracetate, a compound present in the leaves of *A. unedo* [22]. Compound **22** showed an ion [M − H]$^-$ at m/z 177.0917 corresponding to the molecular formula C$_{11}$H$_{14}$O$_2$ and has been identified as *tert*-butylbenzoic acid [36]. Compounds **5**, **13**, and **24** have been reported in Algerian origin honey and have been identified as camphoric acid, di-*tert*-butyl-benzoquinone and di-*tert*-butyl-benzoquinone isomer, respectively [24]. Compounds **6**, **8**, **12–13**, **15–21**, **23–29** have already been reported to be present in different types of honey. Compound **6** showed an ion [M − H]$^-$ at m/z 301.1798 corresponding to the formula C$_{19}$H$_{26}$O$_3$, and was identified as allethrin, an insecticide already found in other honeys [25]. Compounds **8**, **28**, and **29** are phenolic compounds and were identified, respectively, as sakuranin [27], chrysin [40], and kurarinone [41], and were found to be present in the more bitter fraction Z. Compound **12** was identified as 2-hydroxyisophorone already reported in the ST honey of Sardinia and mostly present in sweet fractions. Finally, Compound **15** showed an ion [M − H]$^-$ at m/z 239.0910 corresponding to the molecular formula C$_{13}$H$_{20}$O$_4$. It was identified as unedone, a compound previously reported in strawberry tree honey [9] and more abundantly present in the bitter fraction Z. Thus, Compound **15** could be one of the metabolites particularly responsible for the bitter taste, along with Compounds **8** and **29**. Interestingly, no arbutin was detected in any of the ST fractions.

2.2. Untargeted Metabolomic Analysis of Strawberry Tree Honey Fractions

For the untargeted approach, the LC-ESI/LTQ-Orbitrap-MS chromatograms were pre-treated using the free software Mzmine [42], to compensate for changes in retention time and m/z ratio values between the chromatograms. The pre-treated chromatograms were exported as a data matrix, with the rows relative to the individual samples and the columns relative to the integrated and normalised peak areas obtained through LC-ESI/LTQ-Orbitrap-MS. The numerical values attributed to the variables were pre-treated through logarithmic transformation. Data transformation is intended to remove unwanted systematic behaviour. The exploratory analysis of the samples in terms of similarity or differences was performed using the PCA projection method. The score scatter plot obtained from PCA is shown in Figure 2. The graph shows good discrimination between the sweet fractions coloured in red at the bottom of the plot and the more bitter fractions located in the upper part of the plot. Therefore, it is the second principal component that has a more pronounced influence on the spatial distribution of the samples in the score scatter plot. PCA remains an unsupervised technique, which, therefore, cannot have predictive value and can only provide us preliminary information on the biochemical markers underlying the classification, by reading the corresponding loading scatter plot (Figure 3), in which the variables corresponding to the m/z values are shown. In particular, variables that contribute most to the differentiation of the samples in the score scatter plot and to their location in a specific area of the space can be highlighted.

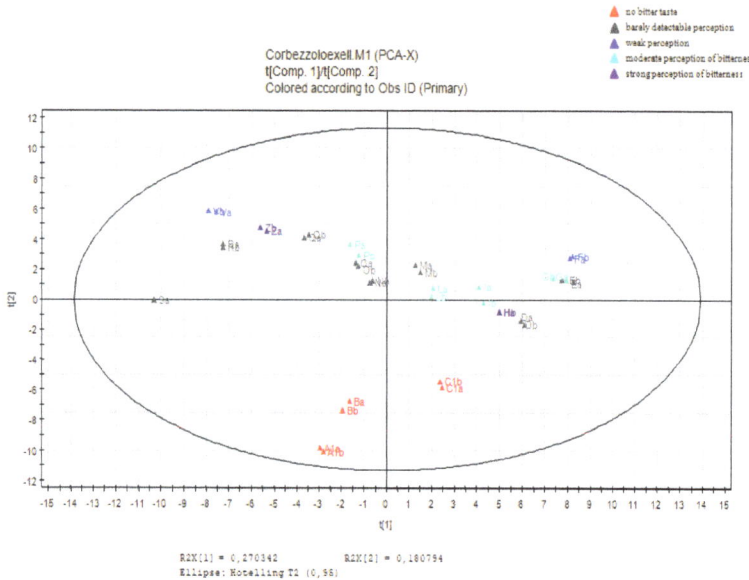

Figure 2. Principal component analysis: score scatter plot obtained from the untargeted analysis of the A-Z fractions.

To classify the samples and to understand which metabolites were most responsible for the bitter taste, the data were statistically processed through another projection technique, the partial least square (PLS) analysis. When the system under consideration is described by a data table and one or more single variable (Y) and the question is what relationship exists between the data block and the single variable (Y), the multivariate technique that is applied is the projection to latent structures by PLS. Therefore, the PLS projection aims to find linear relationships and then to build a plot of the variables that can explain this linear relationship. The model built from the data matrix obtained from the variables, and subsequently transformed and scaled (X), correlated with the Y relative to the perception of bitter taste, was then validated as suggested by Schievano et al. [43] and Stocchero [44], through the analysis of Q2, whose value was higher than 0.5, and through the permutation test, a test through which it is possible to evaluate the randomness and the presence of overfitting in the model.

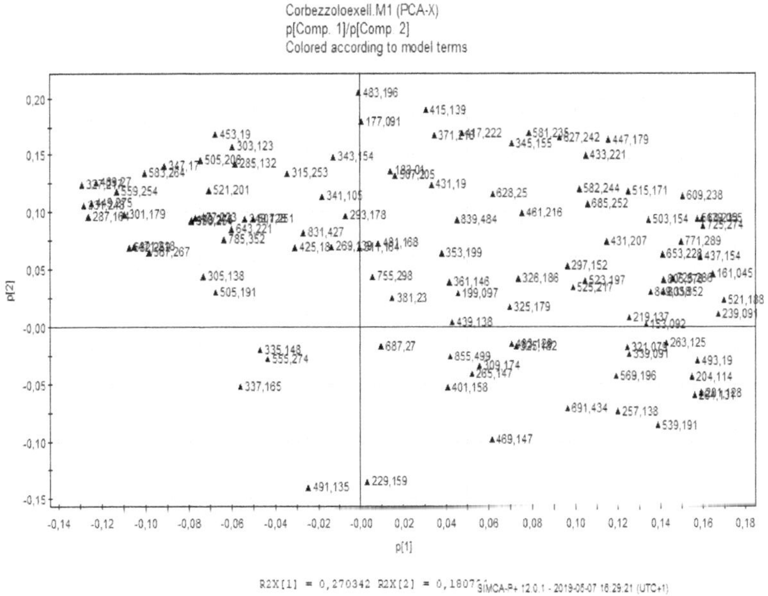

Figure 3. Principal component analysis: loading scatter plot obtained from the untargeted analysis of the A–Z fractions.

As regards the fractions of ST honey, in the space built between t1 and t2 the samples (observations) seem to line up according to a linear relationship that sees in the positions at the top right of the plot the fractions capable of giving the most decisive sensation of bitter taste, and in the lower left the fractions capable of giving the most decisive sensation of sweet taste. The linear relationship can be seen in the score scatter plot (Figure 4). From the analysis of the loading scatter plots (Figure 5), in addition to the construction of a w*c plot, it is possible to evaluate the effect of the variables of the X block on the Y response and, in the specific case, determine the values of m/z observed in the chemical profiling that cause the fractions to have more or less pronounced Y responses (perception of the bitter taste).

From this analysis, the variables found in the loading plot area closest to the Y variable are those that have the highest positive coefficient in the model. From the observation of the LSP, it seems that the influence of *t,t*-ABA, *c,t*-ABA, and homogentisic acid, described as floral markers of ST honey [9], is decidedly significant, being positioned in the central part of the plot and at a fair distance from the variable Y. Other compounds appear to be relevant in the upper right quadrant of the loading scatter plot, namely unedone (m/z 239.0910), sakuranin (m/z 447.1277), and kurarinone (m/z 437.1952).

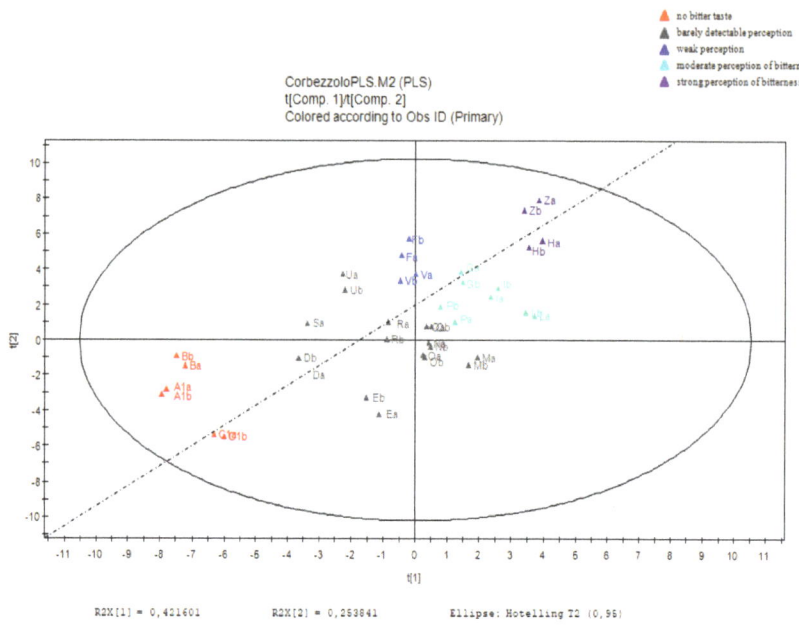

Figure 4. Partial least square: score scatter plot. Untargeted analysis of A–Z fractions.

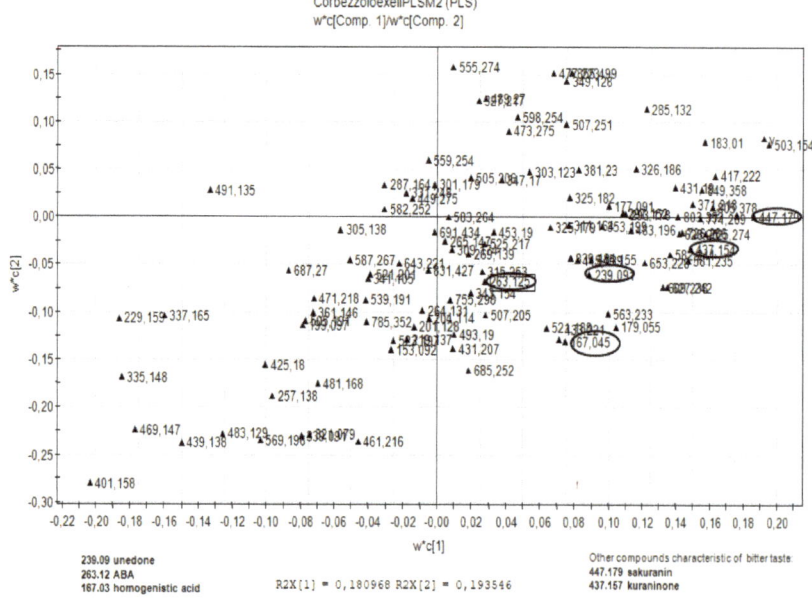

Figure 5. Partial least squares: loading scatter plot. Untargeted analysis of A–Z fractions.

2.3. Quantitative Analysis of Isoprenoid Compounds by HPLC-DAD

(HR) LC-ESI/LTQ-Orbitrap-MS analyses allowed the identification of the four typical ST honey floral markers (homogentisic acid, c,t-ABA, t,t-ABA, and unedone), and thus quantitative HPLC-DAD analysis was used to evaluate whether their variability in the different fractions was connected with the perception of the bitter taste. Homogentisic acid was excluded from this evaluation because it was known that the compound does not have a bitter taste [8]. Table 2 shows the average content of the single isoprenoid compounds (± standard deviation) in the different fractions and expressed as mg of the active ingredient in 1 g of dry fraction. As shown in Table 2, the floral markers c,t-ABA and t,t-ABA are mainly found in the fractions characterised by a less bitter taste, while unedone has been separated mainly in the Z fraction, which was that represented by a more intense sensation of bitterness (Figure S1).

Table 2. Compounds quantified by HPLC-DAD at λ = 262 nm (mg/g, n = 3).

Sample	Bitter Taste [a]	c,t-ABA	t,t-ABA [b]	Unedone [b]
A	0	28.39 ± 1.59	12.36 ± 1.62	nd
B	0	46.14 ± 2.03	91.98 ± 2.05	nd
C	0	11.29 ± 1.05	25.02 ± 1.87	nd
D	1	tr	54.32 ± 2.60	nd
E	1	nd	29.61 ± 1.43	nd
F	2	nd	94.59 ± 1.55	nd
H	3	nd	21.40 ± 1.75	tr
L	3	nd	tr	tr
Z	4	nd	nd	7.10 ± 0.50
LOD (mg/L)		0.4	0.6	0.3
LOQ (mg/L)		1.2	1.9	0.9

[a] Level of bitter perception: (0) no bitter taste; (1) barely detectable; (2) weak, (3) moderate (4) strong; [b] dosed with c,t-ABA calibration curve; nd: not detected (<LOD); traces (<LOQ).

Unedone, an epoxidised derivative of abscisic acid, seems to play a significant role in the bitter taste of ST honey, being positioned in the upper right quadrant of the loading scatter plot. For this reason, a commercial standard of unedone was submitted to sensory analysis. Interestingly, unedone was found to be bitter, and a 300 mg/L solution of unedone gave a level of bitter sensation similar to 10 mg/L quinine. Similarly, c,t-ABA and t,t-ABA were submitted to sensory analysis, but no bitter taste was detected for up to 1000 mg/L solutions. Pydi et al. [45] investigated several abscisic acid (ABA) precursors and metabolites on the T2R4 receptor. It was observed that the structure deeply affects the bitter taste perception. For instance, ABA acts as an antagonist for T2R4, while xanthoxin is an agonist. Interestingly, both unedone and xanthoxin present an epoxide structure. Thus, it can be speculated that unedone acts as an agonist on T2R4 receptor.

In addition to unedone, two other compounds of a flavonoid nature present in very low quantities in ST honey fraction Z were also worthy of interest, namely sakuranin and kurarinone. Indeed, several compounds of flavonoid nature show the property of inducing the perception of bitterness [11]; thus, sakuranin and kurarinone are presumed to be potential sensory biomarkers. Sakuranin is a flavanone, a glucosylated derivative of sakuranetin. Unfortunately, no pure standard of sakuranin was found from suppliers of chemicals, so it was not possible to evaluate its involvement in the bitter perception. However, sakuranetin, the aglycone of sakuranin, was submitted to sensory analysis and no bitter taste was detected for this compound up to 1000 mg/L solutions. Sakuranin, like other flavanones, shows inhibitory activity against acetylcholinesterases [46], enzymes involved in the typical neurodegeneration of Alzheimer's disease. Finally, a pure standard of kurarinone was submitted to sensory analysis and no bitter taste was detected for up to 1000 mg/L solutions also for this flavonoid. Kurarinone has shown good anti-tumour activity against non-small cell lung cancer (NSCLC). The activity has been demonstrated both in vitro and in vivo and appears to be related to the induction of apoptosis in A549 cells [47]. Another property attributed to kurarinone is its antifibrotic activity in the treatment of hepatitis B [48].

To conclude, unedone can be considered responsible for the bitter taste of *A. unedo* honey due to its constant and abundant presence already reported in ST honey (30–

50 mg/kg) [9,49], although other natural compounds can modulate its bitter perception in this unifloral honey.

This developed approach can be useful for future studies on other honey samples to detect specific markers and for food quality control.

3. Materials and Methods

3.1. Chemicals

All the used chemicals were of analytical grade. Standard of homogentistic, acid (±)-2-*cis*, 4-*trans*-abscisic acid, kurarinone, sakuranetin, and quinine sulphate were purchased from Sigma–Aldrich (Milan, Italy). (±)-2-*trans*, 4-*trans*-abscisic acid was purchased from A.G. Scientific, Inc (San Diego, CA, USA). Unedone was purchased from Chem Faces Biochemical Co., Ltd. (Wuhan, China). Methanol, acetonitrile, phosphoric acid 85% w/w, and absolute ethanol were obtained from Sigma–Aldrich (Milan, Italy). Acetonitrile, water, and formic acid of LC-MS grade were purchased from Merck (Darmstadt, Germany). Ultrapure water (18 MΩ cm) was obtained with Milli-Q Advantage A10 System apparatus (Millipore, Milan, Italy).

3.2. Strawberry Tree Honey and Preparation of the Hydrophilic Fractions

Strawberry tree (ST) honey was produced in Sardinia (Italy) in 2018. Unifloral origin was verified by melissopalinological analysis, sensorial evaluation, and LC-DAD evaluation of the typical markers (homogentisic acid, unedone and the two isomers of abscisic acid) [9].

Hydrophilic fractions (HF) of the honey were prepared by dissolving 500 g of the honey in 500 mL of water acidified at pH 4.5 with HCl. The solution was poured on a chromatographic column filled with C18 resin (Sigma–Aldrich) previously activated with ethanol and equilibrated with water. The column charged with honey was washed with water and all the eluted solutions were discharged because no bitter taste was perceived. Elution of the fractions was obtained with increasing quantities of ethanol (H_2O: EtOH in the ratios 95:5, 90:10, 80:20, 70:30, 50:50, 0:100 v/v). In total, 21 fraction were obtained and bitterness was evaluated by testing the solution after ethanol removal by Rotavapor®.

3.3. Sensory Analysis

A pre-trained sensory panel (5 judges) was used to determine the taste of both the fractions from the ST honey and pure compounds. Panellists provided written consent prior to participating indicating that they were not allergic to ST honey (included its main known compounds) or quinine. For establishing the level of perceptions of bitterness or sweetness, solutions of quinine and sucrose were prepared in water at six concentrations in the range 5 to 500 mg/L for quinine and 2 to 200 g/L for sucrose. Recognition threshold was performed according to testing samples in rank order (ISO 8587:2006) and was fixed at 5 mg quinine /L for bitterness and at 5 g sucrose /L for sweetness. For the sensory evaluation of the fractions from the ST honey, a scoring scale graded on a 5-point scale of bitterness perception was used: (0) no bitter taste; (1) barely detectable; (2) weak; (3) moderate; and (4) strong. Both (5)-point scale of bitterness perception and unedone level of bitterness (prepared at 300 mg/L in water) were established by comparison with the six concentrations of quinine. The samples were equilibrated to room temperature (20 °C ± 1) and the analysis was done during daylight. 1 mL of standard solution at proper concentration was applied to the upper surface of the tongue for 15 s then the test solution was expectorated.

3.4. LC-MS/MS Analysis

The electrospray ionisation (ESI) source of a Thermo Scientific LTQ-Orbitrap XL (Thermo Fisher Scientific, Dreieich, Germany) mass spectrometer was tuned in negative ion mode with a standard solution of kaempferol-3-*O*-glucoside (1 µg/mL) introduced at a flow rate of 10 µL/min by a syringe pump. Calibration of the Orbitrap analyser was performed using the standard LTQ calibration mixture composed of caffeine and the peptide MRFA (calibration solution purchased from the manufacturer), dissolved in 50:50% v/v water/acetonitrile solution.

Resolution for the Orbitrap mass analyser was set at 30,000. The mass spectrometric spectra were acquired by full range acquisition covering m/z 120–1400 in LC-MS. The study of the fragmentations was carried out using the data dependent scan experiment mode, by which the most intense [M-H] ions were selected during the LC-MS analysis. The data recorded were processed with Xcalibur 2.0 software (Thermo Fisher Scientific, Dreieich, Germany).

LC/ESI/LIT Orbitrap MS was performed using a Thermo Scientific liquid chromatography system consisting of a quaternary Accela 600 pump and an Accela autosampler, connected to a hybrid linear ion trap (IT) Orbitrap mass spectrometer (Thermo Scientific). LC-ESI-Orbitrap-MS analyses were performed using a Phenomenex Luna (150 mm × 2.1 mm particle size 5 µm) column, eluted with water containing 0.1% formic acid (solvent A) and acetonitrile containing 0.1% formic acid (solvent B). A linear gradient program at a flow rate of 0.200 mL/min was used: 0–35 min, from 5 to 95% (B); then back to 5% (B) for another 5 min. Honey fractions were dissolved in water with a concentration of 1 mg/mL and 5 µL of each samples was injected. The ESI source and MS parameters were the same as those used by D'Urso et al. [50]: capillary voltage: −48 V; tube lens voltage: −176.47 V; capillary temperature: 280 °C; sheath and auxiliary gas flow (N_2), 15 and 5; sweep gas: 0; spray voltage: 5.

3.5. Quantitative Analysis by HPLC-DAD

Quantitative analysis of the strawberry tree honey markers was carried out by a modified HPLC-UV method from Tuberoso et al. [2,9] using an Agilent 1260 Infinity system (Agilent Technologies, Palo Alto, CA, USA), equipped with a binary pump (G-1312C), and a DAD detector (G-4212B). The mobile phase consisted of solvent A (H_2O + 0.1% formic acid) and solvent B (CH_3CN + 0.1% formic acid), while the stationary phase was constituted by a Luna C 18 (150 × 2 mm, 5 µm) column (Phenomenex). The wavelength was set at 262 nm (Figure S1). The gradient started from 5% of B ending at 100% of B in 20 min, at a flow rate of 0.2 mL/min. The injection volume was 10 µL. An interval of 15 min was used to allow the column to equilibrate before injection of the next samples. Calibration curves were built with the external standard method, correlating the area of the peaks vs. the concentration. The commercial standard of (±)-2-*cis*, 4-*trans*-abscisic acid was used in different concentrations (0.1, 0.25, 0.50, 0.75, 1 mg/mL) to construct the calibration curve (y = 12,984x + 9664 R^2 = 0.991). The established method was validated in agreement with the International Conference on Harmonisation of Technical Requirements for Registration of Pharmaceuticals for Human Use (ICH) guidance note which describes validation of analytical methods [51]. Both the precision under conditions of repeatability and intermediate precisions were determined by performing either six injections of standard on the same day or six injections of the same standard on different days, respectively. The limit of detection (LOD) and limit of quantification (LOQ) were calculated on the basis of the data of the regression of the analytical curve and corresponding value are reported in Table 2. The *c,t*-ABA calibration curve was used to dose *t,t*-ABA and unedone as well.

3.6. Multivariate Data Analysis

For the untargeted approach, the chromatograms resulting from the LC-ESI/Orbitrap-MS analysis (negative ion mode) were normalised and aligned using MZmine software [42]. Thanks to the use of this toolbox with normalization of the total raw signal, 253 peaks were detected. After exporting the processed data in tabular format (.cvs file), a further statistical analysis of the data matrix was performed with SIMCA P + software 12.0 (Umetrix AB, Umeå, Sweden) using the PCA method. PCA was performed by applying the area of the peak obtained from LC-MS analysis [52,53] and the data were scaled through the application of Pareto scaling. After PCA, PLS was also applied with the SIMCA P + software. PLS is a regression technique used to relate two sets of data. The Y used for the statistical model was the perception of bitterness returned by the different fractions subjected to sensory analysis. Models were validated by cross-validation techniques and permutation tests such as Hotelling's permutation test and the T^2 test according to standardised good practice to minimise false discoveries and to obtain robust statistical models. The significance was evaluated by measuring the value of Q^2, which was higher than 0.5.

4. Conclusions

The chemometric model obtained by preliminary PCA of LC-ESI/LTQ-Orbitrap-MS data, followed by a PLS regression model was found to be a selective tool to detect the compounds responsible for the perception of bitterness in ST honey. The study allowed unedone to be identified as the main chemical compound responsible for the bitter taste of strawberry tree honey. Taking into account that unedone was characterised for the first time as a new natural product in strawberry tree honey, it is a novelty for further studies on the action on the bitter taste receptor. Further studies on the 25 human T2Rs bitter taste receptors should be developed to better define the action of unedone. Moreover, investigation of the complex matrix of the ST honey could be interesting due to the presence of the bitter compound unedone, an antagonist for the T2R4 receptor (ABA) and sweets compounds (sugars). Finally, 29 chemical compounds were identified and putatively identified in ST honey, using LC-ESI/LTQ-Orbitrap-MS and LC-ESI/LTQ-Orbitrap-MS/MS, and some of them for the first time in this honey.

Supplementary Materials: The following are available online, Table S1. Sensory analysis results (5 judges, mean ± sd) of the bitter perception in the 21 fractions obtained from the fractioning of strawberry tree honey on C18 column. Figure S1. LC-DAD representative chromatograms at λ = 262 nm of fraction B, fraction Z, and pure strawberry tree honey (dilution 1:100 in water w/v). t,t-ABA: (±)-2-*trans*, 4-*trans*-abscisic acid; c,t-ABA: (±)-2-*cis*, 4-*trans*-abscisic acid; U: unedone; HGA: homogentisic acid. Chromatographic conditions are described in the text.

Author Contributions: Conceptualization, P.M. and C.I.G.T.; methodology, P.M. and C.I.G.T.; software, P.M. and G.D.; formal analysis, P.M., G.D., and C.I.G.T.; investigation, P.M., G.D., and C.I.G.T.; resources, P.M. and C.I.G.T.; data curation, P.M., G.D., A.K., and C.I.G.T.; writing—original draft preparation, P.M. and C.I.G.T.; writing—review and editing, C.I.G.T., A.K., and P.M.; visualization, P.M. and C.I.G.T.; supervision, P.M., A.K., and C.I.G.T.; project administration, P.M. and C.I.G.T. All authors have read and agreed to the published version of the manuscript.

Funding: This research received no external funding.

Institutional Review Board Statement: Ethical review and approval were waived for this study, due to the use of food-grade products used in small amount.

Informed Consent Statement: Informed consent was obtained from all subjects involved in the study.

Data Availability Statement: The data presented in this study are available in article and supplementary material.

Conflicts of Interest: The authors declare no conflict of interest.

Sample Availability: Honey sample used for this study is available for limited time from the authors

References

1. Rosa, A.; Tuberoso, C.I.G.; Atzeri, A.; Melis, M.P.; Bifulco, E.; Dessì, M.A. Antioxidant profile of strawberry tree honey and its marker homogentisic acid in several models of oxidative stress. *Food Chem.* **2011**, *129*, 1045–1053. [CrossRef]
2. Tuberoso, C.I.G.; Boban, M.; Bifulco, E.; Budimir, D.; Pirisi, F.M. Antioxidant capacity and vasodilatory properties of Mediterranean food: The case of Cannonau wine, myrtle berries liqueur and strawberry-tree honey. *Food Chem.* **2013**, *140*, 686–691 [CrossRef] [PubMed]
3. Afrin, S.; Forbes-Hernandez, T.Y.; Gasparrini, M.; Bompadre, S.; Quiles, J.L.; Sanna, G.; Spano, N.; Giampieri, F.; Battino, M Strawberry-tree honey induces growth inhibition of human colon cancer cells and increases ROS generation: A comparison with Manuka honey. *Int. J. Mol. Sci.* **2017**, *18*, 613. [CrossRef]

4. Oses, S.M.; Nieto, S.; Rodrigo, S.; Pérez, S.; Rojo, S.; Sancho, M.T.; Fernández-Muiño, M.Á. Authentication of strawberry tree (*Arbutus unedo* L.) honeys from southern Europe based on compositional parameters and biological activities. *Food Biosci.* **2020**, *38*, 100768. [CrossRef]
5. Di Petrillo, A.; Santos-Buelga, C.; Era, B.; González-Paramás, A.M.; Tuberoso, C.; Medda, R.; Pintus, F.; Fais, A. Sardinian honeys as sources of xanthine oxidase and tyrosinase inhibitors. *Food Sci. Biotechnol.* **2018**, *27*, 139–146. [CrossRef] [PubMed]
6. Jurič, A.; Gašić, U.; Brčić Karačonji, I.; Jurica, K.; Milojković-Opsenica, D. The phenolic profile of strawberry tree (*Arbutus unedo* L.) honey. *J. Serb. Chem. Soc.* **2020**, *85*, 1011–1019. [CrossRef]
7. Miguel, M.G.; Faleiro, M.L.; Guerreiro, A.C.; Antunes, M.D. Arbutus unedo L.: Chemical and biological properties. *Molecules* **2014**, *19*, 15799–15823. [CrossRef]
8. Cabras, P.; Angioni, A.; Tuberoso, C.; Floris, I.; Reniero, F.; Ghelli, S. Homogentisic acid: A phenolic acid as a marker of strawberry-tree (Arbutus unedo) honey. *Food Chem.* **1999**, *47*, 4064–4067. [CrossRef]
9. Tuberoso, C.I.G.; Bifulco, E.; Caboni, P.; Cottiglia, F.; Cabras, P.; Floris, I. Floral markers of strawberry tree (Arbutus unedo L.) honey. *J. Agric. Food Chem.* **2010**, *58*, 384–389. [CrossRef] [PubMed]
10. Fierro, F.; Giorgetti, A.; Carloni, P.; Meyerhof, W.; Alfonso-Prieto, M. Dual binding mode of "bitter sugars" to their human bitter taste receptor target. *Sci. Rep.* **2019**, *9*, 8437. [CrossRef]
11. Izawa, K.; Amino, Y.; Kohmura, M.; Ueda, Y.; Kuroda, M. 4.16—Human–Environment Interactions—Taste. In *Comprehensive Natural Products II*, 1st ed.; Liu, H.-W., Mander, L., Eds.; Elsevier: Amsterdam, The Netherlands, 2010; Volume 4, pp. 631–671. ISBN 9780080453828. [CrossRef]
12. Meyerhof, W.; Batram, C.; Kuhn, C.; Brockhoff, A.; Chudoba, E.; Bufe, B.; Appendino, G.; Behrens, M. The molecular receptive ranges of human TAS2R bitter taste receptors. *Chem. Senses* **2010**, *35*, 157–170. [CrossRef]
13. Valdés, A.; Cifuentes, A.; León, C. Foodomics evaluation of bioactive compounds in foods. *Trends Anal. Chem.* **2017**, *96*, 2–13. [CrossRef]
14. Cerulli, A.; Masullo, M.; Montoro, P.; Hosek, J.; Pizza, C.; Piacente, S. Metabolite profiling of "green" extracts of *Corylus avellana* leaves by ^{1}H NMR spectroscopy and multivariate statistical analysis. *J. Pharm. Biomed. Anal.* **2018**, *160*, 168–178. [CrossRef]
15. La Barbera, G.; Capriotti, A.L.; Cavaliere, C.; Montone, C.M.; Piovesana, S.; Samperi, R.; Zenezini Chiozzi, R.; Laganà, A. Liquid chromatography-high resolution mass spectrometry for the analysis of phytochemicals in vegetal-derived food and beverages. *Food Res. Int.* **2017**, *100*, 28–52. [CrossRef]
16. D'Urso, G.; d'Aquino, L.; Pizza, C.; Montoro, P. Integrated mass spectrometric and multivariate data analysis approaches for the discrimination of organic and conventional strawberry (Fragaria ananassa Duch.) crops. *Food Res. Int.* **2015**, *77*, 264–272. [CrossRef]
17. Pascale, R.; Bianco, G.; Cataldi, T.R.I.; Schmitt Kopplin, P.; Bosco, F.; Vignola, L.; Uhl, J.; Lucio, M.; Milella, L. Mass spectrometry-based phytochemical screening for hypoglycemic activity of Fagioli di Sarconi beans (Phaseolus vulgaris L.). *Food Chem.* **2018**, *242*, 497–504. [CrossRef]
18. D'Urso, G.; Montoro, P.; Piacente, S. Detection and comparison of phenolic compounds in different extracts of black currant leaves by liquid chromatography coupled with High-Resolution ESI-LTQ-Orbitrap MS and High-Sensitivity ESI-QTrap MS. *J. Pharm. Biomed. Anal.* **2020**, *179*, 112926. [CrossRef]
19. KNApSAcK Core System. Available online: http://www.knapsackfamily.com/knapsack_core/top.php (accessed on 11 December 2020).
20. Sumner, L.W.; Amberg, A.; Barrett, D.; Beale, M.H.; Beger, R.; Daykin, C.A.; Fan, T.W.; Fiehn, O.; Goodacre, R.; Griffin, J.L.; et al. Proposed minimum reporting standards for chemical analysis—Chemical Analysis Working Group (CAWG) Metabolomics Standards Initiative (MSI). *Metabolomics* **2007**, *3*, 211–221. [CrossRef] [PubMed]
21. Pimpao, R.C.; Dew, T.; Figueira, M.E.; McDougall, G.J.; Stewart, D.; Ferreira, R.B.; Santos, C.N.; Williamson, G. Urinary metabolite profiling identifies novel colonic metabolites and conjugates of phenolics in healthy volunteers. *Mol. Nutr. Food Res.* **2014**, *58*, 1414–1425. [CrossRef] [PubMed]
22. Karikas, G.A.; Giannitsaros, A. Phenolic glucosides from leaves of Arbutus unedo. *Plantes Med. Phytothe.* **1990**, *24*, 27–30.
23. Marmarinos, V.; Paschalakis, P. Photopolymerization Material for Gums Isolation. U.S. Patent 8,334,328, 18 December 2012. Available online: https://patentimages.storage.googleapis.com/77/fb/c3/c3cf3433898be2/CA2677468C.pdf (accessed on 11 December 2020).
24. Ouchemoukh, S.; Amessis-Ouchemoukh, N.; Gomez-Romero, M.; Aboud, F.; Giuseppe, S.; Fernandez-Gutierrez, A.; Segura-Carretero, A. Characterisation of phenolic compounds in Algerian honeys by RP-HPLC coupled to electrospray time-of-flight mass spectrometry. *LWT—Food Sci. Technol.* **2017**, *85 Pt B*, 460–469. [CrossRef]
25. Long, E.Y.; Krupke, C.H. Non-cultivated plants present a season-long route of pesticide exposure for honey bees. *Nat. Commun.* **2016**, *7*, 11629. [CrossRef]
26. Wilkins, A.L.; Lu, Y.; Tan, S.T. Extractives from New Zealand honeys. 4. Linalool derivatives and other components from nodding thistle (Carduus nutans) honey. *J. Agric. Food Chem.* **1993**, *41*, 873–878. [CrossRef]
27. Kaskoniene, V.; Maruska, A.; Kornysova, O.; Charczun, N.; Ligor, M.; Buszewski, B. Quantitative and qualitative determination of phenolic compounds in honey. *Chem. Technol.* **2009**, 74–80.
28. Rouphael, Y.; Bernardi, J.; Cardarelli, M.; Bernardo, L.; Kane, D.; Colla, G.; Lucini, L. Phenolic compounds and sesquiterpene lactones profile in leaves of nineteen artichoke cultivars. *J. Agric. Food Chem.* **2016**, *64*, 8540–8548. [CrossRef]

29. Blunt, J.; Munro, M.G.; Swallow, W. Carbon-13 NMR analysis of tutin and related substances: Application to the identification of minor components of toxic honey. *Aust. J. Chem.* **1979**, *32*, 1339–1343. [CrossRef]
30. Dalla Serra, A.; Franco, M.A.; Mattivi, F.; Ramponi, M.; Vacca, V.; Versini, G. Aroma characterization of Sardinian strawberry tree (Arbutus unedo L.) honey. *Ital. J. Food Sci.* **1999**, *11*, 47–56.
31. Sun, M.; Zhao, L.; Wang, K.; Han, L.; Shan, J.; Wu, L.; Xue, X. Rapid identification of "mad honey" from Tripterygium wilfordii Hook. f. and Macleaya cordata (Willd) R. Br using UHPLC/Q-TOF-MS. *Food Chem.* **2019**, *294*, 67–72. [CrossRef]
32. Ma, L.; Ashworth, D.; Yates, S.R. Simultaneous determination of estrogens and progestogens in honey using high performance liquid chromatography-tandem mass spectrometry. *J. Pharm. Biomed.* **2016**, *131*, 303–308. [CrossRef] [PubMed]
33. Staub Spörri, A.; Jan, P.; Cognard, E.; Ortelli, D.; Edder, P. Comprehensive screening of veterinary drugs in honey by ultra-high-performance liquid chromatography coupled to mass spectrometry. *Food Addit. Contam. A* **2014**, *31*, 806–816. [CrossRef] [PubMed]
34. Bentivenga, G.; D'Auria, M.; Fedeli, P.; Mauriello, G.; Racioppi, R. SPME-GC-MS analysis of volatile organic compounds in honey from Basilicata. Evidence for the presence of pollutants from anthropogenic activities. *Int. J. Food Sci.* **2004**, *39*, 1079–1086. [CrossRef]
35. Češen, M.; Lambropoulou, D.; Laimou-Geraniou, M.; Kosjek, T.; Blaznik, U.; Heath, D.; Heath, E. Determination of bisphenols and related compounds in honey and their migration from selected food contact materials. *J. Agric. Food Chem.* **2016**, *64*, 8866–8875. [CrossRef] [PubMed]
36. Rodríguez-Gonzalo, E.; Domínguez-Alvarez, J.; García-Gómez, D.; García-Jiménez, M.G.; Carabias-Martínez, R. Determination of endocrine disruptors in honey by CZE-MS using restricted access materials for matrix cleanup. *Electrophoresis* **2010**, *31*, 2279–2288. [CrossRef]
37. Sosath, S.; Ott, H.H.; Hecker, E. Irritant principles of the spurge family (Euphorbiaceae). XIII. Oligocyclic and macrocyclic diterpene esters from latexes of some *Euphorbia* species utilized as source plants of honey. *J. Nat. Prod.* **1988**, *51*, 1062–1074. [CrossRef] [PubMed]
38. Duke, C.C.; Tran, V.H.; Duke, R.K.; Abu-Mellal, A.; Plunkett, G.T.; King, D.I.; Hamid, K.; Wilson, K.L.; Barrett, R.L.; Bruhl, J.J. A sedge plant as the source of Kangaroo Island propolis rich in prenylated *p*-coumarate ester and stilbenes. *Phytochemistry* **2017**, *134*, 87–97. [CrossRef]
39. Leyva-Jimenez, F.J.; Lozano-Sanchez, J.; Borras-Linares, I.; Cadiz-Gurrea, M.d.l.L.; Mahmoodi-Khaledi, E. Potential antimicrobial activity of honey phenolic compounds against Gram positive and Gram negative bacteria. *LWT—Food Sci. Technol.* **2019**, *101*, 236–245. [CrossRef]
40. Seraglio, S.K.T.; Valese, A.C.; Daguer, H.; Bergamo, G.; Azevedo, M.S.; Gonzaga, L.V.; Fett, R.; Costa, A.C.O. Development and validation of a LC-ESI-MS/MS method for the determination of phenolic compounds in honeydew honeys with the diluted-and-shoot approach. *Food Res. Int.* **2016**, *87*, 60–67. [CrossRef]
41. Liu, P.; Deng, T.; Ye, C.; Qin, Z.; Hou, X.; Wang, J. Identification of kurarinone by LC/MS and investigation of its thermal stability. *J. Chil. Chem. Soc.* **2009**, *54*, 80–82. [CrossRef]
42. MZmine 2. Available online: http://mzmine.github.io/ (accessed on 11 December 2020).
43. Schievano, E.; Stocchero, M.; Morelato, E.; Facchin, C.; Mammi, S. An NMR-based metabolomic approach to identify the botanical origin of honey. *Metabolomics* **2012**, *8*, 679–690. [CrossRef]
44. Stocchero, M. Relevant and irrelevant predictors in PLS2. *J. Chemom.* **2020**, *34*, e3237. [CrossRef]
45. Pydi, S.P.; Jaggupilli, A.; Nelson, K.M.; Abrams, S.R.; Bhullar, R.P.; Loewen, M.C.; Chelikani, P. Abscisic acid acts as a blocker of the bitter taste G protein-coupled receptor T2R4. *Biochemistry* **2015**, *28*, 2622–2631. [CrossRef]
46. Remya, C.; Dileep, K.V.; Tintu, I.; Variyar, E.J.; Sadasivan, C. Flavanone glycosides as acetylcholinesterase inhibitors: Computational and experimental evidence. *Indian J. Pharm. Sci.* **2014**, *76*, 567–570.
47. Yang, J.; Chen, H.; Wang, Q.; Deng, S.; Huang, M.; Ma, X.; Song, P.; Du, J.; Huang, Y.; Wen, Y.; et al. Inhibitory effect of kurarinone on growth of human non-small cell lung cancer: An experimental study both in vitro and in vivo studies. *Front. Pharm.* **2018**, *9*, 252. [CrossRef] [PubMed]
48. Huai, D. Anti-fibrotic effect of kurarinone in treatment of chronic hepatitis B. *Xiandai Zhenduan Yu Zhiliao* **2014**, *25*, 4641–4642.
49. Deiana, V.; Tuberoso, C.; Satta, A.; Pinna, C.; Camarda, I.; Spano, N.; Ciulu, M.; Floris, I. Relationship between markers of botanical origin in nectar and honey of the strawberry tree (Arbutus unedo) throughout flowering periods in different years and in different geographical areas. *J. Apic. Res.* **2016**, *54*, 342–349. [CrossRef]
50. D'Urso, G.; Maldini, M.; Pintore, G.; d'Aquino, L.; Montoro, P.; Pizza, C. Characterisation of Fragaria vesca fruit from Italy following a metabolomics approach through integrated mass spectrometry techniques. *LWT—Food Sci. Technol.* **2016**, *74*, 387–395. [CrossRef]
51. EMEA. Quality Guidelines: Validation of Analytical Procedures: Text and Methodology (ICH Q2). Available online: http://www.emea.europa.eu/pdfs/human/ich/038195en.pdf (accessed on 11 December 2020).
52. Mari, A.; Montoro, P.; Pizza, C.; Piacente, S. Liquid chromatography tandem mass spectrometry determination of chemical markers and principal component analysis of Vitex agnus-castus L. fruits (Verbenaceae) and derived food supplements. *J. Pharm. Biomed. Anal* **2012**, *70*, 224–230. [CrossRef]
53. Soufi, S.; D'Urso, G.; Pizza, C.; Rezgui, S.; Bettaieb, T.; Montoro, P. Steviol glycosides targeted analysis in leaves of Stevia rebaudiana (Bertoni) from plants cultivated under chilling stress conditions. *Food Chem.* **2016**, *190*, 572–580. [CrossRef]

Electrochemical Determination of the "Furanic Index" in Honey

Severyn Salis [1], Nadia Spano [2,*], Marco Ciulu [3], Ignazio Floris [4], Maria I. Pilo [2] and Gavino Sanna [2]

1. Istituto Zooprofilattico Sperimentale della Sardegna, Via Duca degli Abruzzi 8, 07100 Sassari, Italy; severyn.salis@izs-sardegna.it
2. Dipartimento di Chimica e Farmacia, Università degli Studi di Sassari, Via Vienna 2, 07100 Sassari, Italy; mpilo@uniss.it (M.I.P.); sanna@uniss.it (G.S.)
3. Department of Animal Sciences, University of Göttingen, Kellnerweg 6, 37077 Göttingen, Germany; marco.ciulu@uni-goettingen.de
4. Dipartimento di Agraria, Università degli Studi di Sassari, Viale Italia 39/a, 07100 Sassari, Italy; ifloris@uniss.it
* Correspondence: nspano@uniss.it; Tel.: +39-079-229-569

Abstract: 5-(hydroxymethyl)furan-2-carbaldehyde, better known as hydroxymethylfurfural (HMF), is a well-known freshness parameter of honey: although mostly absent in fresh samples, its concentration tends to increase naturally with aging. However, high quantities of HMF are also found in fresh but adulterated samples or honey subjected to thermal or photochemical stresses. In addition, HMF deserves further consideration due to its potential toxic effects on human health. The processes at the origin of HMF formation in honey and in other foods, containing saccharides and proteins—mainly non-enzymatic browning reactions—can also produce other furanic compounds. Among others, 2-furaldehyde (2F) and 2-furoic acid (2FA) are the most abundant in honey, but also their isomers (i.e., 3-furaldehyde, 3F, and 3-furoic acid, 3FA) have been found in it, although in small quantities. A preliminary characterization of HMF, 2F, 2FA, 3F, and 3FA by cyclic voltammetry (CV) led to hypothesizing the possibility of a comprehensive quantitative determination of all these compounds using a simple and accurate square wave voltammetry (SWV) method. Therefore, a new parameter able to provide indications on quality of honey, named "Furanic Index" (FI), was proposed in this contribution, which is based on the simultaneous reduction of all analytes on an Hg electrode to ca. −1.50 V vs. Saturated Calomel Electrode (SCE). The proposed method, validated, and tested on 10 samples of honeys of different botanical origin and age, is fast and accurate, and, in the case of strawberry tree honey (*Arbutus unedo*), it highlighted the contribution to the FI of the homogentisic acid (HA), i.e., the chemical marker of the floral origin of this honey, which was quantitatively reduced in the working conditions. Excellent agreement between the SWV and Reverse-Phase High-Performance Liquid Chromatography (RP-HPLC) data was observed in all samples considered.

Keywords: HMF; honey; furanic aldehydes; furanic acids; homogentisic acid; cyclic voltammetry; square wave voltammetry; RP-HPLC

1. Introduction

Honey is the sweet natural product that honeybees obtain by conversion of the nectar gathered from flowers. Its composition depends on the geographical, floral, and entomological origins and includes water (15–20%), sugars (80–85% w/w), nitrogenous compounds, enzymes, phenolic and volatile compounds, organic and amino acids, minerals, and vitamins. On the other hand, it is well known that seasonal, environmental, storage, and processing time and conditions may affect honey's composition. Although the nutritional and healing features of this food are appreciated, the possible presence of heavy metals (usually in traces) [1], some alkaloids [2], and reactive organic compounds, such as aldehydes, might imply a possible threat for the health of consumers. The 5-(hydroxymethyl)furan-2-carbaldehyde, better known as hydroxymethylfurfural (HMF), is produced during the

Maillard reaction and, in honey, acidic conditions and the predominance of monosaccharides, as fructose and glucose, promote its formation during aging or processing [3,4]. For these reasons, the amount of HMF in honey is considered a freshness parameter: while it is basically absent in fresh and well-preserved honey, its amount naturally increases with aging and particularly in the presence of other factors such as heat, sunlight, and metal ions, which promote the course of the Maillard's reaction [5–7]. As a consequence, the Codex Alimentarius Standard commission fixed the maximum content of HMF in honey at 40 mg kg^{-1} (80 mg kg^{-1} for honey produced in tropical regions) and the legislative directives of many countries transposed these indications (Council Directive 2001/110/EC of 20 December 2001 relating to honey). The growing attention of the scientific community toward the effects of HMF helped to reveal not only potential adverse effects on human health, such as cytotoxicity, mutagenicity, chromosomal aberrations, and carcinogenicity [8–10], but also positive effects such as antioxidative [11], anti-allergic [12], anti-inflammatory [13], anti-hypoxia [14], anti-sickling [15], and anti-hyperuricemic properties [16]. Hence, the need for accurate methods devoted to the quantitative analysis of HMF in honey is an increasingly relevant target in the field of the quality assurance in hive products and, more in general, for food risk assessment. Three protocols for the determination of HMF in honey are recommended by the International Honey Commission [17]; two of them are spectrophotometric methods (White [18] and Winkler [19]), whereas the last is a reverse-phase high-performance liquid chromatographic (RP-HPLC) method [20]. Both spectrophotometric methods are scarcely sensitive and accurate; in addition, the Winkler method requires the use of the carcinogen *p*-toluidine. On the other hand, the RP-HPLC determination is more accurate than the spectrophotometric ones but it is quite slow. In addition to these, other procedures based on separation techniques, such as RP-HPLC-UV [21] or micellar electrokinetic capillary chromatography (MEKC) [22], electrochemical methods, such as differential pulse polarography (DPP) [23] and square wave voltammetry (SWV) [24,25], or sensors [26,27] were used. Spectroscopic approaches were also reported, including a Winkler-based automated flow injection protocol [28] or a recent method based on the HMF derivatization with the Seliwanoff reagent [29]. Albeit often in minor amounts than HMF, other related species, such as 2-furaldehyde (2F), 3-furaldehyde (3F), 2-furoic acid (2FA), and 3-furoic acid (3FA) may be formed by the degradation processes of honey, hence contributing to the minority composition of reactive organic species present in it. While 2F and 2FA derive, in variable amounts, from the hydrolysis [30] and oxidative degradation of ascorbic acid [31], respectively, the origin of both 3F and 3FA is still debated [32,33]. The literature's contributions devoted to the contemporary determination of HMF and of the minority furanic aldehydes and acids are rare. All—or some—of these compounds have been measured only using HPLC methods in matrices such as honey [32,33] or in fruit juices and drinks [34]. Despite their simplicity, sensitivity, and accuracy, electrochemical methods have never been used to the best of our knowledge in the comprehensive determination of HMF and of the related furanic aldehydes and acids. Hence, the principal goal of this contribution has been to develop, validate, and apply to real samples a rapid, sensitive, and reliable electrochemical method aimed to contemporaneously determine the total amount of HMF, 2F, 3F, 2FA, and 3FA in honey. Figure 1 reports the chemical structures of all these analytes. To do this, a preliminary evaluation of the electrochemical behavior of all analytes has been accomplished using different techniques (i.e., CV and SWV) in different solvent/supporting electrolyte couples and using different working electrodes (WE). Since the results of the preliminary experimental evidences, an SWV method, aimed to the contemporary determination of HMF, 2F, 3F, 2FA, and 3FA, has been developed, and a new freshness parameter for honey, the so-called "Furanic Index" (FI), may therefore be proposed. The method has been optimized, validated in terms of limit of detection (LoD) and of quantification (LoQ), linearity, precision, and trueness, and tested on a number of real samples of unifloral and blossom honeys, which are different also for age and geographical origin.

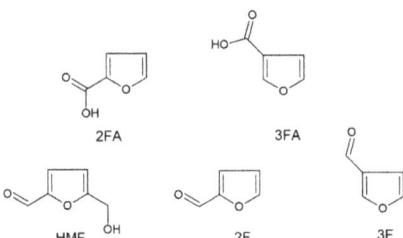

Figure 1. Chemical structures of the furanic acids and aldehydes. 2FA, 2-furoic acid; 3FA, 3-furoic acid; HMF, hydroxymethylfurfural; 2F, 2-furaldehyde; 3F, 3-furaldehyde.

2. Results and Discussion

2.1. Cyclic Voltammetry

2.1.1. Electrochemical Behavior of HMF

The electrochemical behavior of HMF was studied in cyclic voltammetry by varying the nature of the WE (Pt, GC, Au, Hg), the solvent system (solvents: water or methanol; supporting electrolytes: NaH_2PO_4, Na_2HPO_4, LiCl, CH_3COONa, $NaHCO_3$, $LiClO_4$, $[(C_4H_9)_4N]ClO_4$) and the potential scan rate (between 0.05 and 2.0 V s^{-1}). Based on the cathodic nature of the responses of the analyte, a preliminary evaluation has allowed to discard both Pt and Au working electrodes. Hence, only Hg and GC electrodes were further considered in the optimization of the experimental conditions, as summarized below in Table 1.

Table 1. CV operative conditions.

	Operative Parameters
WE	GC disk (ø 2 mm) or Hg hemisphere (ø 1 mm)
RE	SCE
AE	Pt
Potential scan rate	0.10 V s^{-1}
Supporting electrolytes (0.1 mol dm^{-3} in water)	NaH_2PO_4, Na_2HPO_4, LiCl, $LiClO_4$, CH_3COONa, $NaHCO_3$
Supporting electrolytes (0.1 mol dm^{-3} in methanol)	$LiClO_4$, $[(C_4H_9)_4N]ClO_4$
HMF solution	10 cm^3, 1 mmol dm^{-3}

The CV response of an aqueous 1 mmol dm^{-3} solution of HMF showed a well-defined and reproducible irreversible cathodic peak at potentials between -1.10 and -1.55 V vs. SCE, as a function of the nature of both the WE and of the supporting electrolytes (Table 2 and Figure 2).

Table 2. Cathodic peak potential ($E_{p,c}$) and current density at the cathodic peak ($J_{p,c}$) of the irreversible reduction process of HMF by varying the nature of the WE (Hg or GC) and the nature and the pH of the solvent system (a 0.1 mol dm^{-3} water solution of five different supporting electrolytes). HMF concentration: 1 mmol dm^{-3}, AE: Pt, RE: SCE, potential scan rate: 0.10 V s^{-1}.

WE	Hg		GC	
Supporting Electrolytes (pH)	$E_{p,c}$ (V)	$J_{p,c}$ (A mm^{-2})	$E_{p,c}$ (V)	$J_{p,c}$ (A mm^{-2})
NaH_2PO_4 (4.1)	-1.10	-3.21×10^{-3}	-1.35	-7.93×10^{-4}
LiCl (7.0)	-1.48	-1.02×10^{-3}	-1.52	-2.58×10^{-5}
$LiClO_4$ (6.9)	-1.48	-3.54×10^{-5}	-1.52	-3.18×10^{-5}
CH_3COONa (8.8)	-1.46	-4.59×10^{-5}	-1.52	-8.40×10^{-6}
Na_2HPO_4 (9.8)	-1.48	-3.09×10^{-5}	-1.55	-8.82×10^{-6}

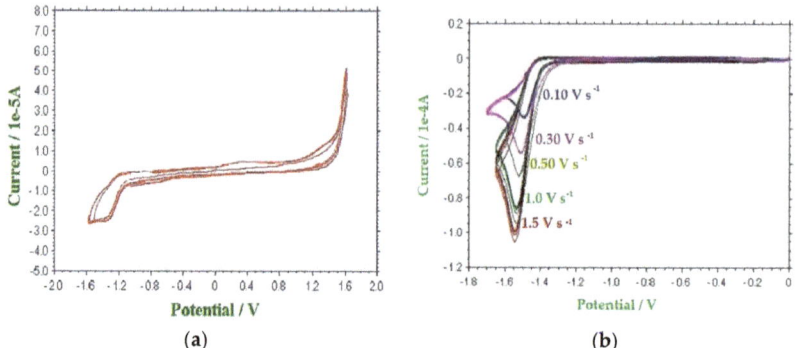

Figure 2. CV responses of HMF (1 mmol dm^{-3}) in a 0.1 mol dm^{-3} water solution of supporting electrolyte. AE: Pt, RE: SCE. (**a**) WE: GC, supporting electrolyte: NaH$_2$PO$_4$, potential scan rate: 0.10 V s^{-1} (**b**) WE: Hg, supporting electrolyte: LiClO$_4$, potential scan rate: between 0.10 and 1.5 V s^{-1}.

No anodic peak was observed in anodic direct scan (WE = GC) or after the inversion of the scan direction after crossing the cathodic peak (Figure 2a). Voltammograms acquired using Hg as WE at scan rates between 0.10 and 1.5 V s^{-1} showed a cathodic shift of $E_{p,c}$ (from −1.48 to −1.54 V vs SCE) and a linear increase of the peak's current (from −3 × 10^{-5} to −1 × 10^{-4} A) (Figure 2b). No backward anodic peak has been observed in this range of scan rates. The linear decreasing of the $i_{p,c}/\sqrt{(\log V_{scan})}$ ratio as a function of the cathodic shift of the $E_{p,c}$ peak suggests a kinetic control on the overall electrochemical process [35].

The replacement of water with methanol as solvent did not cause meaningful changes in the morphology of the main cathodic peak previously observed. Indeed, in these conditions, the CV responses recorded on GC, in cathodic direct scan, showed the same irreversible peak (peak **1** in Figure 3) at about −1.50 V vs. SCE. When the scan direction was inverted (from −2.0 to 2.0 V), three ill-resolved and consecutive irreversible anodic peaks became evident (Figure 3, peaks 2, 3 and 4, respectively, at $E_{p,a}$ between 0.90 and 1.70 V vs. SCE). A further inversion of the direction of the potential scan, performed just after peak 4, caused the appearance of a low and irreversible cathodic peak at $E_{p,c}$ of −1.00 V (peak 5 in Figure 3), which was never observed in the cathodic direct scan (Figure 3).

The voltammetric responses of the solutions of HMF were in agreement with those reported by other authors who studied this analyte's behavior in aqueous solutions [25] as well as the reduction processes that involve the carbonylic group in 2F [35,36]. All the authors attributed the irreversible cathodic peak at about −1.50 V to the two-electron reduction of the aldehydic group to the corresponding alcohol, that, in the case of HMF, produces the 2,5-dihydroxymethylfurane. In addition, the meaningful cathodic shift of this irreversible reduction peak, observed passing by a solvent system at pH = 4.1 (supporting electrolyte: NaH$_2$PO$_4$) to that at pH = 9.8 (supporting electrolyte: Na$_2$HPO$_4$), is consistent with the observations of Ganesan et al. [36], because low pH values assist the reduction process. On the other hand, the most anodic peak, observed at 1.68 V vs. SCE in the direct anodic scan on GC in the methanol/LiClO$_4$ solvent system, was previously attributed to the irreversible oxidation of HMF to 2,5-diformylfurane (1.63 V vs. SCE, [25]), which was likely reducible, in the reverse cathodic scan, to HMF (cathodic peak at −1.00 V vs. SCE). Since the reduction of carbonylic group of HMF is the main (often the unique) peak and the most reproducible one present in the different solutions, we decided to further investigate it to find the optimal conditions for its quantitative determination in honey samples.

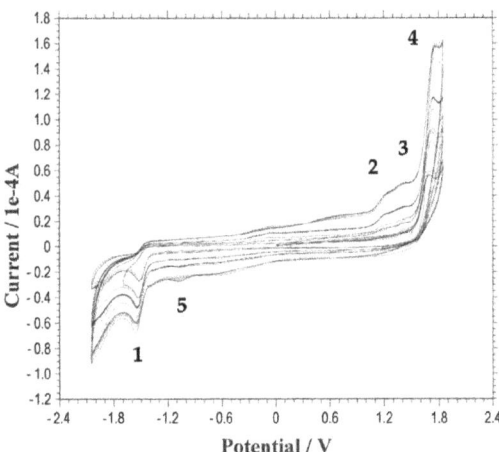

Figure 3. CV responses of HMF (1 mmol dm^{-3}) in a 0.1 mol dm^{-3} of LiClO$_4$ in methanol. WE: GC; AE: Pt; RE: SCE, potential scan rate: 0.10 V s^{-1}, red line; 0.30 V s^{-1}, blue line, 0.50 V s^{-1}, green line; 1.0 V s^{-1}, brown line.

2.1.2. Electrochemical Behavior of Furanic Aldehydes and Acids

The electrochemical behavior of 2F, 3F, 2FA, and 3FA was studied in CV in the same conditions used for HMF. Table 3 reports the potentials of the cathodic peak measured using Hg and GC as working electrodes, and LiCl, LiClO$_4$, CH$_3$COONa, and Na$_2$HPO$_4$ as supporting electrolytes in water. All the species showed an irreversible cathodic reduction peak at potential between −1.46 and −1.84 V vs. SCE, using both the working electrodes, when LiCl and LiClO$_4$ were used as supporting electrolytes. These peaks can be tentatively ascribed to the reductions of 2F and 3F to give the corresponding alcohols (i.e., 2- and 3-hydroxymethylfurane, respectively) and the reductions of 2FA and 3FA toward 2F and 3F, respectively. On the other hand, the cathodic shift of the reduction process caused by the use of alkaline supporting electrolytes brought the $E_{p,c}$ of 2FA and 3FA out of the available cathodic window. This happened when the GC working electrode was used in the presence of CH$_3$COONa as supporting electrolyte and, for both Hg and GC electrodes, when Na$_2$HPO$_4$ was used. Hence, the possibility to measure at the same time all these analytes in honey is achievable only using neutral supporting electrolytes such as LiCl or LiClO$_4$.

Table 3. Cathodic peak potentials $E_{p,c}$ (V vs. SCE) for HMF, 2F, 3F, 2FA, and 3FA measured in a CV direct cathodic scan on a 0.1 mol dm^{-3} aqueous solution of four different supporting electrolytes. Analyte concentration: 1 mmol dm^{-3}, potential scan rate: 0.10 V s^{-1}, AE: Pt, RE: SCE.

	Supporting Electrolyte							
	LiCl		LiClO$_4$		CH$_3$COONa		Na$_2$HPO$_4$	
	WE ($E_{p,c}$, V)		WE ($E_{p,c}$, V)		WE ($E_{p,c}$, V)		WE ($E_{p,c}$, V)	
Analyte	Hg	GC	Hg	GC	Hg	GC	Hg	GC
HMF	−1.48	−1.54	−1.48	−1.52	−1.46	−1.52	−1.48	−1.55
2F	−1.49	−1.54	−1.50	−1.55	−1.48	−1.54	−1.47	−1.49
3F	−1.68	−1.76	−1.65	−1.76	−1.67	−1.74	−1.64	−1.69
2FA	−1.46	−1.78	−1.48	−1.84	−1.64	nd	nd	nd
3FA	−1.48	−1.76	−1.49	−1.77	−1.61	nd	nd	nd

nd: no reduction process has been observed within the cathodic window available for the WE/solvent system couple.

2.2. Square Wave Voltammetry

Method Optimization

Preliminary CV experiments allowed to ascertain the feasibility in a contemporary determination of HMF and of the principal furanic aldehydes and acids in honey. Unfortunately, CV is an unsuitable electrochemical technique for an accurate and sensitive determination of trace compounds. On the other hand, the amount of the most abundant of these analytes in fresh honey, the HMF, is typically of few tens of mg kg^{-1}, i.e., well below the concentration used for the preliminary CV tests. Hence, the quantitative method to be assessed had to be able to reach LoQ levels at least of few mg kg^{-1}, and the best choice to quantify trace amounts of electroactive organic analytes was the square wave voltammetry (SWV). The method has been optimized as a function of the general parameters, such as the nature of the working electrode (Hg or GC) and of the supporting electrolyte (LiCl or LiClO$_4$), as well as of specific SWV parameters, such as the pulse height (values between 2.5 and 250 mV), the frequency (between 1.5 and 150 Hz), and the modulation amplitude (between 1 and 40 mV). The optimized SWV parameters are pulse height, 4 mV; frequency, 15 Hz; modulation amplitude, 25 mV. Table 4 shows the optimization of both WE and supporting electrolyte in the development of the SWV method. HMF was chosen as a molecular benchmark, since it is the most abundant furanic compound present in honey.

Table 4. Cathodic peak potentials $E_{p,c}$ (V vs. SCE) and cathodic peak current densities $J_{p,c}$ (µA mm^{-2}) in the SWV determination of HMF (concentration: 16 µmol dm^{-3}) by varying the nature of WE (Hg or GC) and that of the supporting electrolyte (LiCl or LiClO$_4$, both 0.1 mol dm^{-3} in water). Potential scan rate: 0.10 V s^{-1}; cathodic window of potential: between 0 and -2.00 V vs. SCE; AE: Pt; RE: SCE; pulse height: 4 mV; frequency: 15 Hz; modulation amplitude: 25 mV.

WE		Hg		GC	
Supporting Electrolyte	$E_{p,c}$ (V)	$J_{p,c}$ (µA mm^{-2})	$E_{p,c}$ (V)	$J_{p,c}$ (µA mm^{-2})	
LiCl	-1.43	-0.152	nd	nd	
LiClO$_4$	-1.45	-0.072	-1.53	-0.164	

Despite the higher sensitivity of the GC electrode, when LiClO$_4$ was the supporting electrolyte, no response was obtained using LiCl. On the other hand, the Hg electrode could detect the benchmark HMF content using both the supporting electrolytes. In these cases, the density of current measured in the water/LiCl solvent system was roughly double the amount measured in water/LiClO$_4$. Based on the results here obtained, the SWV method was tested also in the presence of 2F, 3F, 2FA, and 3FA. Additionally in this case, two WEs and two supporting electrolytes were used in these experiments. Table 5 reports the electrochemical parameters obtained for all the analytes in this cycle of measurements.

Data reported in Table 5 confirmed the preliminary results obtained for HMF. The $E_{p,c}$ values for the five analytes measured with Hg electrode were more reproducible and closer to each other in LiCl rather than in LiClO$_4$. In addition, it was impossible to detect the cathodic peak of 3FA with GC electrode, when LiClO$_4$ was used as supporting electrolyte. Furthermore, the current densities measured for all analytes, when Hg was the working electrode, were much higher than those measured by GC electrode, and the same happened switching from LiClO$_4$ to LiCl as the supporting electrolyte. Hence, Hg as the working electrode and LiCl as the supporting electrolyte were the best choices for the SWV determination of HMF, 2F, 3F, 2FA, and 3FA in honey. If the extreme closeness of all the $E_{p,c}$ values made it almost impossible to discriminate the amount of each analyte, individually, as clearly demonstrated by Figure 4, it was however possible—and easy—to quantify the amounts of all these analytes at once, expressing them in terms of an equivalent amount of HMF. This observation allowed defining the "Furanic Index" (FI) as a freshness-related parameter for honey that is able to evaluate the overall amount of HMF, 2F, 3F, 2FA, and 3FA, as expressed in mg of HMF per kg of honey. From a quantitative viewpoint, the expression of FI is:

$$FI\ (mg\ dm^{-3}) = C_{HMF} + C_{2F} \cdot (HMF/2F) + C_{3F} \cdot (HMF/3F) + C_{2FA} \cdot (HMF/2FA) + C_{3FA} \cdot (HMF/3FA)$$

where $C_{analyte}$ is the amount (in mg dm^{-3}) of each furanic compound, and the HMF/analyte ratio refers to the relevant molecular weights.

Table 5. Cathodic peak potentials $E_{p,c}$ (V vs. SCE) and cathodic peak current densities $J_{p,c}$ (µA mm^{-2}) in the SWV determination of HMF, 2F, 3F, 2FA, and 3FA (concentration: 1 mmol dm^{-3} each) by varying the nature of WE (Hg or GC) and that of the supporting electrolyte (LiCl or LiClO$_4$, both 0.1 mol dm^{-3} in water). Potential scan rate: 0.10 V s^{-1}; cathodic window of potential: between 0 and −2.0 V vs. SCE; AE: Pt; RE: SCE; pulse height: 4 mV; frequency: 15 Hz; modulation amplitude: 25 mV.

	Supporting Electrolyte							
	LiCl				LiClO$_4$			
	WE = Hg		WE = GC		WE = Hg		WE = GC	
Analyte	$E_{p,c}$ (V vs. SCE)	$J_{p,c}$ (µA mm^{-2})	$E_{p,c}$ (V vs. SCE)	$J_{p,c}$ (µA mm^{-2})	$E_{p,c}$ (V vs. SCE)	$J_{p,c}$ (µA mm^{-2})	$E_{p,c}$ (V vs. SCE)	$J_{p,c}$ (µA mm^{-2})
HMF	−1.43	−77.1	−1.47	−13.7	−1.45	−54.9	−1.48	−10.4
2F	−1.47	−33.4	−1.49	−4.55	−1.47	−2.98	−1.51	−0.48
3F	−1.61	−13.1	−1.70	−1.16	−1.62	−2.53	−1.70	−3.34
2FA	−1.42	−21.3	−1.70	−1.46	−1.43	−1.69	−1.84	−0.32
3FA	−1.44	−14.0	−1.70	−1.96	−1.44	−9.57	nd	nd

nd: no reduction process has been observed within the cathodic window available for the WE/solvent system couple.

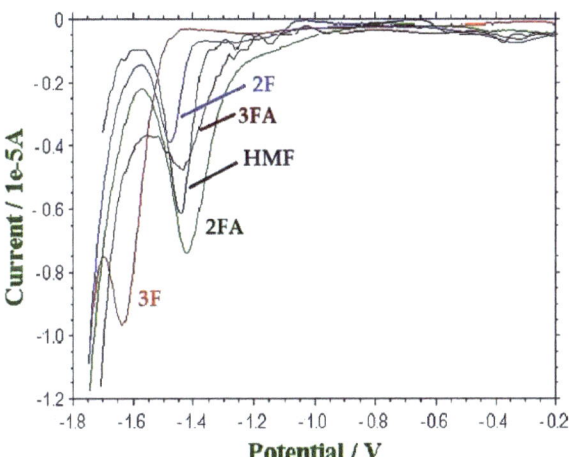

Figure 4. SW voltammograms of: HMF (0.08 mmol dm^{-3}), 2F (0.10 mmol dm^{-3}), 3F (0.62 mmol dm^{-3}), 2FA (0.36 mmol dm^{-3}), and 3FA (0.18 mmol dm^{-3}) in 0.1 mol dm^{-3} LiCl aqueous solution, WE: Hg; AE: Pt; RE: SCE. Potential scan rate: 0.10 V s^{-1}; pulse height: 4 mV; frequency: 15 Hz; modulation amplitude: 25 mV.

2.3. Validation

Validation of the proposed method has been accomplished in terms of LoD, LoQ, linearity, precision, and trueness. Table 6 reports the features describing the performances of the method proposed.

2.3.1. LoD and LoQ

These parameters were calculated, according to IUPAC recommendation, through the ULA1 method [37]. Four different 0.1 mol dm^{-3} LiCl aqueous solutions containing HMF in concentrations of 3 mg dm^{-3}, 5 mg dm^{-3}, 7 mg dm^{-3}, and 10 mg dm^{-3}, respectively, were prepared and analyzed by means of SWV in triplicate. The LoD obtained in this way is 0.6 mg dm^{-3}. This amount is significantly higher than that (i.e., 0.0021 mg dm^{-3}) recently measured by Sahli et al. [25], but in this case, the approach used for the calculation was not described. On the other hand, it is well known that in the past, it has been possible to associate to the general expression "detection limit" values spanning over more than three orders of magnitude [38]. Hence, even without arguing on the overall reliability of very low LoDs reported in the literature, it is well known that the application of the ULA methods, albeit providing sometimes LoDs higher than those obtained by more "optimistic" approaches [33], is one of the best choices to avoid both type-1 and type-2 decision errors. For the ULA1 approach, LoQ is three times the LoD value, i.e., 1.8 mg dm^{-3}.

2.3.2. Linearity

Linearity was measured over a concentration range between LoQ and 200 mg dm^{-3}. This is the typical interval of amounts observed for HMF in honeys of different geographical origin and age. As shown in Table 6, very good correlation coefficients (R^2) were observed in this case. In addition, no "hidden" deviations from linearity have been evidenced by the output of the graphical analysis performed on the residuals of the regression line.

Table 6. Validation parameters for the SWV determination of the furanic index (FI) in honey.

LoD	LoQ	Linearity	Repeatability [3]	Intermediate Precision [4]	Trueness [5]
		Concentration range: 2–200 mg dm^{-3}	RSD% (FI, mg kg^{-1})	RSD% (FI, mg kg^{-1})	Two tails t-test ($p = 0.95$)
0.6 mg dm$^{-3,\,1}$	1.8 mg dm$^{-3,\,1}$	Y = (a ± s$_a$)X + (b ± s$_b$) a = 1.07 × 10^{-7} s$_a$ = 0.01 × 10^{-7}	9.9 (4.8) 4.4 (22)	3.0 (160)	$t_{tab} = 2.57$ $t_{exp} < 2.19$
1.2 mg kg$^{-1,\,2}$	3.6 mg kg$^{-1,\,2}$	b = 2 × 10^{-7} s$_b$ = 1 × 10^{-7} $R^2 = 0.9991$	2.3 (120) 0.72 (306)		

The LoD value was calculated according to [37]. [1] LoD calculated on a 50% solution of honey in a 0.1 mol dm^{-3} LiCl/water solvent system; [2] LoD calculated on pure honey; [3] parameter evaluated by analyzing four different honey samples five times in the same analytical session; [4] parameter evaluated by analyzing five times in five different analytical sessions over five months a honey sample exhibiting an intermediate FI with respect the concentration range measured; [5] evaluated by means of comparison with an independent analytical method (i.e., the RP-HPLC method, [33]) made on five different honey samples.

2.3.3. Precision

This parameter was evaluated on real samples in terms of repeatability and intermediate precision, which are both expressed as the percent relative standard deviation (RSD%). Repeatability was obtained by analyzing five times, in the same analytical session, four honey samples with a concentration range between 4.8 and 306 mg kg^{-1}. On the other hand, intermediate precision was evaluated by analyzing five times, in five different analytical sessions over five months, a honey sample showing an intermediate value of FI. As a function of the FI in the honey samples, the RSD% of repeatability ranged between 0.72% (FI of 306 mg kg^{-1}) and 9.9% (FI of 4.8 mg kg^{-1}), whereas the RSD% of the intermediate precision was of 3.0%, which was measured on a honey showing an FI of 160 mg kg^{-1}. All precision parameters were acceptable according to the Horwitz's theory [39].

2.3.4. Trueness

Given the absence of any Certified Reference Materials, but in consideration that an independent analytical method is currently available [33], trueness was evaluated through the comparison of results obtained from five real samples by both the proposed method and the RP-HPLC method previously published by this research group, which is able to singularly quantify HMF, 2F, 3F, 2FA, and 3FA in the same chromatographic run. The comparison of the experimental t (between 0.55 and 2.19) with the tabulated t (2.57 for 5 degrees of freedom, $p = 0.95$ in a two-tail t-test) allowed concluding that the proposed method was bias-free.

2.4. SWV Determination of the Furanic Index in Honeys
2.4.1. Method Application

The proposed method was tested on ten honey samples of different botanical origin (three from blossom, one from eucalyptus, one from thistle, and five from strawberry tree) and of different age (between 2014 and 2019). Figure 5 reports a typical SW voltammogram for one real sample.

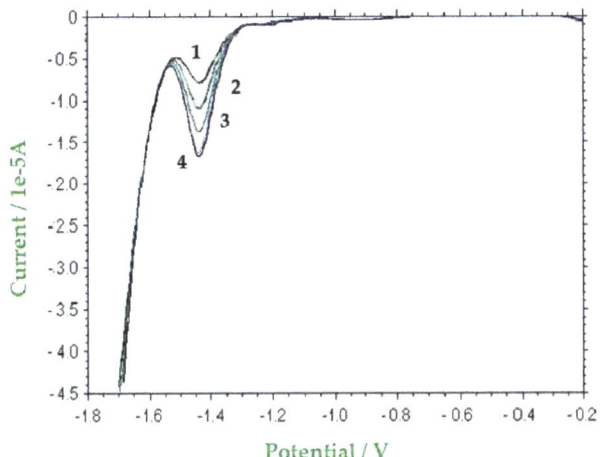

Figure 5. SW voltammograms of (1) a pure strawberry tree honey (50% w/w) in a 0.1 mol dm^{-3} LiCl solution in water; (2) the same solution 1, after the addition of 0.75 mg of HMF; (3) the same solution 1, after the addition of 1.50 mg of HMF; (4) the same solution 1, after the addition of 2.25 mg of HMF. WE: Hg; AE: Pt; RE: SCE. Potential scan rate: 0.10 V s^{-1}; pulse height: 4 mV; frequency: 15 Hz; modulation amplitude: 25 mV.

2.4.2. Furanic Index in Honeys from Different Botanical Origin and Age: Comparison with HPLC Data

Table 7 reports the amounts of FI measured in these samples. For comparison purposes, the last column of Table 7 also reported the sum of all analytes (i.e., HMF, 2F, 3F, 2FA, and 3FA) measured in the same samples by means of a literature RP-HPLC method [33], which are all expressed in equivalent amounts of HMF. It is possible to observe that the FI measured on blossom, eucalyptus, and thistle honey is never statistically different (two tails t-test, $p = 0.95$) from the amount measured by the RP-HPLC method. This fact is relevant, as it shows that the quicker SWV determination of the total amount of the furanic compounds is effective in the assessment of the freshness of the honey samples. In particular, RP-HPLC data substantiated the expected large predominance of the amount of HMF on those of the other analytes. As a matter of fact, HMF constitutes on average 93% of the total furanic species in these honeys. 2F and 2FA were sometimes quantified, at concentrations never higher than 3.0 mg kg^{-1} and 4.0 mg kg^{-1}, respectively, while 3F and 3FA were always below the relevant LoQs (i.e., 0.3 mg kg^{-1} and 0.09 mg kg^{-1}, respectively). The furanic indexes measured in these samples were always congruent to a recently produced and well-preserved honey: indeed, the measured FI ranged between 4.8 ± 0.4 mg kg^{-1} (B3 honey) and 22.0 ± 0.5 mg kg^{-1} (E honey).

Table 7. Furanic index (FI) in ten honeys of different floral origin and age. Comparison with RP-HPLC data.

Sample	Year of Production	Floral Origin	SWV FI (mg kg^{-1})	RP-HPLC [1] ΣHMF + Fs + FAs (mg kg^{-1})
B1	2018	Blossom	14.2 ± 0.6	13.3 ± 0.2
B2	2019	Blossom	12.0 ± 0.5	10.4 ± 0.2
B3	2019	Blossom	4.8 ± 0.4	3.7 ± 0.1
T	2018	Thistle	19.4 ± 0.8	17.4 ± 0.4
E	2019	Eucalyptus	22.0 ± 0.5	19.5 ± 0.3
S1	2015	Strawberry tree	192 ± 1	79 ± 1
S2	2016	Strawberry tree	245 ± 2	70.4 ± 0.8
S3	2018	Strawberry tree	120 ± 2	52 ± 1
S4	2016	Strawberry tree	160 ± 2	67 ± 2
S5	2014	Strawberry tree	306 ± 2	160 ± 2

All data are expressed as average ± standard deviation, n = 3. [1] Sum of the concentrations of HMF, of 2F, of 3F, of 2FA and of 3FA, expressed as mg of HMF on kg of honey.

On the other hand, a very different situation was exhibited by the five strawberry tree honeys: in comparison to the RP-HPLC data, the amount of the FI is much higher (from 191% to 350%) than the sum of furanic species measured by RP-HPLC. The fact that this result only occurred for the samples of this botanical origin—irrespectively of the age of the honey led to supposing that in this honey, a compound normally absent in the other honeys was present, which was reduced at the same potential of all the analytes considered. As a matter of fact, the strawberry tree honey is rich in 2,5-dihydroxyphenylacetic acid, homogentisic acid (HA), and L-tyrosine catabolite [40], and this compound has been found until now only in this honey [41,42]. The electrochemical behavior of HA was studied in the same conditions used for the characterization of all the furanic compounds, and the voltammetric evidences acquired both in CV and in SWV confirmed that it is irreversibly reduced in the cathodic direct scan, at −1.48 V vs. SCE. Figure 6 shows a SW voltammogram of a solution 0.1 mol dm^{-3} of HA.

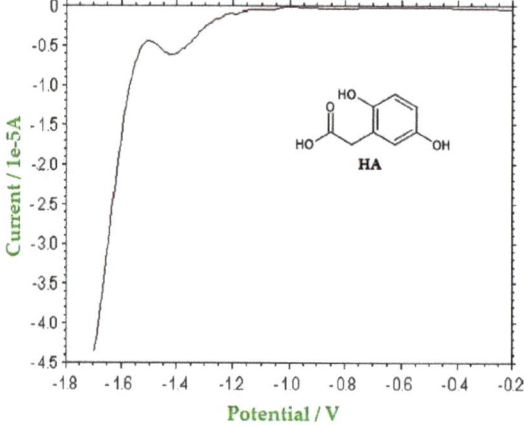

Figure 6. SW voltammogram of HA (0.1 mmol dm^{-3}) in 0.1 mol dm^{-3} LiCl aqueous solution, WE Hg; AE: Pt; RE: SCE. Potential scan rate: 0.10 V s^{-1}; pulse height: 4 mV; frequency: 15 Hz; modulation amplitude: 25 mV.

To verify the contribution of this molecule to the FI measured in strawberry tree honeys, HA was quantified in all the samples of this botanical origin by means of another RP-HPLC literature method [42]. The amounts of HA ranged between 92.5 ± 0.2 mg kg^{-1} and 241.2 ± 0.5 mg kg^{-1}, as shown in Table 8. This table also reports the theoretical amount of reducible species present in the strawberry tree honey, as measured by RP-HPLC, and expressed in equivalent HMF concentration, column RP-HPLC ΣHMF+Fs+FAs+HA. It is interesting to note that—for honey samples S1 and S3—this amount is not statistically different from that measured by SWV (criteria: two tails t-test, $p = 0.95$), whereas for the remaining samples, these data exhibit a slight bias (between −3.16%, sample S5 and −6.43%, sample S4) in comparison to the RP-HPLC one. However, this bias is well within the acceptability range for this level of concentration, as suggested by the AOAC International in its Peer Verified Methods Programs [43].

Table 8. RP-HPLC determination of homogentisic acid (HA) in the strawberry tree honeys and comparison among results obtained by means of both RP-HPLC and SWV methods.

Sample	RP-HPLC [1] HA (mg kg^{-1})	RP-HPLC [2] ΣHMF + Fs + FAs (mg kg^{-1})	RP-HPLC [3] ΣHMF + Fs + Fas + HA (mg kg^{-1})	SWV ΣHMF + Fs + Fas + HA (mg kg^{-1})
S1	150.0 ± 0.1	79 ± 1	194 ± 1	192 ± 1
S2	241.2 ± 0.5	70.4 ± 0.8	256 ± 1	245 ± 2
S3	92.5 ± 0.2	52 ± 1	123 ± 1	120 ± 2
S4	135.3 ± 0.3	67 ± 2	171 ± 2	160 ± 2
S5	202.8 ± 0.2	160 ± 2	316 ± 2	306 ± 2

All data are expressed as average ± standard deviation, $n = 3$. [1] Data obtained according [42]; [2] sum of the concentrations of HMF, 2F, 3F, 2FA, and 3FA, obtained according to [33] and expressed as mg of HMF on kg of honey; [3] sum of the concentrations of HMF, 2F, 3F, 2FA, 3FA, and HA, obtained according to [33,42] and expressed as mg of HMF on kg of honey.

3. Materials and Methods

3.1. Honey Samples

The study was carried out on 10 honey samples of different botanical origin: strawberry tree (*Arbutus unedo*), eucalyptus (*Eucalyptus camaldulensis*), thistle (*Galactites tomentosus*), and blossom. Some of these honeys were directly provided by beekeepers; the others were commercial samples. All the samples were produced in Italy, almost all in the Sardinia Island, between 2014 and 2019. The floral origin was established on the basis of the indications of the producers and verified by melissopalynological analysis [44]. All samples were stored in the dark at 4 °C until SWV analysis.

3.2. SWV Determination of FI in Honey Samples

Ca. 5 g (exactly weighted on an analytical balance) of homogenized honey was diluted to 10 cm^3 by a 0.1 mol dm^{-3} LiCl aqueous solution and transferred in a polarographic cell in glass equipped with a four-hole gas-tight cover. The solution was carefully deaerated with Ar for at least 15 min; hence, a gentle flow of Ar ensured an inert blanket over the solution during the tests. Quantification was performed by internal calibration, by means of the multiple additions method (i.e., with three consecutive spikings of HMF, respectively 50%, 100%, and 150% of the amount of the furanic index evaluated by measurements performed by a literature HPLC method [33]). Each analytical evaluation was replicated at least three times.

3.3. Chemicals and Reagents

2-furoic acid, >98%; 3-furoic acid, ≥98%; 2-furaldehyde, >99%; 3-furaldehyde, ≥97%; 5-hydroxymethylfurfural, ≥99%; 2,5-dihydroxyphenylacetic acid, ≥99%; mercury, 99.9995%; methanol anhydrous, >99.8%; methanol HPLC grade; 1.0000 mol dm^{-3} H$_2$SO$_4$ solution in water and type I water were purchased from Merck, Milan, Italy. In addition, lithium perchlorate, 99.99%; lithium chloride, 99.98%; sodium hydrogen carbonate, >99.7%; sodium acetate, >99.9%; disodium hydrogen phosphate, 99.99%; sodium dihydrogen phosphate, 99.99%; tetrabutylammonium perchlorate, >99%, all used as supporting electrolytes were analytical grade Merck reagents. 99.999% Ar gas was used for degassing the solvent systems used in all electrochemical experiments and was purchased from Sapio, Monza, Italy.

3.4. Equipment

All the electrochemical measurements were performed using a CHI-650 electrochemical workstation (CH Instruments, Austin, TX, USA) interfaced with a computer with the specific software CHI-650, in a conventional three-electrodes voltammetric cell in glass, equipped with a gas-tight four-hole cover, positioned inside a Faraday cage. Four different working electrodes (WE) were used in this study: a 2 mm diameter glassy carbon (GC) disk, a 2 mm diameter Pt disk, a ca. 1 mm diameter Au hemisphere electrode, and a ca. 1 mm diameter Hg/Au amalgam hemisphere electrode, respectively. A 5 mm Pt disk was always the auxiliary electrode (AE), and a saturated calomel electrode (SCE) was the reference electrode (RE); all potential values given were referred. Before use, all of the electrode surfaces of WE and AE were polished with alumina powder (0.3 µm of diameter) and rinsed with ultrapure water. All the experiments were performed at room temperature in an Ar saturated solution.

The HPLC equipment used for confirmatory purposes was a Series 200 apparatus (Perkin Elmer, Milan, Italy) formed by a binary pump, by an UV–vis variable wavelength detector, and by a sampling valve equipped with a 20 mm^3 sample loop. Separation was accomplished on an Alltima C18 column 250 mm × 4.6 mm, 5 µm particle size (Alltech, Sedriano, Italy) fitted with a guard cartridge packed with the same stationary phase. Data were processed by a Turbochrom Workstation Software (Perkin Elmer, Milan, Italy).

An Ultra-turrax mixer model T18 (IKA, Staufen, Germany) was used to homogenize honey samples before analysis.

4. Conclusions

Although HMF has been a well-recognized freshness index of honeys for years, it is possible that the aging and/or imperfect storage conditions allow the formation also of different furanic compounds, such as furaldehydes and furoic acids. In order to provide a thorough evaluation of any reactive furanic-based compounds formed in honeys by degradation of this matrix, a rapid and accurate SWV method has been developed, optimized, and validated to measure the amount of HMF as well as of 2-furaldehyde, 2F, 3-furaldehyde, 3F, 2-furoic acid, 2FA, and 3-furoic acid, 3FA. All these analytes underwent an irreversible reduction on Hg electrode at potentials close to −1.50 V vs. SCE. Hence, a new chemical index (i.e., the so-called "Furanic Index", FI) was proposed for a quick and accurate assessment of the freshness of honeys. FI expresses, as mg of HMF per kg, the total amount of HMF, 2F, 3F, 2FA, and 3FA found in honey. The method exhibits satisfactory LoD and LoQ values, excellent linearity inside the operative range of concentration (i.e., between 2 and 200 mg dm^{-3}). The method was tested on ten real honey samples that were different in age and floral origin. An excellent accuracy has been demonstrated for blossom as well as eucalyptus and thistle honeys, while for strawberry tree samples, homogentisic acid, the chemical marker of this floral origin, also contributed to FI as it was quantitatively reduced in the analytical conditions. However, RP-HPLC determination of HA allows evaluating the freshness of these samples by FI. So, it is possible to believe that the proposed method could be advantageously used for screening purposes due its simplicity and its speed.

Author Contributions: Conceptualization, N.S. and G.S.; methodology, N.S. and G.S.; validation, N.S., S.S. and G.S.; formal analysis, N.S. and S.S.; investigation, S.S.; data curation, N.S. and S.S.; writing—original draft preparation, N.S. and G.S.; writing—review and editing, N.S., S.S., M.C., I.F., M.I.P. and G.S.; supervision, N.S., M.I.P. and G.S.; funding acquisition, N.S., M.I.P. and G.S. All authors have read and agreed to the published version of the manuscript.

Funding: This research was funded by Università degli Studi di Sassari ("Fondo di Ateneo per la ricerca 2019").

Institutional Review Board Statement: Not applicable.

Informed Consent Statement: Not applicable.

Data Availability Statement: The data used to support the findings of this study are available from the corresponding author upon request.

Acknowledgments: Authors gratefully acknowledge Università degli Studi di Sassari for financial support ("Fondo di Ateneo per la ricerca 2019").

Conflicts of Interest: The authors declare no conflict of interest. The funders had no role in the design of the study; in the collection, analyses, or interpretation of data; in the writing of the manuscript, or in the decision to publish the results.

References

1. Sanna, G.; Pilo, M.I.; Piu, P.C.; Tapparo, A.; Seeber, R. Determination of heavy metals in honey by anodic stripping voltammetry at microelectrodes. *Anal. Chim. Acta* **2000**, *415*, 165–173. [CrossRef]
2. Islam, N.; Khalil, I.; Islam, A.; Gan, S.H. Toxic compounds in honey. *J. Appl. Toxicol.* **2014**, *34*, 733–742. [CrossRef] [PubMed]
3. Kuster, B.F.M. 5-Hydroxymethylfurfural (HMF). A review focusing on its manufacture. *Starch Stäerke* **1990**, *42*, 314–321. [CrossRef]
4. Markowicz, D.B.; Monaro, E.; Siguemoto, E.; Séfora, M.; Valdez, B. Maillard reaction products in processed foods: Pros and cons. In *Food Industrial Processes—Methods and Equipment*, 1st ed.; Valdez, B., Ed.; InTech: Rijeka, Croatia, 2012; pp. 281–300.
5. Khalil, M.I.; Sulaiman, S.A.; Gan, S.H. High 5-hydroxymethylfurfural concentrations are found in Malaysian honey samples stored for more than 1 year. *Food Chem. Toxicol.* **2010**, *48*, 2388–2392. [CrossRef] [PubMed]
6. Fallico, B.; Zappala, M.; Arena, E.; Verzera, A. Effects of conditioning on HMF content in unifloral honeys. *Food Chem.* **2004**, *85*, 305–313. [CrossRef]
7. Islam, A.; Khalil, I.; Islam, N.; Moniruzzaman, M.; Mottalib, A.; Sulaiman, S.A.; Gan, S.H. Physicochemical and antioxidant properties of Bangladeshi honeys stored for more than 1 year. *BMC Complement. Altern. Med.* **2012**, *12*, 177. [CrossRef]
8. Glatt, H.; Schneider, H.; Liu, Y. V79-hCYP2E1-hSULT1A1, a cell line for the sensitive detection of genotoxic effects induced by carbohydrate pyrolysis products and other food-borne chemicals. *Mutat. Res. Genet.* **2005**, *580*, 41–52. [CrossRef]
9. Lee, Y.C.; Shlyankevich, M.; Jeong, H.K.; Douglas, J.S.; Surh, Y.J. Bioactivation of 5-hydroxymethyl-2-furaldehyde to an electrophilic and mutagenic allylic sulfuric acid ester. *Biochem. Biophys. Res. Commun.* **1995**, *209*, 996–1002. [CrossRef]
10. Monien, B.H.; Engst, W.; Barknowitz, G.; Seidel, A.; Glatt, H. Mutagenicity of 5-hydroxymethylfurfural in V79 cells expressing human SULT1A1: Identification and mass spectrometric quantification of DNA adducts formed. *Chem. Res. Toxicol.* **2012**, *25*, 1484–1492. [CrossRef]
11. Zhao, L.; Chen, J.; Su, J.; Li, L.; Zhang, X.; Xu, Z.; Chen, T. In vitro antioxidant and antiproliferative activities of 5-hydroxymethylfurfural. *J. Agric. Food Chem.* **2013**, *61*, 10604–10611. [CrossRef]
12. Yamada, P.; Nemoto, M.; Shigemori, H.; Yokota, S.; Isoda, H. Isolation of 5-(hydroxymethyl) furfural from lycium chinense and its inhibitory effect on the chemical mediator release by basophilic cells. *Planta Med.* **2011**, *77*, 434–440. [CrossRef]
13. Kitts, D.D.; Chen, X.-M.; Jing, H. Demonstration of antioxidant and anti-inflammatory bioactivities from sugar–amino acid Maillard reaction products. *J. Agric. Food Chem.* **2012**, *60*, 6718–6727. [CrossRef] [PubMed]
14. Li, M.-M.; Wu, L.-Y.; Zhao, T.; Xiong, L.; Huang, X.; Liu, Z.-H.; Fan, X.-L.; Xiao, C.-R.; Gao, Y.; Ma, Y.-B.; et al. The protective role of 5-HMF against hypoxic injury. *Cell Stress Chaperones* **2011**, *16*, 267–273. [CrossRef] [PubMed]
15. Abdulmalik, O.; Safo, M.K.; Chen, Q.; Yang, J.; Brugnara, C.; Ohene-Frempong, K.; Abraham, D.J.; Asakura, T. 5-hydroxymethyl-2-furfural modifies intracellular sickle haemoglobin and inhibits sickling of red blood cells. *Br. J. Haematol.* **2005**, *128*, 552–561. [CrossRef] [PubMed]
16. Lin, S.-M.; Wu, J.-Y.; Su, C.; Ferng, S.; Lo, C.-Y.; Chiou, R.Y.-Y. Identification and mode of action of 5-hydroxymethyl-2-furfural (5-HMF) and 1-methyl-1,2,3,4-tetrahydro-β-carboline-3-carboxylic acid (MTCA) as potent xanthine oxidase inhibitors in vinegars. *J. Agric. Food Chem.* **2012**, *60*, 9856–9862. [CrossRef]
17. Bogdanov, S.; Lüllmann, C.; Martin, P.; von der Ohe, W.; Russmann, H.; Vorwhol, G.; Persano Oddo, L.; Sabatini, A.G.; Marcazzan, G.L.; Piro, R.; et al. Honey quality and international regulatory standards: Review by the International Honey Commission. *Bee World* **1999**, *80*, 61–69. [CrossRef]
18. White, J., Jr. Spectrophotometric method for hydroxymethylfurfural in honey. *J. Assoc. Off. Anal. Chem.* **1979**, *62*, 509–514. [CrossRef]
19. Winkler, O. Beitrag zum Nachweis und zur Bestimmung von Oxymethylfurfurol in Honig und Kunsthonig. *Z. Lebensm. Forsch. A* **1955**, *102*, 161–167. [CrossRef]
20. Jeuring, H.J.; Kuppers, F.J. High performance liquid chromatography of furfural and hydroxymethylfurfural in spirits and honey. *J. Assoc. Off. Anal. Chem.* **1980**, *63*, 1215–1218. [CrossRef]

21. Spano, N.; Casula, L.; Panzanelli, A.; Pilo, M.I.; Piu, P.C.; Scanu, R.; Tapparo, A.; Sanna, G. An RP-HPLC determination of 5-hydroxymethylfurfural in honey. The case of strawberry tree honey. *Talanta* **2006**, *68*, 1390–1395. [CrossRef]
22. Rizelio, V.M.; Gonzaga, L.V.; Borges, G.d.S.C.; Micke, G.A.; Fett, R.; Costa, A.C.O. Development of a fast MECK method for determination of 5-HMF in honey samples. *Food Chem.* **2012**, *133*, 1640–1645. [CrossRef]
23. Reyes-Salas, E.O.; Manzanilla-Cano, J.A.; Barceló-Quintal, M.H.; Juárez-Mendoza, D.; Reyes-Salas, M. Direct electrochemical determination of hydroxymethylfurfural (HMF) and its application to honey samples. *Anal. Lett.* **2006**, *39*, 161–171. [CrossRef]
24. Shamsipur, M.; Beigi, A.A.M.; Teymouri, M.; Tash, S.A.; Samimi, V. Electrocatalytic Application of Girard's Reagent T to Simultaneous Determination of Furaldehydes in Pharmaceutical and Food Matrices by Highly Sensitive Voltammetric Methods. *Electroanalysis* **2010**, *22*, 1314–1322. [CrossRef]
25. Salhi, I.; Samet, Y.; Trabelsi, M. Direct electrochemical determination of very low levels of 5-hydroxymethylfurfural in natural honey by cyclic and square wave voltammetric techniques. *J. Electroanal. Chem.* **2020**, *873*, 114326. [CrossRef]
26. Luo, K.; Chen, H.; Zhou, Q.; Yan, Z.; Su, Z.; Li, K. A Sensitive and Visual Molecularly Imprinted Fluorescent Sensor Incorporating CaF_2 Quantum Dots and b-cyclodextrins for 5-hydroxymethylfurfural Detection. *Anal. Chim. Acta* **2020**, *1124*, 113–120. [CrossRef]
27. Alonso-Lomillo, M.A.; Javier del Campo, F.; Javier Munõz Pascual, F. Preliminary Contribution to the Quantification of HMF in Honey by Electrochemical Biosensor Chips. *Electroanalysis* **2006**, *18*, 2435–2440. [CrossRef]
28. De la Iglesia, F.; Lázaro, F.; Puchades, R.; Maquieira, A. Automatic determination of 5-hydroxymethylfurfural (5-HMF) by a flow injection method. *Food Chem.* **1997**, *60*, 245–250. [CrossRef]
29. Besir, A.; Yazici, F.; Mortas, M.; Gul, O. A Novel Spectrophotometric Method Based on Seliwanoff Test to Determine 5-(Hydroxymethyl) Furfural (HMF) in Honey: Development, in House Validation and Application. *LWT Food Sci. Technol.* **2021**, *139*, 110602. [CrossRef]
30. Bonn, G.; Bobleter, O. Determination of the Hydrothermal Degradation Products of D-(U-14C) Glucose and D-(U-14C) Fructose by TLC. *J. Radioanal. Chem.* **1983**, *79*, 171–177. [CrossRef]
31. Sawamura, M.; Takemoto, K.; Matsuzaki, Y.; Ukeda, H.; Kusunose, H. Identification of Two Degradation Products from Aqueous Dehydroascorbic Acid. *J. Agric. Food Chem.* **1994**, *42*, 1200–1203. [CrossRef]
32. Nozal, M.J.; Bernal, J.L.; Toribio, L.; Jiménez, J.J.; Martin, M.T. High-performance liquid chromatographic determination of methyl anthranilate, hydroxymethylfurfural and related compounds in honey. *J. Chromatogr. A* **2001**, *917*, 95–103. [CrossRef]
33. Spano, N.; Ciulu, M.; Floris, I.; Panzanelli, A.; Pilo, M.I.; Piu, P.C.; Salis, S.; Sanna, G. A direct RP-HPLC method for the determination of furanic aldehydes and acids in honey. *Talanta* **2009**, *78*, 310–314. [CrossRef]
34. Yuan, J.-P.; Chen, F. Separation and identification of furanic compounds in fruit juices and drinks by high-performance liquid chromatography photodiode array detection. *J. Agric. Food Chem.* **1998**, *46*, 1286–1291. [CrossRef]
35. Sackz, A.A.; de Oliveira, M.F.; Okumura, L.L.; Stradiotto, N.R. Voltammetric behavior of 2-furaldehyde reduction in ethanol using glassy carbon electrode. *Eclètica Quìmica* **2002**, *27*, 141–151. [CrossRef]
36. Ganesan, M.; Kottaisamy, M.; Thangavelu, S.; Manisankar, P. Voltammetric studies of the reduction of furfural on glassy carbon electrode in aqueous solutions. *Bull. Electrochem.* **1992**, *8*, 424–426.
37. Mocak, J.; Bond, A.M.; Mitchell, S.; Scollary, G. A statistical overview of standard (IUPAC and ACS) and new procedures for determining the limits of detection and quantification: Application to voltammetric and stripping techniques. *Pure Appl. Chem.* **1997**, *69*, 297–328. [CrossRef]
38. Currie, L.A. Detection: International Update, and Some Emerging Di-Lemmas Involving Calibration, the Blank, and Multiple Detection Decisions. *Chemom. Intel. Laborat. Syst.* **1997**, *37*, 151–181. [CrossRef]
39. Horwitz, W. Evaluation of Analytical Methods Used for Regulation of Foods and Drugs. *Anal. Chem.* **1982**, *54*, 67–76. [CrossRef]
40. Fernández-Cañón, J.M.; Granadino, B.; De Bernabé, D.B.; Renedo, M.; Fernández-Ruiz, E.; Peñalva, M.A.; De Córdoba, S.R. The Molecular Basis of Alkaptonuria. *Nat. Genet.* **1996**, *14*, 19–24. [CrossRef]
41. Cabras, P.; Angioni, A.; Tuberoso, C.; Floris, I.; Reniero, F.; Guillou, C.; Ghelli, S. Homogentisic acid: A phenolic acid as a marker of Strawberry-Tree (Arbutus unedo) honey. *J. Agric. Food Chem.* **1999**, *47*, 4064–4067. [CrossRef]
42. Scanu, R.; Spano, N.; Panzanelli, A.; Pilo, M.I.; Piu, P.C.; Sanna, G.; Tapparo, A. Direct chromatographic methods for the rapid determination of homogentisic acid in strawberry tree (*Arbutus unedo* L.) honey. *J. Chromatogr. A* **2005**, *1090*, 76–80. [CrossRef] [PubMed]
43. AOAC International. *AOAC Peer Verified Methods Programs: Manual on Policies and Procedures*; Association of Official Analytical Chemists: Arlington, TX, USA, 1998.
44. Louveaux, J.; Maurizio, A.; Vorwhol, G. Methods of melissopalynology. *Bee World* **1978**, *59*, 139–157. [CrossRef]

Article

Multi-Elemental Analysis as a Tool to Ascertain the Safety and the Origin of Beehive Products: Development, Validation, and Application of an ICP-MS Method on Four Unifloral Honeys Produced in Sardinia, Italy

Andrea Mara [1], Sara Deidda [1], Marco Caredda [2], Marco Ciulu [1], Mario Deroma [3], Emanuele Farinini [4], Ignazio Floris [3], Ilaria Langasco [1], Riccardo Leardi [4], Maria I. Pilo [1], Nadia Spano [1] and Gavino Sanna [1,*]

1. Dipartimento di Chimica e Farmacia, Università degli Studi di Sassari, Via Vienna 2, I-07100 Sassari, Italy; a.mara@studenti.uniss.it (A.M.); saradeidda96@tiscali.it (S.D.); marcociulu@yahoo.it (M.C.); ilangasco@uniss.it (I.L.); mpilo@uniss.it (M.I.P.); nspano@uniss.it (N.S.)
2. AGRIS Sardegna, Loc. Bonassai S.S. 291 Km 18.6, I-07100 Sassari, Italy; mcaredda@agrisricerca.it
3. Dipartimento di Agraria, Università degli Studi di Sassari, Viale Italia 39/a, I-07100 Sassari, Italy; mderoma@uniss.it (M.D.); ifloris@uniss.it (I.F.)
4. Dipartimento di Farmacia, Università di Genova, Viale Cembrano 4, I-16148 Genova, Italy; farinini@difar.unige.it (E.F.); leardi@difar.unige.it (R.L.)
* Correspondence: sanna@uniss.it; Tel.: +39-079-229-500

Abstract: Despite unifloral honeys from Sardinia, Italy, being appreciated worldwide for their peculiar organoleptic features, their elemental signature has only partly been investigated. Hence, the principal aim of this study was to measure the concentration of trace and toxic elements (i.e., Ag, As, Ba, Be, Bi, Cd, Co, Cr, Cu, Fe, Hg, Li, Mn, Mo, Ni, Pb, Sb, Sn, Sr, Te, Tl, V, and Zn) in four unifloral honeys produced in Sardinia. For this purpose, an original ICP-MS method was developed, fully validated, and applied on unifloral honeys from asphodel, eucalyptus, strawberry tree, and thistle. Particular attention was paid to the method's development: factorial design was applied for the optimization of the acid microwave digestion, whereas the instrumental parameters were tuned to minimize the polyatomic interferences. Most of the analytes' concentration ranged between the relevant LoDs and few mg kg^{-1}, while toxic elements were present in negligible amounts. The elemental signatures of asphodel and thistle honeys were measured for the first time, whereas those of eucalyptus and strawberry tree honeys suggested a geographical differentiation if compared with the literature. Chemometric analysis allowed for the botanical discrimination of honeys through their elemental signature, whereas linear discriminant analysis provided an accuracy level of 87.1%.

Keywords: honey; ICP-MS; trace elements; toxic elements; botanical origin; geographical origin; asphodel; eucalyptus; strawberry tree; thistle

1. Introduction

Honey is an ancient natural and functional food used for centuries in the traditional medicine of many cultures. Beyond monosaccharides, such as glucose and fructose, which it is composed of up to 80% w/w, honey is a very complex matrix containing many bioactive compounds such as proteins, amino acids and enzymes, organic acids, polyphenols, flavonoids, vitamins, and inorganic elements [1]. This combination makes honey an outstanding functional food with many health-promoting activities and antimicrobial, antiviral, antifungal, anticancer, and antidiabetic properties [2–6].

Sardinia, Italy, is the second-largest island in the Mediterranean Sea. Its extension and distance from the European and African continental shelves characterize the endemic flora and fauna. Furthermore, due to the paucity of population and industries, the poor anthropogenic pressure is considered by consumers as an intrinsic guarantee of the quality of agri-food products, albeit sometimes this is not supported enough by scientific data. For these reasons, Sardinian honey is a typical product universally appreciated and recognized for its quality and for its peculiar organoleptic features.

Beekeeping is a significant sector of the regional agriculture, since Sardinia produces more than the 11% of Italian honey, and Italy is the fourth largest producer in the European Union [7]. Besides the multiflora ones, the production of Sardinian honey is mainly focused on four uniflora varieties: asphodel (*Asphodelus* spp.), eucalyptus (*Eucalyptus* spp.), strawberry tree (*Arbutus unedo* L.), and thistle (*Galactites tomentosa*) [8,9].

Given the importance of the sector for the economy of Sardinia, the traceability of their uniflora honeys should be ascertained with analytical methods aimed toward food authentication. Among the possible methods, those based on elemental metabolomics [10] provide great achievements in both botanical [11,12] and geographical classification [11,13]. The elemental metabolomics approach involves the determination of trace elements to achieve an elemental signature [10]. Hence, although they are not directly related with most of the therapeutical and nutraceutical properties of honey [14–17], trace elements are nevertheless of great scientific interest.

First, the health-promoting properties of honey must be coupled with the highest level of food safety and, consequently, the concentration of toxic elements should be negligible. Thus, honey is considered a valuable bioindicator of environmental pollution [18–21] because its chemical composition is strictly related to the environmental quality of the area next to the beehive [22,23]. In fact, the concentration of some potentially toxic elements has been found to be higher in hive matrices from industrial and urban areas with respect to what was measured in uncontaminated areas [18,24–27]. For these reasons, the concern for the potential presence of toxic elements in honey moved the European Union to set maximum levels for Hg (0.01 mg kg^{-1}) [28] and for Pb (0.1 mg kg^{-1}) [29].

Finally, the elemental signature of honey has been frequently used for its authentication and traceability because foraged plants tend to accumulate specific elements in the nectar, allowing for classification according to their botanical origin [30,31]. In addition, soil composition affects element availability in honey, allowing, in this case, their georeferentiation [31,32].

Despite trace elements being determined in honeys from many countries, such as Bulgaria [12], Kosovo [20], Ethiopia [24], Turkey [26], Italy [18,27,32,33], Poland and Greece [31], Hungary [34], and New Zealand [35], there are few contributions to the literature reporting on the amounts of trace elements in uniflora honeys produced in Sardinia. The literature is also poor regarding studies reporting on the elemental composition of the honeys of asphodel, eucalyptus, strawberry tree, and thistle produced out of Sardinia. Among them, eucalyptus honey was the most investigated [36–39] and data from Tunisia [36], Argentina [37], and Italy [38] have been reported in the literature. In contrast, the elemental composition of strawberry tree honey has been reported only for those coming from Croatia [40], whereas, to the best of our knowledge, no study has dealt with a trace element characterization of either asphodel or thistle honeys.

From an analytical viewpoint, microwave-assisted digestion [41,42] and inductively coupled plasma spectrometry (ICP-MS) [41,43] are the preferred techniques for sample pre-treatment and for elemental analysis, respectively. Among the former ones, microwave-assisted digestion ensures high efficiency and good performances [44–47]. On the other hand, ICP-MS allows to achieve great results in trace and ultra-trace analysis and to perform a reliable investigation for food authentication and traceability [13,41]. Regardless of the tool chosen for the elemental analysis, the data obtained should be elaborated using a multivariate approach, essential for classification purposes. Principal component analysis (PCA) [36], cluster analysis (CA) [32], discriminant analysis (DA) [35], partial least squares (PLS) [31], and self-organizing maps (SOMs) [12] provide great results in exploratory analysis and honeys classification.

This research group has been active for decades in the assessment and validation of original analytical methods applied to beehive matrices to ascertain their quality [48–55] as well as origin [56,57]. Hence, the principal aim of this research was to ascertain the elemental signature of the most renowned unifloral honeys of Sardinia as a reliable tool to ensure their healthiness and guarantee their origin. Therefore, an original ICP-MS method able to simultaneously measure the total amount of 23 elements of potential health concern (i.e., Ag, As, Ba, Be, Bi, Cd, Co, Cr, Cu, Fe, Hg, Li, Mn, Mo, Ni, Pb, Sb, Sn, Sr, Te, Tl, V, and Zn) was developed and thoroughly validated. Finally, the proposed method was applied to a large sampling of honeys from asphodel, eucalyptus, strawberry tree, and thistle.

2. Results and Discussion

2.1. Sample Pre-Treatment

To avoid the matrix effect and maximize the signal-to-noise ratio, great attention was paid to the optimization of the sample pre-treatment step [44–47]. Although the sample amounts and the instrumental conditions depend on the specific features of the microwave system, according to the recent method proposed by Astolfi et al. [42], HNO_3 and H_2O_2 were chosen as the acidic/oxidizing mixture due the fact of their proven capability to minimize the formation of the interfering polyatomic ions in the plasma and their efficiency in organic matter decomposition. According to Muller et al. [45], a simply 2^2 full factorial design was applied to improve the digestion efficiency and the composition of the oxidizing mixture. The optimization allowed for a reduction in the HNO_3 volume required for sample digestion, increasing the amount of the greenest hydrogen peroxide and the mass of sample digested, which is an important achievement for trace element analysis. These results were also achieved thanks to the versatile performance of the ultraWAVE digestion system. The optimized method is reported in Table 1.

Table 1. UltraWAVE SRC microwave digestion operational program and conditions.

Step		Time (min)	Temperature (°C)	
1st	Heating	25	240	Initial pressure: 40 bar
2nd	Holding	10	240	Final temperature: <40 °C
3rd	Cooling	~30	<40	Pressure release rate: 8 bar/min
				Rack: 15 positions
				Vessel: volume 15 cm^3, PTFE
				Sample amount: 0.7 g of honey
				Reagents: 0.5 cm^3 HNO_3, 3 cm^3 of H_2O_2, and 4 cm^3 H_2O

First, honey samples were heated until 40 °C and then homogenized by an Ultraturrax. Next, approximately 0.7 g of honey, exactly weighted on an analytical balance (± 0.0001 g uncertainty), was treated with 0.5 cm^3 of HNO_3, 3 cm^3 of H_2O_2, and 4 cm^3 of type I water inside a PTFE vessel of the SCR mineralization system. After the digestion, samples were diluted to a final volume of 15 cm^3 and filtered through a 0.22 μm nylon filter. Finally, they were stored at 4 °C in the dark until the analysis. The typical amount of residual acidity found in digested samples was 0.2 mol dm^{-3}, whereas the efficiency of organic matter decomposition (EOMD%) was generally higher than 94%.

2.2. Validation of the ICP-MS Method

The validation of the ICP-MS method proposed was accomplished in terms of the limit of detection (LoD), limit of quantification (LoQ), linearity, precision, and trueness. Table 2 reports the validation parameters of the ICP-MS method for the determination of the total amount of 23 trace elements in honey.

Table 2. Validation parameters of the ICP-MS method aimed at the determination of 23 trace elements in unifloral honeys.

Element	LoD [a] (µg kg^{-1})	LoQ [a] (µg kg^{-1})	Calibration Range [b] (µg dm^{-3})	Repeatability [c] CV (%)	Intermediate Precision [d] CV (%)	Trueness, Recovery (% ± s [e])
Ag	5	17	0.1–50	4	8	106 ± 5
As	2	7	0.1–50	4	6	92 ± 1
Ba	20	70	1–250	1	3	90 ± 20
Be	0.4	1.3	0.02–50	4	13	103 ± 1
Bi	0.1	0.3	0.005–50	4	12	85 ± 7
Cd	0.3	1.0	0.01–50	4	8	117 ± 1
Co	0.3	1.0	0.01–50	8	9	99 ± 1
Cr	7	23	0.1–50	4	5	97 ± 1
Cu	20	70	1–100	4	11	107 ± 4
Fe	30	100	1–100	4	17	105 ± 15
Hg	6	20	0.1–50	4	13	130 ± 10
Li	2	7	0.1–500	4	14	96 ± 1
Mn	8	27	0.1–500	3	3	107 ± 4
Mo	0.7	2.3	0.04–50	4	8	94 ± 1
Ni	3	10	0.1–100	6	18	95 ± 2
Pb	3	10	0.1–250	4	6	92 ± 1
Sb	0.7	2.3	0.04–50	4	9	115 ± 1
Sn	2.1	6.9	0.1–50	5	5	103 ± 2
Sr	3	10	0.1–100	3	7	103 ± 1
Te	1.2	3.9	0.04–50	4	21	108 ± 3
Tl	0.04	0.13	0.005–50	7	5	96 ± 1
V	0.2	0.7	0.01–50	4	6	94 ± 2
Zn	40	130	1–500	12	16	101 ± 1

[a] The LoD and LoQ values were measured according to [58]. [b] Instrumental calibration range. [c] Evaluated on the sample replicates in the same analytical session (n = 3). [d] Evaluated on the sample replicates within one month (n = 3). [e] Standard deviation.

The LoDs and LoQs of the method were calculated according to Currie [58] on 30 measurements of method blanks obtained in different analytical sessions. The LoDs were generally below 10 µg dm^{-3} for all of the elements except for Ba, Cu, Fe, and Zn, which were between 10 and 40 µg dm^{-3}.

Because of the large variability of the analyte concentrations, great attention was paid to the instrument calibration phase. Although the linearity of the ICP-MS could be extended over several orders of magnitude of concentration, in this case, this parameter was explored for each element only within the relevant operative range of concentration. For each analyte, the calibration function was the result of a linear regression of six standard solutions, three for each extreme point of the calibration range. To evaluate the experimental error and verify the linearity of the calibration function, three different solutions at an analyte concentration equal to the central point of the calibration range were analyzed. For each analyte, the difference between the experimental and predicted values was not significant (t-test, α = 0.05). Hence, keeping in consideration the random distribution of the residuals around the mean value as well as the very high coefficient of determination (R^2 always above 0.999), the linearity of the calibration function was successfully ascertained.

Precision, measured as the coefficient of variation (CV%), was assessed in terms of repeatability and intermediate precision. For this purpose, a batch of honey samples was replicated in the same way and in different analytical sessions. Table 2 reports the CV%s calculated at the analyte's average concentration. For elements in which the concentration in the honey was below the LoD, precision was measured by spiking the samples with standard solutions of analyte. Repeatability exhibited CV% between 1% (Ba) and 12% (Zn), while intermediate precision ranged between 3% (Ba and Mn) and 21% (Te). All precision parameters were in the range of CV% defined by the Horwitz's theory [59]; hence, the overall precision level of the method was acceptable.

Due to the lack of certified reference material (CRM) for the determination of trace elements in honey, trueness was evaluated by recovery tests. Hence, three aliquots of each sample were spiked with increasing amounts of a standard solution containing all analytes. All aliquots then underwent the whole analytical procedure described in Section 2.1. Furthermore, an additional aliquot for each honey sample was analyzed without any spiking. Recoveries ranged between 85% (Bi) and 130% (Hg). Quantitative recoveries (t-test, $\alpha = 0.05$) were observed for most of the elements, except for As, Li, Mo, Pb, and V (underestimation bias) and for Be, Cd, and Sb (overestimation bias). Nevertheless, the recoveries obtained in these cases were also acceptable according to the AOAC guidelines [60]. In summary, the proposed method was fully and successfully validated, and the parameters here considered were comparable or better than those reported in previous works [31,35,36,41,42].

2.3. Honey Analysis

Table 3 summarizes, in terms of both the mean value and range, the amount of toxic and trace elements belonging to the four unifloral Sardinian honeys, while the full data set is available in the Supplementary Materials (Table S1).

Table 3. Mean and range concentrations (in µg kg^{-1}) of 23 toxic and trace elements in the unifloral honey samples of asphodel, eucalyptus, strawberry tree, and thistle.

Element	Asphodel (n = 33) Asphodel spp.		Eucalyptus (n = 30) Eucalyptus spp.		Strawberry Tree (n = 31) Arbutus unedo L.		Thistle (n = 39) Galactites tomentosa	
	Mean	Range	Mean	Range	Mean	Range	Mean	Range
Ag	<5	<5–≤17	<5	<5–<5	<5	<5–≤17	<5	<5–<5
As	≤3	<2–9	≤6	<2–19	<2	<2–≤7	≤4	<2–≤7
Ba	180	<20–1600	340	80–690	550	≤70–2600	300	≤70–1800
Be	<0.4	<0.4–<0.4	<0.4	<0.4–<0.4	<0.4	<0.4–<0.4	<0.4	<0.4–<0.4
Bi	≤0.2	<0.1–1.6	≤0.2	<0.1–1.2	≤0.2	<0.1–0.8	≤0.6	<0.1–10.7
Cd	≤0.7	<0.3–2.5	1.7	<0.3–9.2	≤0.3	<0.3–≤1	1.4	<0.3–5.2
Co	2.5	≤1–10.2	7.6	≤1–109	1.8	<0.3–9.7	5.1	≤1–15.5
Cr	≤11	<7–24	≤17	<7–26	≤15	<7–24	≤8	<7–24
Cu	90	<20–250	170	≤70–630	90	<20–250	180	≤70–1030
Fe	160	≤100–660	570	≤100–1600	180	≤100–630	340	120–820
Hg	<6	<6–<6	<6	<6–<6	<6	<6–≤10	<6	<6–≤20
Li	<2	<2–14	11	<2–30	280	<2–8500	≤5	<2–22
Mn	190	40–770	2600	140–5100	330	≤27–4900	330	40–3300
Mo	≤1.8	<0.7–3.6	≤2.3	<0.7–3.9	1.8	<0.7–3.8	≤2.1	<0.7–3.7
Ni	19	<3–170	22	<3–122	12	<3–33	24	≤10–220
Pb	23	<3–400	16	<3–95	≤10	<3–90	≤9	<3–30
Sb	≤0.8	<0.7–≤2.3	≤1.3	<0.7–4.7	≤0.9	<0.7–≤2.3	≤0.8	<0.7–2.8
Sn	44	<2.1–210	30	≤6.9–110	43	<2.1–200	46	<7.1–240
Sr	38	<3–174	180	20–290	140	22–350	98	18–420
Te	≤1.5	<1.2–6.8	<1.2	<1.2–<1.2	<1.2	<1.2–<1.2	<1.2	<1.2–≤3.9
Tl	≤0.13	<0.04–0.4	0.3	<0.04–2.2	≤0.13	<0.04–1.4	0.18	<0.04–1.3
V	≤0.4	<0.2–≤0.7	4.1	≤0.7–12.8	≤0.6	<0.2–1.8	1.3	≤0.7–5.6
Zn	550	≤130–1400	660	330–1400	400	<40–1200	800	300–2000
Total (mg kg−1)	1.30	0.56–2.70	2.20	0.74–6.60	2.10	0.76–10.30	4.70	1.00–8.40

Each sample was analyzed twice. Italics represent data below the LoD; underlined values represent data below the LoQ.

Table 4 shows the results of the amounts of toxic and trace elements reported for unifloral eucalyptus honeys produced in different countries.

Table 4. Mean concentration and range [a] (both in µg kg^{-1}) of toxic and trace elements in unifloral Eucalyptus honeys from different geographical origins.

Element	Tunisia (n = 3) [36]	Argentina (n = 1) [37]	Italy (n = 1) [38]	Unknown (n = 1) [39]	Sardinia (n = 29) (This Work)
Ag					<5; <5–<5
As	19.08	<10	5.99	3.33	≤6; <2–19
Ba					340; 80–690
Be		<10			*<0.4; <0.4–<0.4*
Bi					*≤0.2; <0.1–1.2*
Cd	<0.01	<10	0.592	0.70	1.7; <0.3–9.2
Co		10			7.6; ≤1–109
Cr	130	<10	1.50	2.73	≤17; <7–26
Cu	800	120	219	140	170; ≤70–630
Fe	7100	3380	1008	914	570; ≤100–1600
Hg			<0.75		*<6; <6–<6*
Li					11; <2–30
Mn	1250	8840	1009	1976	2600; 140–5100
Mo					≤2.3; <0.7–3.9
Ni	220	50	11.3	8.04	22; <3–122
Pb	250	10	5.00	141	16; <3–95
Se	130	10	5.60		
Sb	100				≤1.3; <0.7–4.7
Sn				7.85	30; ≤6.9–110
Sr					180; 20–290
Ti	610				
Te					*<1.2; <1.2–<1.2*
Tl		<10			0.3; <0.04–2.2
U		<10			
V	50	<10		3.36	4.1; ≤0.7–12.8
Zn	2060	550	791	414	660; 330–1400

[a] Data are presented in the "average; range" format. n = number of honey samples analyzed. Italics represent data below the LoD; underlined values represent data below the LoQ.

The determination of As, Cd, Cr, Cu, Fe, Mn, Ni, Pb, and Zn was performed in all studies, and the amounts of these elements in honeys from different geographical origins normally spanned over one or two orders of magnitude. The different concentrations of the most abundant elements (i.e., Cr, Cu, Fe, Ni, Pb, and Zn, significantly higher in the samples from Tunisia, or the very high concentration of Mn in the Argentinian honey) allowed to envisage the feasibility of a discrimination of these honeys according to geographical origin. However, due to the paucity of the samples analyzed in the literature [36–39], no definitive conclusion could be obtained by this comparison.

The comparison of the data set obtained from the determination of 23 elements in nine samples of strawberry tree from Croatia [40] was more reliable. In this case, 17 out of 23 elements were measured in both countries. Meaningful differences in the average concentrations of many trace elements (i.e., Co, Cu, Mo, Ni, Pb, Sb, Sn, and V) appeared to support the possibility to achieve a differentiation according to the different geographical origin. The average distributions reported in Figure 1 highlights the differences between Croatian and Sardinian strawberry tree honeys.

Figure 1. Average distributions of strawberry tree honeys from Croatia and Sardinia. Elemental concentrations are in µg kg^{-1}.

2.4. Chemometric Analysis

In order to evaluate the suitability of trace elements to ascertain the origin of Sardinian honeys, principal component analysis (PCA) and linear discriminant analysis (LDA) were performed for exploratory and classification purposes, respectively.

The original data set, constituted by 133 samples and 18 variables (Ag, Be, Hg, Sb, and Te were removed because they were almost never quantified) were randomly divided into training (85 samples) and validation (48 samples) sets for internal validation.

The logarithmic transformation (log10) was used to reduce the skewness of the probability density distribution present in the original data. However, the information embodied in the loadings has changed and so it could lead to dangerous misunderstandings [61].

From the loading plot (Figure 2a), PC1 explained mainly the global concentration of the trace elements characterized by positive loadings on PC1 (Fe, V, Mo, and Mn). On the other hand, PC2 explained the contrast between the two clusters of correlated trace elements: Li, Sr, and Ba at positive loadings and Co, Cd, Cu, Zn, and Ni at negative loadings. Couples of elements, such as Mn and V, Cd and Co, and, mainly, Cu and Zn, were strongly correlated among them. Finally, Sn, Cr, Pb, and Bi were the less significant variables, as suggested by their overall low expression in both PC1 and PC2. Looking at the score plot in the plane PC1–PC2 (Figure 2b), with the objects colored according to the different botanical origin, it was noticed that the four classes were generally characterized by a specific location on the plane: asphodel honey samples (1) can be found at negative scores of both PC1 and PC2; eucalyptus samples (2) at positive scores of both PC1 and PC2; strawberry tree samples (3) at negative score of PC1 and positive scores of PC2 and, finally, thistle samples (4) can be found at positive scores of PC1 and negative scores of PC2. The information on the within-class variability of the four classes was smoothed by the logarithmic transformation, as well as the correlations between the variables. The high variability can be confirmed from the width of the ranges reported in Table 3. Furthermore, an analysis of the correlations was performed before and after the logarithmic transformation (Table S2) to ensure a correct interpretation of the variables. The coefficients highlight a strong correlation between Fe–Mn–V and Cu–Ni–Zn, confirmed by the loading plot, in contrast, the correlations between Ba–Sr–Li and Cd–Co were emphasized by the data pre-treatment.

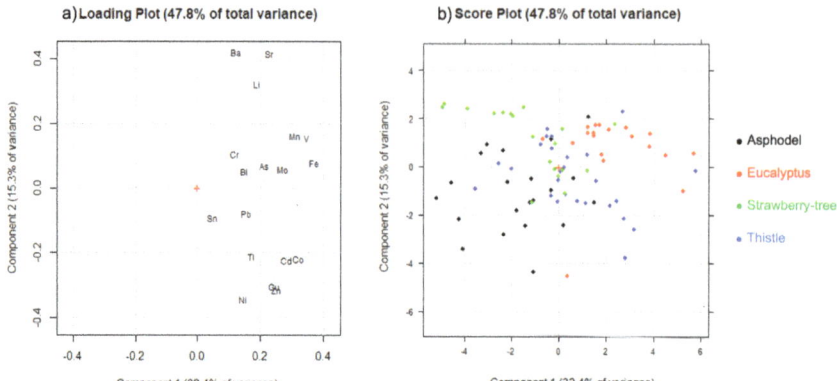

Figure 2. PCA performed on 85 unifloral honey samples and 18 trace elements: (a) loading plot; (b) score plot. Objects were colored according to the different botanical origin of the samples.

For classification purposes, linear discriminant analysis (LDA) was used. This method is based on the description of data by means of probability density distributions under two hypotheses: (i) probability distributions are multivariate normal within all the classes; (ii) dispersion and correlation structures are the same within all the classes. Fortunately, the method is quite robust against slight deviations from these hypotheses [62].

In this case, after the pre-processing, it was possible to use LDA because the classes were similar in size and orientation and, therefore, it was possible to assume that they had similar variance and covariance matrix.

The results obtained in cross-validation and prediction are reported in Table 5. The botanical origin with the lowest percentage of correct prediction was asphodel (61.9%), whose samples were mainly confused with thistle ($n = 3$) and strawberry tree ($n = 5$), while the best results were obtained for eucalyptus (84.2%). A permutation test was performed to compare the performance obtained with the correct assignment with a series of random permutation of the classes (in this case, 100 permutations were calculated). Figure S3 shows how the model works, since in no event in the classification were the results obtained randomly. In prediction, a correct classification percentage of 87.1% was achieved, quite in agreement with the results obtained in the CV. By observing the confusion matrix (Table 6), it is possible to look at the samples characterized by the wrong assignment, highlighting that asphodel and thistle were relatively less accurate but still acceptable: strawberry tree was twice erroneously classified as thistle and once as asphodel; eucalyptus was erroneously re-classified as both thistle and asphodel; a sample of asphodel was wrongly classified as a thistle.

Table 5. Results of the LDA, % of correct classified samples in the CV and prediction.

% Correctly Classified	Asphodel	Eucalyptus	Strawberry Tree	Thistle	Total
Cross-validation	61.9	84.2	78.9	76.9	75.5
Prediction [a]	91.7	81.5	75.0	100.0	87.1

[a] Internal validation.

Table 6. Confusion matrix in prediction.

	Asphodel	Eucalyptus	Strawberry Tree	Thistle
Asphodel	11	0	0	1
Eucalyptus	1	9	0	1
Strawberry tree	1	0	9	2
Thistle	0	0	0	13

The level of discrimination achieved in this contribution is less than those obtained by this research group using chemometric approaches based on untargeted physical and chemical data [49] or FT-ATR spectroscopic data [50].

Several factors could affect the ascertainment of the botanical origin. First, Sardinian unifloral honeys are typically underrepresented from a melissopalynological point of view since the pollen spectrum is often largely overlapping due to both accompanying and rare pollens [8]. This highlights a partly common botanical origin, especially for spring productions of asphodel and thistle and, secondly, also for summer and autumn production of eucalyptus and strawberry tree, respectively. Finally, a last possible factor, could be related to the geographical origin of some honeys, especially for those coming from the mining areas (i.e., central and southwestern Sardinia), where the natural characteristics of soil composition may have influenced the concentration of some elements.

In addition, a second LDA was unsuccessfully used in the attempt to achieve a geographical classification of Sardinian samples belonging to the same floral origin (data not reported). Probably, the reason for this failure was related both to the difficulty to localize the exact position of the hives [8] and to the extreme geochemical and pedological variability of the soil of Sardinia [63].

Summarizing, the elemental signature provided rather robust descriptors for the botanical discrimination of Sardinian unifloral honeys, whereas the intra-regional geographical discrimination was currently compromised and should be investigated by acquiring additional data. Hence, efforts will be paid to carefully solve these issues. Moreover, additional samples of both eucalyptus and strawberry tree honeys produced outside Sardinia will be gathered in order to achieve a reliable geographical discrimination.

3. Materials and Methods

3.1. Honey Samples

The honeys were mainly gathered in the last year directly from local beekeepers and, in a minor amount, purchased in Sardinian markets. The collection, summarized in Figure 3, consisted of 133 samples of the four most renowned unifloral varieties of honey: asphodel ($n = 33$), eucalyptus ($n = 30$), strawberry tree ($n = 31$), and thistle ($n = 39$). Samples were stored in the dark at 4 °C until the analysis. The botanical origin of each sample, primarily based on information directly provided by beekeepers, has been confirmed by the melissopalynological analysis, which provided data within the relevant ranges measured by Floris et al. [8].

Briefly, Sardinia is mainly hilly, therefore, 60% ($n = 80$) of the honeys are from the hills, 36% ($n = 48$) from the plains, and only 4% ($n = 5$) from the mountains. The production areas are divided into rural areas 50% (low urbanization degree, $n = 67$), slightly urbanized areas 45% (medium urbanization degree, $n = 62$) and urbanized 3% (high urbanization degree, $n = 4$) [64]. The odd distribution of the samples in the region, as reported in Figure 3, is representative of the different production areas and the distribution of the botanical sources.

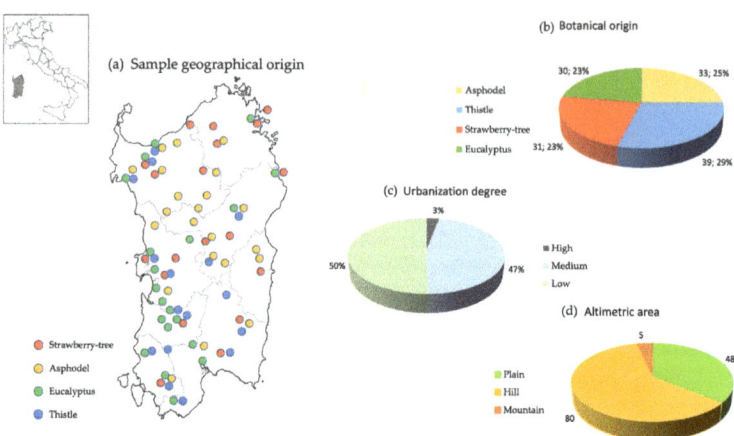

Figure 3. Information about Sardinian honey collection (n = 133). (**a**) Sample geographical origin; (**b**) Botanic source; (**c**) Urbanization degree; (**d**) Altimetric area.

3.2. Reagents and Standard Solutions

In all the analytical phases, type I water (resistivity > 18 MΩ cm^{-1}), produced by means of a MilliQ plus System (Millipore, Vimodrone, Italy) was used. Nitric acid (67–69% *w/w*, NORMATON® for ultra-trace analysis) and hydrogen peroxide (30% *w/w*, NORMATON® for ultra-trace metal analysis) were from VWR (Milan, Italy). The multi-element standard periodic table mix 1 for ICP (TraceCert®, 33 elements, 10 mg dm^{-3}) and the internal standard solution (3% HNO$_3$ *v/v* aqueous solution containing 1000 mg dm^{-3} of Rh) were from Sigma-Aldrich (St. Louis, MI, USA), while the elemental standards of Hg (10 mg dm^{-3}), Mo (10 mg dm^{-3}), Sb (1000 mg dm^{-3}), and Sn (100 mg dm^{-3}) were from Carlo Erba (Milan, Italy). NexION KED Setup Solution (1% HCl *v/v* aqueous solution containing Co 10 µg dm^{-3} and Ce 1 µg dm^{-3}) and NexION Setup Solution (1% HNO$_3$ *v/v* aqueous solution containing 1 µg dm^{-3} each of Be, Ce, Fe, In, Li, Mg, Pb, and U) were from Perkin Elmer (Milano, Italy). A standard solution of NaOH 0.5 mol dm^{-3}, and the potassium dichromate, the ammonium iron (II) sulphate, and the sulfuric acid (96% *v/v*) used for the titrations were from Sigma-Aldrich.

3.3. Instrumentation

The multi-element analysis was performed on a NexION 300X ICP-MS spectrometer, equipped with a S10 autosampler, a glass concentric nebulizer, a glass cyclonic spray chamber, and a kinetic energy discrimination (KED) collision cell, all produced by Perkin Elmer (Milan, Italy). Samples were mineralized by means of a microwave single reaction chamber (SCR) system (ultraWAVE™, Milestone, Sorisole, Italy) equipped with fifteen polytetrafluoroethylene (PTFE) vessels (volume: 15 cm^3 each). An Ultraturrax mixer model T18 (IKA, Staufen, Germany) was used to homogenize the honey samples before analysis. The residual acidity and the dissolved organic carbon (DOC) in the digested sample were determined by acid–base titration procedure and by the Walkley-Black method [65] using a Thermo Scientific Orion 950 titrator, whereas the total organic carbon (TOC) was determined by means of a CHN analyzer Leco 628. Nylon filters (pore diameter: 0.22 µm), polypropylene (PP) metal-free tubes, and polyethylene (PE) flasks were from VWR (Milan, Italy).

3.4. ICP-MS Method Assessment, Quality Control and Assurance

The minimization of the dissolved organic carbon (DOC) in digested honey samples is an essential prerequisite to make a reliable ICP-MS analysis. Failing in this, high residual amounts of organic substances from a partial oxidation of the saccharides contained in honey could cause the formation of molecular ions in the plasma that may interfere with the determination of very important elements such as ^{52}Cr, ^{63}Cu, and ^{75}As. Beyond this specific issue, the need to avoid the interference of molecular ions of any origin on the determination of the chosen isotopes of the analytes suggested to operate in kinetic energy discrimination (KED) mode for 15 elements out of 23. Therefore, the He flow has been carefully optimized for each element, to find the best compromise between the minimization of the polyatomic ion interferences and the maximization of the instrumental signal [66–69]. As a result, ^{75}As, ^{138}Ba, ^{111}Cd, ^{59}Co, ^{52}Cr, ^{63}Cu, ^{57}Fe, ^{55}Mn, ^{60}Ni, ^{121}Sb, ^{120}Sn, ^{88}Sr, ^{120}Te, ^{51}V, and ^{66}Zn were analyzed in KED mode using a He flow rate between 3 and 4 cm^3 min^{-1}, while ^{107}Ag, ^{9}Be, ^{209}Bi, ^{7}Li, ^{202}Hg, ^{98}Mo, ^{208}Pb, and ^{205}Tl were analyzed in normal mode. The optimized instrumental parameters and the elemental settings used for each analyte are reported in Table S3.

The matrix effect was determined comparing the slopes of the calibration function obtained in the absence (2% HNO$_3$ v/v aqueous solutions) or presence (digested honey samples spiked with known amounts of analyte) of the matrix [70]. The data obtained accounted for a substantial absence of any matrix effect for all the analytes considered in the whole calibration range, hence quantification has been accomplished by means of external calibration. The samples, blank-diluted when necessary to lead analyte concentrations within the relevant calibration range, were analyzed in duplicate. Data obtained as a result of a triplicate ICP-MS measurement, were blank-corrected. To compensate for any signal instability, a solution of Rh (10 µg dm^{-3}) was used as an internal standard, while the reliability of the measurements was ensured by analyzing one blank and a standard solution containing each analyte (50 µg dm^{-3}) every 10 samples. Memory effects between consecutive samples were eliminated by interposing a washing cycle of 60 s with a 2% HNO$_3$ v/v aqueous solution.

3.5. Optimization of the Composition of the Acidic/Oxidizing Mixture

After some preliminary evaluations, the digestion program (Table 1; pression, temperature and time) and water volume of the oxidizing mixture (4 cm^3) were kept constant, while the factors considered in the design were the sample amount (0.5–1.0 g), and the ratio between the volumes of HNO$_3$ and H$_2$O$_2$ (0.5/3 cm^3–2/1.5 cm^3). As a result, a two-level 2^2 full factorial design was applied to improve the composition of the oxidizing mixture and minimize the dissolved organic carbon in digested sample.

To estimate the experimental error and validate the regression model, a duplicate of the central point was added to the factorial design. The responses of the design were the residual acidity in the digested sample and the efficiency of organic matter decomposition [71] (EOMD%), which was calculated with Equation (1):

$$\text{EOMD\%} = [(\text{TOC} - \text{DOC})/\text{TOC}]\% \qquad (1)$$

where TOC is the total organic carbon, in mg kg^{-1}; DOC is the dissolved organic carbon, in mg kg^{-1}.

The experimental matrix obtained with the two-level full factorial design is reported in Table 7 as well as the experimental plan.

Table 7. Two-level 2^2 full factorial experimental design with center point.

Experiment	Sample Amount (g)	Ratio HNO$_3$/H$_2$O$_2$	X_1 [a]	X_2 [a]	Residual Acidity (mol dm^{-3})	EOMD%
1	0.50	0.17	−1	−1	0.39	99.7
2	1.00	0.17	+1	−1	0.06	87.2
3	0.50	1.33	−1	+1	1.41	97.7
4	1.00	1.33	+1	+1	0.16	99.1
5	0.75	0.56	0	0	0.48	95.8
6	0.75	0.56	0	0	0.39	96.4

[a] X_1 and X_2 represent the sample amount and the ratio HNO$_3$/H$_2$O$_2$, respectively.

The results obtained with the full factorial design are reported in Table 7. Multilinear regression (MLR) provided the coefficients for Equation (2) for both the responses, residual acidity (Y_1) and EOMD% (Y_2), respectively, which are reported in Table 8.

$$Y_n = b_0 + b_1 \times 1 + b_2 \times 2 + b_{12} \times 1 \times 2 \quad (2)$$

where b_0 is a constant, b_1 and b_2 are the coefficients of the main effects of the factors X_1 and X_2, whereas b_{12} is the coefficient of their interaction.

Table 8. Values of MLR coefficients and their significance levels for both the design's responses.

Coefficient	Residual Acidity		EOMD%	
	Coefficient Value	Significance [a]	Coefficient Value	Significance [a]
b_0	0.49	***	95.9	***
b_1	−0.41	**	−2.8	*
b_2	0.27	*	2.5	*
b_{12}	−0.25	*	3.5	**

[a] Blank, not statistically significant; * $p < 0.05$, ** $p < 0.01$, and *** $p < 0.001$.

As reported in Table 8, all the coefficients were significant for both the responses, including the coefficient of the interactions (b_{12}) especially for EOMD%. From the experiments at the central point, the experimental error can be estimated, and the standard deviation for Y_1 and Y_2 were 0.06 and 0.42, respectively. The predicted values for the residual acidity and for EOMD% at the center point were 0.5 ± 0.1 and 96 ± 1. Since the difference values between the experimental and predicted values are not significant (t-test, $\alpha = 0.05$), the model can be accepted.

The effects of the sample amount and the ratio between the oxidizing compounds on the residual acidity and the EOMD% of the digested samples are shown in the contour plots in Figure 4.

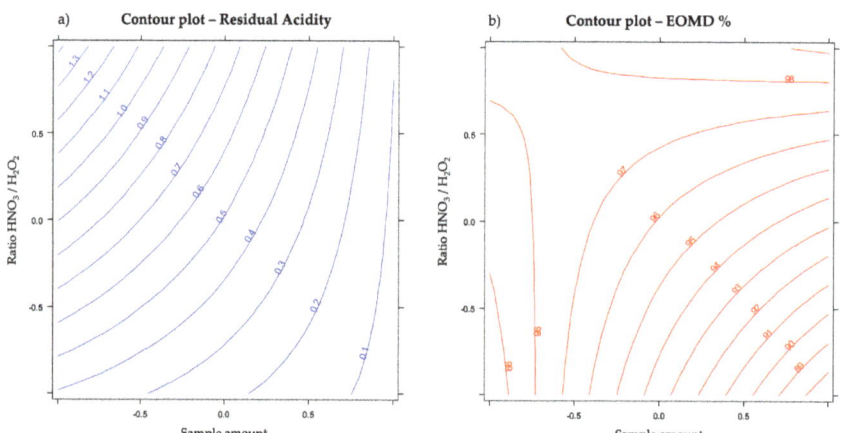

Figure 4. Contour plots for the effects on the design's responses: (a) residual acidity; (b) EOMD%.

From Figure 4 it is possible to understand the interaction between the two variables:

Y_1. Residual Acidity: the effect of the ratio HNO_3/H_2O_2 (X_2) on the residual acidity is present only at the lower sample amount (X_1) where its increase leads to higher acidity, while at higher amount var. X_2 has no effect; on the other hand, at higher ratios (X_2), an increase in the sample amount (X_1) led to a higher decrease in the response, greater than the decrease occurring at lower ratios (X_2).

Y_2. EOMD%: the effect of the ratio HNO_3/H_2O_2 (X_2) on the EOMD is present only at the higher sample amount (X_1) where its increase led to a higher response, while at the lower sample amount var. X_2 had no effect; on the contrary, at lower ratios (X_2), an increase in the sample amount (X_1) led to a decrease in the response, while at higher ratios, the sample amount had no effect.

It is evident that a good compromise between the two responses provides a sample amount of approximately 0.7 g of honey and a ratio HNO_3/H_2O_2 less than 0.5. Therefore, in order to minimize the consumption of nitric acid for the benefit of the greenest hydrogen peroxide, the acceptable composition of the acidic/oxidizing mixture is 0.5 cm^3 HNO_3, 3 cm^3 of H_2O_2 and 4 cm^3 of H_2O.

3.6. Statistical Analysis

A two-tail t-test at $\alpha = 0.05$ was used in ascertaining the existence of matrix effect as well as in the trueness evaluation. Principal components analysis, LDA, and MLR were performed by means of the R-based software Chemometric Agile Tool (CAT) developed by the Italian group of Chemometrics [72].

4. Conclusions

For the first time, the concentration of 23 trace elements were measured in a very large sampling of the most renowned unifloral honeys from Sardinia, Italy, using an original and validated ICP-MS method. Special attention was paid to the development of the acid microwave digestion procedure as well as the optimization of instrumental parameters to improve the efficiency of the organic matter decomposition and to minimize the polyatomic interferences, respectively. Among the most abundant elements (i.e., Ba, Mn, Fe, Cu, and Zn), only Mn was measured in all of the samples, whereas the others ranged from the relevant LoDs to a few mg kg^{-1}. Toxic elements were almost always below the amounts of potential health concern, hence confirming a very good level of food safety for Sardinian honeys. Since no elemental signatures were reported in the literature for asphodel and thistle honeys, a meta-analysis was carried out only on eucalyptus and strawberry tree honeys and it highlighted the possibility of a geographical discrimination thanks to their elemental signature, mainly for strawberry tree honeys from Croatia and Sardinia. Finally, through the elemental signature of the four unifloral honeys here considered, a good classification based on the botanical origin was accomplished by means of linear discrimination analysis.

Supplementary Materials: The following supporting information can be downloaded at: https://www.mdpi.com/article/10.3390/molecules27062009/s1, Figure S1: Scree plot; Figure S2: Influence plot; Figure S3: Permutation test. Table S1: Whole data set of the determination of 23 trace elements in 133 Sardinian unifloral honeys (Excel file); Table S2: Correlation coefficients; Table S3: Instrumental parameters and elemental settings used for the ICP-MS determination of 23 trace elements in unifloral honeys.

Author Contributions: Conceptualization, A.M. and G.S.; methodology, A.M., I.F., M.D. and G.S.; validation, A.M. and G.S.; formal analysis, A.M., M.C. (Marco Caredda) and M.C. (Marco Ciulu); investigation, A.M., I.L., S.D. and M.D.; resources, I.F. and G.S.; data curation, A.M., M.C. (Marco Caredda) and G.S.; writing—original draft preparation, A.M., E.F. and G.S.; writing—review and editing, A.M., M.C. (Marco Caredda), M.C. (Marco Ciulu), I.F., R.L., M.I.P., N.S. and G.S.; visualization, A.M., M.C. (Marco Caredda), M.C. (Marco Ciulu), E.F., R.L. and G.S.; supervision, G.S.; project administration, G.S.; funding acquisition, G.S. All authors have read and agreed to the published version of the manuscript.

Funding: This research was funded by the Università degli Studi di Sassari ("Fondo di Ateneo per la ricerca 2020").

Institutional Review Board Statement: Not applicable.

Data Availability Statement: The data presented in this study are reported in the Supplementary Materials.

Acknowledgments: The authors gratefully thank all the beekeepers, who provided the samples.

Conflicts of Interest: The authors declare no conflict of interest.

Sample Availability: Honey samples are, in part, available from the authors.

References

1. Bogdanov, S.; Jurendic, T.; Sieber, R.; Gallmann, P. Honey for Nutrition and Health: A Review. *J. Am. Coll. Nutr.* **2008**, *27*, 677–689. [CrossRef] [PubMed]
2. Cianciosi, D.; Forbes-Hernández, T.Y.; Afrin, S.; Gasparrini, M.; Reboredo-Rodriguez, P.; Manna, P.P.; Zhang, J.; Bravo Lamas, L.; Martínez Flórez, S.; Agudo Toyos, P.; et al. Phenolic Compounds in Honey and Their Associated Health Benefits: A Review. *Molecules* **2018**, *23*, 2322. [CrossRef] [PubMed]
3. De-Melo, A.A.M.; De Almeida-Muradian, L.B.; Sancho, M.T.; Pascual-Maté, A. Composition and properties of *Apis mellifera* honey: A review. *J. Apic. Res.* **2017**, *57*, 5–37. [CrossRef]
4. Rahman, M.M.; Alam, N.; Fatima, N.; Shahjalal, H.M.; Gan, S.H.; Khalil, I. Chemical composition and biological properties of aromatic compounds in honey: An overview. *J. Food Biochem.* **2017**, *41*, e12405. [CrossRef]
5. Da Silva, P.M.; Gauche, C.; Gonzaga, L.V.; Costa, A.C.O.; Fett, R. Honey: Chemical composition, stability and authenticity. *Food Chem.* **2016**, *196*, 309–323. [CrossRef] [PubMed]
6. Solayman, M.; Islam, M.A.; Paul, S.; Ali, Y.; Khalil, M.I.; Alam, N.; Gan, S.H. Physicochemical Properties, Minerals, Trace Elements, and Heavy Metals in Honey of Different Origins: A Comprehensive Review. *Compr. Rev. Food Sci. Food Saf.* **2016**, *15*, 219–233. [CrossRef] [PubMed]
7. Istituto di Servizi per il Mercato Agricolo Alimentare (Ismea). Il settore apistico: Analisi di mercato. Roma. 2019. Available online: https://www.ismeamercati.it/flex/cm/pages/ServeBLOB.php/L/IT/IDPagina/9685 (accessed on 16 February 2022).
8. Floris, I.; Satta, A.; Ruiu, L. Honeys of Sardinia (Italy). *J. Apic. Res.* **2007**, *46*, 198–209. [CrossRef]
9. Floris, I.; Pusceddu, M.; Satta, A. The Sardinian Bitter Honey: From Ancient Healing Use to Recent Findings. *Antioxidants* **2021**, *10*, 506. [CrossRef]
10. Zhang, P.; Georgiou, C.A.; Brusic, V. Elemental metabolomics. *Brief. Bioinform.* **2017**, *19*, 524–536. [CrossRef]
11. Danezis, G.P.; Georgiou, C.A. Elemental metabolomics: Food elemental assessment could reveal geographical origin. *Curr. Opin. Food Sci.* **2022**, *44*, 100812. [CrossRef]
12. Voyslavov, T.; Mladenova, E.; Balkanska, R. A New Approach for Determination of the Botanical Origin of Monofloral Bee Honey, Combining Mineral Content, Physicochemical Parameters, and Self-Organizing Maps. *Molecules* **2021**, *26*, 7219. [CrossRef] [PubMed]
13. Drivelos, S.A.; Georgiou, C.A. Multi-element and multi-isotope-ratio analysis to determine the geographical origin of foods in the European Union. *Trends Anal. Chem.* **2012**, *40*, 38–51. [CrossRef]
14. Afrin, S.; Haneefa, S.M.; Fernandez-Cabezudo, M.J.; Giampieri, F.; Al-Ramadi, B.K.; Battino, M. Therapeutic and preventive properties of honey and its bioactive compounds in cancer: An evidence-based review. *Nutr. Res. Rev.* **2019**, *33*, 50–76. [CrossRef]
15. Viuda-Martos, M.; Ruiz-Navajas, Y.; Fernández-López, J.; Pérez-Álvarez, J.A. Functional properties of honey, propolis, and royal jelly. *J. Food Sci.* **2008**, *73*, R117–R124. [CrossRef] [PubMed]
16. Cornara, L.; Biagi, M.; Xiao, J.; Burlando, B. Therapeutic Properties of Bioactive Compounds from Different Honeybee Products. *Front. Pharmacol.* **2017**, *8*, 412. [CrossRef]
17. Denisow, B.; Denisow-Pietrzyk, M. Biological and therapeutic properties of bee pollen: A review. *J. Sci. Food Agric.* **2016**, *96*, 4303–4309. [CrossRef]
18. Satta, A.; Verdinelli, M.; Ruiu, L.; Buffa, F.; Salis, S.; Sassu, A.; Floris, I. Combination of beehive matrices analysis and ant biodiversity to study heavy metal pollution impact in a post-mining area (Sardinia, Italy). *Environ. Sci. Pollut. Res.* **2012**, *19*, 3977–3988. [CrossRef]
19. Goretti, E.; Pallottini, M.; Rossi, R.; La Porta, G.; Gardi, T.; Goga, B.C.; Elia, A.C.; Galletti, M.; Moroni, B.; Petroselli, C.; et al. Heavy metal bioaccumulation in honey bee matrix, an indicator to assess the contamination level in terrestrial environments. *Environ. Pollut.* **2019**, *256*, 113388. [CrossRef]
20. Kastrati, G.; Paçarizi, M.; Sopaj, F.; Tašev, K.; Stafilov, T.; Mustafa, M. Investigation of Concentration and Distribution of Elements in Three Environmental Compartments in the Region of Mitrovica, Kosovo: Soil, Honey and Bee Pollen. *Int. J. Environ. Res. Public Health* **2021**, *18*, 2269. [CrossRef]
21. Rashed, M.N.; El-Haty, M.T.A.; Mohamed, S.M. Bee honey as environmental indicator for pollution with heavy metals. *Toxicol. Environ. Chem.* **2009**, *91*, 389–403. [CrossRef]
22. Bargańska, Ż.; Ślebioda, M.; Namieśnik, J. Honey bees and their products: Bioindicators of environmental contamination. *Crit. Rev. Environ. Sci. Technol.* **2016**, *46*, 235–248. [CrossRef]
23. Ćirić, J.; Spirić, D.; Baltić, T.; Lazić, I.B.; Trbović, D.; Parunović, N.; Petronijević, R.; Đorđević, V. Honey Bees and Their Products as Indicators of Environmental Element Deposition. *Biol. Trace Elem. Res.* **2020**, *199*, 2312–2319. [CrossRef] [PubMed]
24. Yayinie, M.; Atlabachew, M. Multi-element Analysis of Honey from Amhara Region-Ethiopia for Quality, Bioindicator of Environmental Pollution, and Geographical Origin Discrimination. *Biol. Trace Elem. Res.* **2022**, 1–15. [CrossRef] [PubMed]

5. Fodor, P.; Molnar, E. Honey as an environmental indicator: Effect of sample preparation on trace element determination by ICP-AES. *Mikrochim. Acta* **1993**, *112*, 113–118. [CrossRef]
6. Tuzen, M.; Silici, S.; Mendil, D.; Soylak, M. Trace element levels in honeys from different regions of Turkey. *Food Chem.* **2007**, *103*, 325–330. [CrossRef]
7. Perna, A.M.; Grassi, G.; Gambacorta, E.; Simonetti, A. Minerals content in Basilicata region (southern Italy) honeys from areas with different anthropic impact. *Int. J. Food Sci. Technol.* **2021**, *56*, 4465–4472. [CrossRef]
8. Commission Regulation (EU) 2018/73 of 16 January 2018 Amending Annexes II and III to Regulation (EC) No 396/2005 of the European Parliament and of the Council as Regards Maximum Residue Levels for Mercury Compounds in or on Certain Products. Available online: https://eur-lex.europa.eu/eli/reg/2018/73/oj (accessed on 16 February 2022).
9. Commission Regulation (EU) 2015/1005 of 25 June 2015 Amending Regulation (EC) No 1881/2006 as Regards Maximum Levels of Lead in Certain Foodstuffs. Available online: https://eur-lex.europa.eu/eli/reg/2015/1005/oj (accessed on 16 February 2022).
10. Bogdanov, S.; Haldimann, M.; Luginbühl, W.; Gallmann, P. Minerals in honey: Environmental, geographical and botanical aspects. *J. Apic. Res.* **2007**, *46*, 269–275. [CrossRef]
11. Drivelos, S.A.; Danezis, G.P.; Halagarda, M.; Popek, S.; Georgiou, C.A. Geographical origin and botanical type honey authentication through elemental metabolomics via chemometrics. *Food Chem.* **2020**, *338*, 127936. [CrossRef]
12. Quinto, M.; Miedico, O.; Spadaccino, G.; Paglia, G.; Mangiacotti, M.; Li, D.; Centonze, D.; Chiaravalle, A.E. Characterization, chemometric evaluation, and human health-related aspects of essential and toxic elements in Italian honey samples by inductively coupled plasma mass spectrometry. *Environ. Sci. Pollut. Res.* **2016**, *23*, 25374–25384. [CrossRef]
13. Pisani, A.; Protano, G.; Riccobono, F. Minor and trace elements in different honey types produced in Siena County (Italy). *Food Chem.* **2008**, *107*, 1553–1560. [CrossRef]
14. Czipa, N.; Andrási, D.; Kovács, B. Determination of essential and toxic elements in Hungarian honeys. *Food Chem.* **2015**, *175*, 536–542. [CrossRef] [PubMed]
15. Grainger, M.N.C.; Klaus, H.; Hewitt, N.; French, A.D. Investigation of inorganic elemental content of honey from regions of North Island, New Zealand. *Food Chem.* **2021**, *361*, 130110. [CrossRef]
16. Di Bella, G.; Potortì, A.G.; Beltifa, A.; Ben Mansour, H.; Nava, V.; Turco, V.L. Discrimination of Tunisian Honey by Mineral and Trace Element Chemometrics Profiling. *Foods* **2021**, *10*, 724. [CrossRef] [PubMed]
17. Conti, M.E.; Finoia, M.G.; Fontana, L.; Mele, G.; Botrè, F.; Iavicoli, I. Characterization of Argentine honeys on the basis of their mineral content and some typical quality parameters. *Chem. Cent. J.* **2014**, *8*, 44. [CrossRef] [PubMed]
18. Forte, G.; D'Ilio, S.; Caroli, S. Honey as a Candidate Reference Material for Trace Elements. *J. AOAC Int.* **2001**, *84*, 1972–1975. [CrossRef]
19. Caroli, S.; Forte, G.; Alessandrelli, M.; Cresti, R.; Spagnoli, M.; D'Ilio, S.; Pauwels, J.; Kramer, G. A pilot study for the production of a certified reference material for trace elements in honey. *Microchem. J.* **2000**, *67*, 227–233. [CrossRef]
20. Lovaković, B.T.; Lazarus, M.; Karačonji, I.B.; Jurica, K.; Semren, T.Ž.; Lušić, D.; Brajenović, N.; Pelaić, Z.; Pizent, A. Multi-elemental composition and antioxidant properties of strawberry tree (*Arbutus unedo* L.) honey from the coastal region of Croatia: Risk-benefit analysis. *J. Trace Elem. Med. Biol.* **2018**, *45*, 85–92. [CrossRef]
21. Pohl, P.; Bielawska-Pohl, A.; Dzimitrowicz, A.; Jamroz, P.; Welna, M.; Lesniewicz, A.; Szymczycha-Madeja, A. Recent achievements in element analysis of bee honeys by atomic and mass spectrometry methods. *Trends Anal. Chem.* **2017**, *93*, 67–77. [CrossRef]
22. Astolfi, M.L.; Conti, M.E.; Marconi, E.; Massimi, L.; Canepari, S. Effectiveness of Different Sample Treatments for the Elemental Characterization of Bees and Beehive Products. *Molecules* **2020**, *25*, 4263. [CrossRef]
23. Caroli, S. Determination of essential and potentially toxic trace elements in honey by inductively coupled plasma-based techniques. *Talanta* **1999**, *50*, 327–336. [CrossRef]
24. Sadowska, M.; Hyk, W.; Ruszczyńska, A.; Roszak, J.; Mycka, A.; Krasnodębska-Ostręga, B. Statistical evaluation of the effect of sample preparation procedure on the results of determinations of selected elements in environmental samples. Honey bees as a case study. *Chemosphere* **2021**, *279*, 130572. [CrossRef] [PubMed]
25. Muller, E.I.; Müller, C.C.; de Souza, J.P.; Muller, A.L.; Enders, M.S.; Doneda, M.; Frohlich, A.C.; Iop, G.D.; Anschau, K.F. Green microwave-assisted wet digestion method of carbohydrate-rich foods with hydrogen peroxide using single reaction chamber and further elemental determination using ICP-OES and ICP-MS. *Microchem. J.* **2017**, *134*, 257–261. [CrossRef]
26. Leme, A.B.P.; Bianchi, S.; Carneiro, R.L.; Nogueira, A.R.A. Optimization of Sample Preparation in the Determination of Minerals and Trace Elements in Honey by ICP-MS. *Food Anal. Methods* **2013**, *7*, 1009–1015. [CrossRef]
27. Oliveira, S.; Alves, C.N.; Morte, E.S.B.; Junior, A.D.F.S.; Araujo, R.G.O.; Santos, D.C.M.B. Determination of essential and potentially toxic elements and their estimation of bioaccessibility in honeys. *Microchem. J.* **2019**, *151*, 104221. [CrossRef]
28. Spano, N.; Ciulu, M.; Floris, I.; Panzanelli, A.; Pilo, M.I.; Piu, P.C.; Scanu, R.; Sanna, G. Chemical characterization of a traditional honey-based Sardinian product: *Abbamele*. *Food Chem.* **2008**, *108*, 81–85. [CrossRef]
29. Spano, N.; Casula, L.; Panzanelli, A.; Pilo, M.I.; Piu, P.C.; Scanu, R.; Tapparo, A.; Sanna, G. An RP-HPLC determination of 5-hydroxymethylfurfural in honey: The case of strawberry tree honey. *Talanta* **2006**, *68*, 1390–1395. [CrossRef]
30. Salis, S.; Spano, N.; Ciulu, M.; Floris, I.; Pilo, M.; Sanna, G. Electrochemical Determination of the "Furanic Index" in Honey. *Molecules* **2021**, *26*, 4115. [CrossRef]
31. Sanna, G.; Pilo, M.I.; Piu, P.C.; Tapparo, A.; Seeber, R. Determination of heavy metals in honey by anodic stripping voltammetry at microelectrodes. *Anal. Chim. Acta* **2000**, *415*, 165–173. [CrossRef]

52. Ciulu, M.; Solinas, S.; Floris, I.; Panzanelli, A.; Pilo, M.I.; Piu, P.C.; Spano, N.; Sanna, G. RP-HPLC determination of water-soluble vitamins in honey. *Talanta* **2011**, *83*, 924–929. [CrossRef]
53. Spano, N.; Piras, I.; Ciulu, M.; Floris, I.; Panzanelli, A.; Pilo, M.I.; Piu, P.C.; Sanna, G. Reversed-Phase Liquid Chromatographic Profile of Free Amino Acids in Strawberry-tree (*Arbutus unedo* L.) Honey. *J. AOAC Int.* **2009**, *92*, S73–S84. [CrossRef]
54. Ciulu, M.; Floris, I.; Nurchi, V.M.; Panzanelli, A.; Pilo, M.I.; Spano, N.; Sanna, G. A Possible Freshness Marker for Royal Jelly: Formation of 5-Hydroxymethyl-2-furaldehyde as a Function of Storage Temperature and Time. *J. Agric. Food Chem.* **2015**, *63*, 4190–4195. [CrossRef] [PubMed]
55. Ciulu, M.; Floris, I.; Nurchi, V.M.; Panzanelli, A.; Pilo, M.I.; Spano, N.; Sanna, G. HPLC determination of pantothenic acid in royal jelly. *Anal. Methods* **2013**, *5*, 6682–6685. [CrossRef]
56. Ciulu, M.; Serra, R.; Caredda, M.; Salis, S.; Floris, I.; Pilo, M.I.; Spano, N.; Panzanelli, A.; Sanna, G. Chemometric treatment of simple physical and chemical data for the discrimination of unifloral honeys. *Talanta* **2018**, *190*, 382–390. [CrossRef]
57. Ciulu, M.; Oertel, E.; Serra, R.; Farre, R.; Spano, N.; Caredda, M.; Malfatti, L.; Sanna, G. Classification of Unifloral Honeys from SARDINIA (Italy) by ATR-FTIR Spectroscopy and Random Forest. *Molecules* **2020**, *26*, 88. [CrossRef] [PubMed]
58. Currie, L.A. Detection and quantification limits: Origins and historical overview. *Anal. Chim. Acta* **1999**, *391*, 127–134. [CrossRef]
59. Horwitz, W. Evaluation of analytical methods used for regulation of foods and drugs. *Anal. Chem.* **1982**, *54*, 67–76. [CrossRef]
60. Association of Official Agricultural Chemists (AOAC). Guidelines for Standard Method Performance Requirements, AOAC International. 2016. Available online: https://www.aoac.org/wp-content/uploads/2019/08/app_f.pdf (accessed on 16 February 2022).
61. Oliveri, P.; Malegori, C.; Simonetti, R.; Casale, M. The impact of signal pre-processing on the final interpretation of analytical outcomes—A tutorial. *Anal. Chim. Acta* **2018**, *1058*, 9–17. [CrossRef]
62. Oliveri, P.; Malegori, C.; Mustorgi, E.; Casale, M. Qualitative pattern recognition in chemistry: Theoretical background and practical guidelines. *Microchem. J.* **2020**, *162*, 105725. [CrossRef]
63. Aru, A.; Baldaccini, P.; Vacca, A.; Delogu, G.; Dessena, M.A.; Melis, R.T.; Vacca, S.; Madrau, S. Carta dei Suoli Della Sardegna. 1991, Cagliari. Available online: http://www.sardegnaportalesuolo.it/cartografia/carte-dei-suoli/carta-dei-suoli-della-sardegna-scala-1250000.html (accessed on 16 February 2022).
64. Istituto Nazionale di Statistica (Istat). Principali Statistiche Geografiche sui Comuni. Roma. 2022. Available online: https://www.istat.it/it/archivio/156224 (accessed on 16 February 2022).
65. Food and Agriculture Organization of the United Nations (FAO). Standard Operating Procedure for Soil Organic Carbon Walkley-Black Method: Titration and Colorimetric Method. 2020, Rome. Available online: https://www.fao.org/publications/card/en/c/CA7471EN/ (accessed on 16 February 2022).
66. Spanu, A.; Langasco, I.; Valente, M.; Deroma, M.A.; Spano, N.; Barracu, F.; Pilo, M.I.; Sanna, G. Tuning of the Amount of Se in Rice (*Oryza sativa*) Grain by Varying the Nature of the Irrigation Method: Development of an ICP-MS Analytical Protocol, Validation and Application to 26 Different Rice Genotypes. *Molecules* **2020**, *25*, 1861. [CrossRef]
67. Spanu, A.; Valente, M.; Langasco, I.; Leardi, R.; Orlandoni, A.M.; Ciulu, M.; Deroma, M.A.; Spano, N.; Barracu, F.; Pilo, M.I.; et al. Effect of the irrigation method and genotype on the bioaccumulation of toxic and trace elements in rice. *Sci. Total Environ.* **2020**, *748*, 142484. [CrossRef]
68. Mara, A.; Langasco, I.; Deidda, S.; Caredda, M.; Meloni, P.; Deroma, M.; Pilo, M.I.; Spano, N.; Sanna, G. ICP-MS Determination of 23 Elements of Potential Health Concern in Liquids of e-Cigarettes. Method Development, Validation, and Application to 37 Real Samples. *Molecules* **2021**, *26*, 6680. [CrossRef] [PubMed]
69. Langasco, I.; Barracu, F.; Deroma, M.A.; López-Sánchez, J.F.; Mara, A.; Meloni, P.; Pilo, M.I.; Estrugo, S.; Sanna, G.; Spano, N.; et al. Assessment and validation of ICP-MS and IC-ICP-MS methods for the determination of total, extracted and speciated arsenic. Application to samples from a soil-rice system at varying the irrigation method. *J. Environ. Manag.* **2021**, *302*, 114105. [CrossRef] [PubMed]
70. de Oliveira, F.A.; de Abreu, A.T.; Nascimento, N.D.O.; Froes, R.; Antonini, Y.; Nalini, H.A.; de Lena, J.C. Evaluation of matrix effect on the determination of rare earth elements and As, Bi, Cd, Pb, Se and In in honey and pollen of native Brazilian bees (*Tetragonisca angustula*—Jataí) by Q-ICP-MS. *Talanta* **2017**, *162*, 488–494. [CrossRef] [PubMed]
71. Bizzi, C.A.; Pedrotti, M.F.; Silva, J.S.; Barin, J.S.; Nóbrega, J.A.; Flores, E.M.M. Microwave-assisted digestion methods: Towards greener approaches for plasma-based analytical techniques. *J. Anal. At. Spectrom.* **2017**, *32*, 1448–1466. [CrossRef]
72. Leardi, R.; Melzi, C.; Polotti, G. CAT (Chemometric Agile Software). Available online: http://gruppochemiometria.it/index.php/software (accessed on 16 February 2022).

Article

Effectiveness of Different Analytical Methods for the Characterization of Propolis: A Case of Study in Northern Italy

Radmila Pavlovic [1], Gigliola Borgonovo [1,2], Valeria Leoni [1], Luca Giupponi [1,*], Giulia Ceciliani [1], Stefano Sala [1], Angela Bassoli [1,2] and Annamaria Giorgi [1,3]

1. Centre of Applied Studies for the Sustainable Management and Protection of Mountain Areas (CRC Ge.S.Di.Mont.), University of Milan, Via Morino 8, 25048 Edolo (BS), Italy; radmila.pavlovic1@unimi.it (R.P.); gigliola.borgonovo@unimi.it (G.B.); valeria.leoni@unimi.it (V.L.); giulia.ceciliani@unimi.it (G.C.); stefano.sala1@unimi.it (S.S.); angela.bassoli@unimi.it (A.B.); anna.giorgi@unimi.it (A.G.)
2. Department of Food, Environmental and Nutritional Sciences (DEFENS), University of Milan, Via Celoria 2, 20133 Milan, Italy
3. Department of Agricultural and Environmental Sciences - Production, Landscape, Agroenergy (DISAA), Via Celoria 2, 20133 Milan, Italy
* Correspondence: luca.giupponi@unimi.it

Received: 20 December 2019; Accepted: 21 January 2020; Published: 23 January 2020

Abstract: Propolis is used as folk medicine due to its spectrum of alleged biological and pharmaceutical properties and it is a complex matrix not still totally characterized. Two batches of propolis coming from two different environments (plains of Po Valley and the hilly Ligurian–Piedmont Apennines) of Northern Italy were characterized using different analytical methods: Spectrophotometric analysis of phenols, flavones and flavonols, and DPPH radical scavenging activity, HPLC, NMR, HSPME and GC–MS and HPLC–MS Orbitrap. Balsam and moisture content were also considered. No statistical differences were found at the spectrophotometric analysis; balsam content did not vary significantly. The most interesting findings were in the VOCs composition, with the Po Valley samples containing compounds of the resins from leaf buds of *Populus nigra* L. The hills (Appennines) samples were indeed characterize by the presence of phenolic glycerides already found in mountain environments. HPLC–Q-Exactive-Orbitrap®–MS analysis is crucial in appropriate recognition of evaluate number of metabolites, but also NMR itself could give more detailed information especially when isomeric compounds should be identified. It is necessary a standardized evaluation to protect and valorize this production and more research on propolis characterization using different analytical techniques.

Keywords: propolis; poplar; HPLC–Q-Exactive-Orbitrap®–MS analysis; phenolic glycerides

1. Introduction

Propolis, or bee glue, is a natural wax-like resinous substance found in beehives where it is used by honeybees as cement and to seal cracks or open spaces [1]. It is used by bees as well to prevent contamination inside the hive by bacteria, viruses or parasites because of its antiseptic effect; as well as to cover intruders who died inside the hive in order to avoid their decomposition [2]. Many comparison studies have now validated the theory that propolis is collected by honeybees from tree buds or other botanical sources in the North Temperate Zone, which extends from the Tropic of Cancer to the Arctic Circle [3]. The best sources of propolis are species of poplar, willow, birch, elm, alder, beech, conifer, and horse-chestnut trees [3]. Its color varies from green to brown and reddish, and the characteristic of each different type of propolis is dependent of some factors as e.g., plant source and edaphoclimatic conditions [4]. The word propolis derives from Hellenistic Ancient Greek *pro-* for or in defence, and *polis* city. There are records suggesting the use of it by ancient Egyptians, Persians, and Romans [5].

Recent studies confirmed the many properties of propolis: research over the last two to three decades has further exposed the wide potential of propolis, particularly its biological applications. Applications like anti-carcinogenic [6], anti-protozoan [7], anti-inflammatory [8], antioxidant [9], immunestimulating [10], antiviral [11], anti-tumor [10], hepato-protective [12], antifungal [13], and antibacterial activity [14], so it has been the subject of increasing scientific interest due to its diverse range of bio-medical properties.

Modern herbalists recommend this bee product for its beneficial properties to increase the natural resistance of human organisms [14]. Today propolis is currently used as a popular remedy and is available in the form of capsules (either in pure form or combined with aloe gel and rosa canina or pollen), as an extract (hydroalcholic or glycolic), as a mouthwash (combined with melissa, sage, mallow andyor rosemary), in throat lozenges, creams, and in powder form (to be used in gargles or for internal use once dissolved in water). It is also available commercially as purified product in which the wax has been removed. Propolis is also claimed to be useful in cosmetics and as a constituent of health foods. Current opinion is that the use of standardized preparations of propolis is safe and less toxic than many synthetic medicines [15].

It is therefore to mention that still there are not quality standard for this product [16]. A European Community report states that, the economic value of propolis is difficult to measure because it has no legal definition and is not a registered product (Evaluation of CAP measures for the apiculture sector, final report, July 2013). Scientific research regarding its chemical composition and biological activity started only about 30 years ago [17]. Since bees use the natural available vegetation to create propolis, there is a high variability in the composition.

It should be kept in mind, that bees collect their products for own benefit in the first place, and human beings are taking advantages of their hard work. If we take care to place the apiary in a location rich in food and material sources they need, then we have healthy bees contributing to pollination and biodiversity maintaining and other secondary aspects derived from these [16]. The role of bees in marginal ecosystems and agroecosystems is very important, and it is also essential to discriminate if the location of the apiary can influence the characteristics of this important bee product. Various factors give rise to the chemical complexity of propolis, for example, phyto-geographical origin, time of collection, and type of bees foraging. Complex chemical composition of propolis is the most important reason for many of the analytical challenges [18]. For what concerns phyto-geographical origin, chromatographic fingerprints are a valuable analytical method to identify different parts of plants [19].

Propolis is one of the most fascinating bee products, for sure a key factor of the success of the important macro-organism of the beehive, and its chemical complexities pose a great challenge to understanding content and percentage uniformity, and the connected biological activity. Complex chemical composition especially polarity of constituents makes it difficult to apply a single analytical technique vis a vis standardization even in today's era of very advanced techniques like High Performance Liquid Chromatography (HPLC), Liquid Chromatography associated with Mass Spectrometry (LC–MS/MS), Liquid Chromatography-High Resolution Mass Spectrometry (LC–HRMS), Gas Chromatography associated with Mass Spectrometry and Solid Phase Microextraction (SPME–GC–MS), and Nuclear Magnetic Resonance (NMR). Propolis is prepared primarily as alcoholic extract and therefore the maceration in alcohol is the most common extraction method used also for experimental purposes. Spectrophotometry, especially the Folin–Ciocalteu method, is the most widely used for the routine determination of total content of phenols and certain groups of flavonoids in propolis. However, other spectrophotometry methodologies have also been widely applied. For example, 1,1-diphenyl-2-picrylhydrazyl (DPPH) is one of the widely used method in evaluation of the antioxidant activity of propolis [20].

The quantification of individual compounds shows a significant discrepancy in the results reported in the bibliography about total phenolic content [21]. Very often, phenolic acids and individual compounds are not quantified. This is mainly due to the difference in the reference standards chosen for the construction of the calibration curves necessary to express the quantitative result [22].

Chromatographic methods, especially HPLC, are used for the separation and quantification of the specific constituent compounds of the phenolic profile, although they are not recommended as routine procedures due to their high cost [23,24] and, of course, to the complexity of the propolis matrix.

Advanced techniques as HPLC–high resolution mass spectrometry (HRMS) accompanied with a metabolomic approach could be a sufficiently descriptive method as it is able to detect the biomarkers that could be used as indicators of authentication. For the in-depth characterization of propolis recently developed analytical platforms based on NMR technique has been proved suitable for defining some a whole series of isomeric compounds found in propolis [25].

In addition to phenolics, another important class of propolis constituents is represented by volatile compounds [26–28]. Solid phase microextraction (SPME) represents a reliable tool for the analysis of volatile organic compounds [29,30] and eliminates most drawbacks to extracting organics, including high cost and excessive preparation time. SPME is a simple and fast modern tool used to characterize the volatile fraction of medicinal plants [31] and foods [29] and offers a valid alternative to HD for gas chromatographic analysis of essential oils from different sources [31]. For what concern propolis, SPME coupled with GC–MS can avoid the loss and degradation of volatile constituents that happen instead with HD (Hydro Distillation), very often used for the characterization of propolis volatiles [32,33].

Taking into account above elaborated considerations, the primary aim of this study was indeed to characterize two propolis produced in a hilly (Ligurian–Piedmont Apennines) and plain areas (Po Valley) of Northern Italy using diverse analytical approaches starting from basic ones (spectrophotometric analysis) to reach those more advanced such as HPLC, LC–HRMS, NMR, and SPME–GC–MS in order to evaluate the information that each of those method can provide for the characterization of propolis.

2. Results

2.1. Balsam and Moisture Content, Total Phenols, Flavones and Flavonols Content, and Scavenging Activity

In Table 1 is reported the composition of propolis in balsam and the moisture content. Antioxidant content and the DPPH radical scavenging activity are also reported, measured as described in materials and methods. The results are expressed in mg/g and agree with previous research [21]. Balsam content was found higher in the hills' samples (75.92 ± 4.92%), while in the plains samples was 63.94 ± 12.86%) but not significantly different according the statistical analysis. Very similar mean value (0.74 ± 0.38% for the hills batch and 0.69 ± 0.52% for the six samples collected in the plains) was found for the moisture content.

The total phenols, calculated as Gallic Acid Equivalents (mg GAE/g), the total Flavones and flavonols were determined using aluminum chloride and expressed as quercetine equivalent (mg QE/g) and the DPPH radical scavenging activity did not vary significantly between the hills and plains batches: the mean value of total flavones and flavonols was 32.14 ± 4.38 mg/g for the hills samples and 26.91 ±4.31 mg/g for the plains, the mean value of total phenol was 242.42 ± 11.67 for the hill samples and 236.32 ± 40.92 mg/g for the plains and this results were reflected by a very similar DPPH radical scavenging activity mean values (45.01 ± 1.39 for the hills and 46.44 ± 0.96 for the plains).

Table 1. Balsam and moisture content total antioxidant compounds and overall scavenging activity in raw propolis samples from hills (Ligurian–Piedmont Apennines) and plains (Po Valley).

Propolis Composition and Activity	Hills	Plains	Statistical Evaluation			
	Average ± SD	Average ± SD	t-Value	DF	p-Value	Sign.
Balsam content (% w/w)	75.92 ± 4.92	63.94 ± 12.86	2.131	6.433	0.074	ns
Moisture content (%)	0.74 ± 0.38	0.69 ± 0.52	0.195	9.233	0.850	ns
total Phenols (mg GAE/g)	242.42 ± 11.67	236.32 ± 40.92	0.351	5.808	0.738	ns
Total Flavones and Flavonols (mg QE/g)	32.14 ± 4.38	26.91 ± 4.31	2.082	9.998	0.064	ns
DPPH radical scavenging activity (%)	45.01 ± 1.39	46.44 ± 0.96	−2.082	8.855	0.068	ns

SD, standard deviation; ns, not significant; DS, degrees of freedom.

2.2. High-Performance Liquid Chromatography (HPLC) Analysis

The HPLC analysis was performed in order to obtain the preliminary phenolic/flavonoids profile, using three different UV wavelength that were used for better identification of the compounds. For example, pinocembrin was detected on 375 nm, phenolic acids caffeic, m-coumaric and ferulic were monitored on 325 nm while p-coumaric and trans-cinammic acids along with chrysin were registered at 295 nm (Figure 1). The quantity did not result significantly different for caffeic acid, chrysin and pinocembrin (Table 2). In hill samples, instead, it was found a considerably higher quantity of p-coumaric acid, ferulic acid, *m*-coumaric acid while the amount of *trans*-cinnamic acid was higher in the plain samples batch.

Table 2. The content of phenolic acids /flavonoids (mg/g) evaluated by HPLC-UV in alcoholic extract of propolis samples from hills (Ligurian–Piedmont Apennines) and plains (Po Valley).

Phenolic Acids/Flavonoids	Hills	Plains	Statistical Evaluation
	Average ± SD	Average ± SD	p-Value
caffeic acid	4.37 ± 0.53	4.21 ± 0.80	ns
p-coumaric acid	6.97 ± 2.12	1.40 ± 0.37	0.0001
ferulic acid	7.41 ± 2.22	1.64 ± 0.30	0.0013
m-coumaric acid	3.72 ± 0.27	2.87 ± 0.66	0.0150
trans-cinnamic acid	3.42 ± 0.21	4.48 ± 1.01	0.0428
pinocembrin	19.06 ± 6.27	17.90 ± 4.20	ns
chrysin	33.62 ± 3.49	35.64 ± 12.71	ns

SD, standard deviation; ns, not significant.

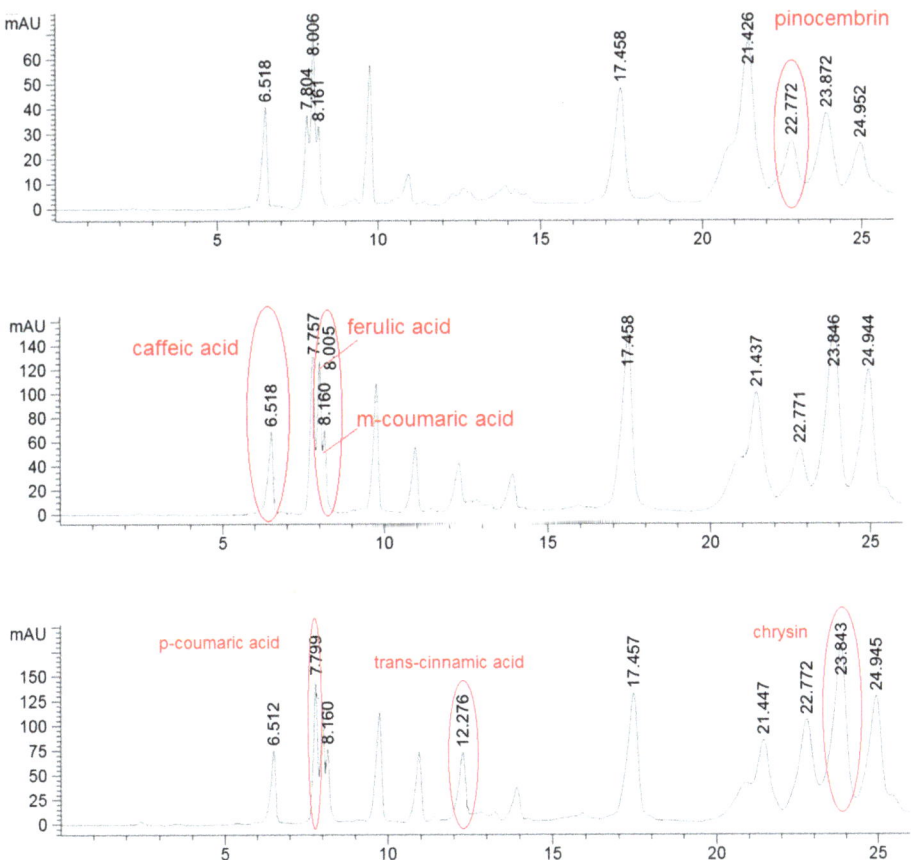

Figure 1. Detection of pinocembrin in one hills sample at 375 nm; detection of caffeic acid, ferulic acid and *m*-coumaric acid in one hills sample at 325 nm; detection of *p*-coumaric acid, *trans*-cinnamic acid and chrysin in one hills sample at 295 nm.

2.3. Propolis Volatile Compounds

For what concerns the volatiles composition, a higher quantity of volatile compounds (VOCs) was found in the plains batches than in the hills (total VOCs of 415 µg/g for the hills and 502 µg/g for the plains), with a corresponding higher content of terpenes and terpenoids (85 µg/g on average for plains samples and 65 for hills samples). Hills and plains samples contained the same compounds, that varied only quantitatively for 18 compounds on the 60 compounds recognized. In the plains it was found a significantly higher quantity of methyl-acetate, 2,4-dimethyl-1-heptene, methyl propanoate, and benzaldehyde. The quantity was significantly higher β-Linalool, cinnamaldehyde, α-copaen-11-ol, aceto-cinnamone, cinnamyl-alcohol, and finally α-eudesmol and β-eudesmol. In the hills, instead, was found to be higher the hydrocarbons 2-butenal, 2-methyl- and 2-butenal, 3-methyl- as well as for two unknown sesquiterpenes (called sesquiterpene_3 and sesquiterpene_4). The PCA sub lining these differences is in Figure 2.

Some of the compound mentioned above as characteristic of one or both the locations were also higher of 1% of total VOCs (Table 3). β-Linalool and cinnamyl alcohol were in fact present in a percentage higher than 1% only in plains samples. The most important volatiles with an amount that exceeded 1% of total VOCs for both Appennines and Plains were cinnamaldehyde, β-eudesmol and δ-cadinene. Aliphatic and aromatics alcohols, carbonyl compounds and aliphatic acids have been characterized among non-terpenes volatiles in a fraction of 280 µg/g for hills samples and 350 µg/g for the plains, while some compounds were not identified (about 70 µg/g for both hills and plains samples). A substantial amount of acids was found in the samples from both locations: Acetic acid, 2-methyl butanoic acid, 2-butenoic acid and 2-methyl propanoic acid and propanoic acid and α-methyl crotonic acid. The aromatic compounds such as benzaldehyde, benzyl acetate, benzyl alcohol and phenethyl alcohol constituted the significant amount of non-terpenoids VOCs fraction (Table 3).

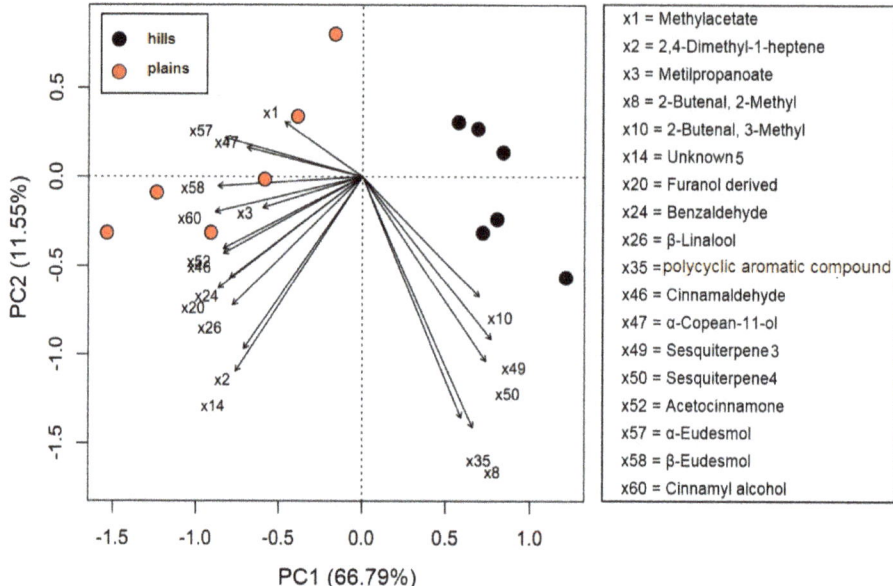

Figure 2. PCA biplot of volatile compounds (VOCs) of propolis samples collected in hills (Ligurian–Piedmont Apennines) and plains (Po Valley).

Table 3. Volatile compounds identified in raw propolis samples from hills (Ligurian–Piedmont Apennines) and plains (Po Valley).

RT [a]	Compounds	Hills Mean[b] ± SD [c]	% [d]	Plains Mean [b] ± SD [c]	% [d]	t-Value	DF	p-Value	Signif. Code
2.33	methyl-acetate	22.85 ± 5.11	5.52	34.23 ± 4.85	6.75	−3.9538	9.9717	0.002729	**
2.93	2,4–dimethyl-1-heptene	6.55 ± 0.83	1.58	10.41 ± 3.25	2.06	−2.8173	5.6514	0.03251	*
3.23	methyl-propanoate	0.69 ± 0.16	0.17	1.10 ± 0.28	0.22	−3.1264	8.1668	0.01373	*
6.00	α-pinene	0.55 ± 0.31	0.13	0.87 ± 0.46	0.17	−1.448	8.7216	0.1826	ns
7.23	3-buten-2-ol, 2-methyl-	1.33 ± 0.41	0.32	1.30 ± 0.36	0.26	0.14423	9.793	0.8882	ns
7.77	camphene	1.92 ± 0.55	0.46	3.53 ± 1.81	0.70	−2.0744	5.9194	0.08402	ns
8.98	esanal	2.96 ± 0.73	0.71	3.86 ± 1.59	0.76	−1.2607	7.039	0.2476	ns
9.24	2-butenal, 2-methyl-	1.15 ± 0.32	0.28	0.67 ± 0.13	0.13	3.4176	6.6211	0.01217	*
11.03	unknown_1	4.49 ± 0.52	1.08	4.40 ± 1.33	0.87	0.15257	6.4759	0.8834	ns
14.34	2-butenal, 3-methyl-	3.76 ± 0.85	0.91	2.68 ± 0.46	0.53	2.7426	7.6617	0.0264	*
14.76	unknown_2	7.04 ± 2.87	1.70	6.09 ± 1.75	1.20	0.69089	8.2804	0.5085	ns
16.70	unknown_3	19.67 ± 3.08	4.75	15.30 ± 6.80	3.02	1.4333	6.9764	0.195	ns
16.95	unknown_4	5.71 ± 2.21	1.38	5.48 ± 1.72	1.08	0.20213	9.4253	0.8433	ns
19.04	unknown_5	8.61 ± 1.21	2.08	12.20 ± 2.99	2.41	−2.7322	6.5943	0.03101	*
20.98	nonanal	1.92 ± 0.49	0.46	2.18 ± 0.35	0.43	−1.0457	9.071	0.3228	ns
21.32	benzene, 1-methoxy-2-methyl-	0.34 ± 0.12	0.08	0.93 ± 0.84	0.18	−1.7233	5.2184	0.143	ns
21.43	tetradecane	0.53 ± 0.13	0.13	0.65 ± 0.23	0.13	−1.1373	7.9121	0.2887	ns
21.74	2-octenal	0.50 ± 0.19	0.12	0.53 ± 0.20	0.10	−0.22144	9.9454	0.8292	ns
22.17	acetic acid	45.28 ± 6.83	10.93	57.80 ± 17.65	11.42	−1.6215	6.465	0.1525	ns
22.76	terpene_1	0.73 ± 0.26	0.18	2.46 ± 1.02	0.49	−4.0402	5.6519	0.007701	**
22.82	trans-linalool oxide	1.74 ± 0.50	0.42	1.61 ± 0.44	0.32	0.46595	9.8211	0.6514	ns
23.33	α-copaene	2.29 ± 0.94	0.55	1.98 ± 0.50	0.39	0.69802	7.6553	0.5058	ns
23.61	(+)-camphor	1.44 ± 0.30	0.35	2.89 ± 1.62	0.57	−2.1594	5.3419	0.07974	ns
23.84	benzaldehyde	11.11 ± 2.69	2.68	18.05 ± 4.59	3.57	−3.1941	8.0829	0.01255	*
24.23	propanoic acid	6.34 ± 1.30	1.53	9.04 ± 4.21	1.79	−1.5016	5.938	0.1844	ns
24.66	β-linalool	3.06 ± 1.34	0.74	6.70 ± 3.19	1.32	−2.5733	6.7075	0.03821	*
24.91	2-methyl-propanoic acid	19.88 ± 3.52	4.80	27.67 ± 11.59	5.47	−1.5762	5.9158	0.1668	ns
25.69	sesquiterpene_1	0.44 ± 0.08	0.11	0.42 ± 0.09	0.08	0.49589	9.7629	0.6309	ns
25.99	β-cyclocitral	3.80 ± 1.16	0.92	3.37 ± 1.41	0.67	0.57503	9.6507	0.5784	ns
26.17	unknown_6	16.14 ± 1.92	3.90	15.86 ± 4.33	3.13	0.14412	6.8858	0.8895	ns
27.03	2-methyl-butanoic acid	20.72 ± 3.90	5.00	28.79 ± 11.02	5.69	−1.6923	6.2338	0.1397	ns
27.35	2-butenoic acid	5.39 ± 1.52	1.30	10.50 ± 4.88	2.08	−2.4501	5.9572	0.05	ns
27.51	sesquiterpene_2	2.38 ± 0.82	0.57	1.72 ± 0.57	0.34	1.6229	8.9525	0.1392	ns
28.19	benzyl-acetate	8.76 ± 2.10	2.12	8.09 ± 2.47	1.60	0.506	9.7442	0.6241	ns

Table 3. Cont.

RT [a]	Compounds	Hills		Plains		t-Value	DF	p-Value	Signif. Code [d]
		Mean [b] ± SD [c]	% [d]	Mean [b] ± SD [c]	% [d]				
28.29	polycyclic aromatic compound	3.55 ± 0.54	0.86	2.69 ± 0.59	0.53	2.6531	9.9141	0.02435	*
28.84	δ-cadinene	13.63 ± 3.88	3.29	12.45 ± 3.63	2.46	0.54221	9.9557	0.5996	ns
28.99	unknown_7	3.25 ± 0.61	0.79	4.26 ± 1.89	0.84	−1.2451	6.0414	0.2592	ns
29.07	unknown_8	2.51 ± 1.40	0.61	3.46 ± 1.23	0.68	−1.2557	9.8345	0.2343	ns
29.25	unknown_9	1.50 ± 0.25	0.36	1.72 ± 0.59	0.34	−0.85269	6.6761	0.4234	ns
29.55	pentanoic acid, 4-methyl-	1.42 ± 0.56	0.34	1.27 ± 0.45	0.25	0.50394	9.5007	0.6258	ns
29.81	unknown_10	1.24 ± 0.51	0.30	1.57 ± 0.45	0.31	−1.1763	9.8542	0.2671	ns
30.18	calamenene	4.97 ± 1.45	1.20	4.01 ± 1.39	0.79	1.1844	9.9827	0.2637	ns
30.28	α-methyl crotonoic acid	43.69 ± 6.73	10.55	59.28 ± 16.62	11.72	−2.1297	6.5967	0.07309	ns
30.90	benzyl-alcohol	48.04 ± 10.13	11.60	38.98 ± 10.75	7.70	1.5022	9.9647	0.1641	ns
31.54	phenethyl-alcohol	24.41 ± 3.79	5.90	31.67 ± 7.52	6.26	−2.1124	7.3894	0.07045	ns
32.81	cinnamaldehyde	4.98 ± 0.89	1.20	9.57 ± 2.63	1.89	−4.0482	6.1243	0.006458	**
32.88	α-copaen-11-ol	0.74 ± 0.48	0.18	1.50 ± 0.48	0.30	−2.7538	9.9998	0.02034	*
33.04	octanoic acid	0.38 ± 0.11	0.09	0.54 ± 0.20	0.11	−1.7018	7.941	0.1275	ns
33.08	sesquiterpene_3	1.16 ± 0.43	0.28	0.22 ± 0.07	0.04	5.3763	5.2743	0.002547	**
33.15	sesquiterpene_4	1.69 ± 0.50	0.41	0.62 ± 0.22	0.12	4.8115	6.8895	0.002028	**
33.39	guaiol	0.39 ± 0.46	0.09	0.37 ± 0.12	0.07	0.12043	5.6469	0.9083	ns
33.56	acetocinnamone	1.14 ± 0.21	0.27	2.22 ± 0.74	0.44	−3.4706	5.8256	0.01394	*
34.11	unknown_11	0.78 ± 0.28	0.19	0.88 ± 0.23	0.17	−0.70911	9.6282	0.4951	ns
34.22	Sesquiterpene_5	3.04 ± 1.68	0.73	3.26 ± 0.55	0.64	−0.30368	6.0741	0.7715	ns
34.30	Sesquiterpene_6	1.19 ± 0.68	0.29	1.91 ± 0.58	0.38	−1.965	9.7885	0.07841	ns
34.40	Sesquiterpene_7	1.88 ± 0.99	0.45	1.97 ± 0.30	0.39	−0.21625	5.9323	0.836	ns
34.79	α-eudesmol	2.05 ± 0.93	0.49	4.49 ± 0.72	0.89	−5.105	9.399	0.00056	**
34.90	β-eudesmol	4.21 ± 0.68	1.02	8.42 ± 1.32	1.66	−6.9394	7.5091	0.000161	**
35.16	α-Copaen-11-ol	0.52 ± 0.47	0.13	0.79 ± 0.34	0.16	−1.1297	9.1588	0.2873	ns
35.60	cinnamyl alcohol	1.64 ± 0.35	0.40	5.39 ± 1.90	1.06	−4.7549	5.3445	0.004271	**

[a] RT, retention time (min); [b] Mean, mean value ($n = 6$); data are expressed in µg/g; [c] SD, standard deviation ($n = 6$); [d] %—percentage of total VOCs; [d] Signif. Code, * $p < 0.05$; ** $p < 0.01$; ns, not significant.

2.4. Nuclear Magnetic Resonance (NMR)

The NMR spectra of samples were recorded on a Bruker Avance spectrometer with proton operating frequency 600.13 MHz with a 5mm TBI probe. The spectra were performed at 300 K using 16 K of TD (time domain), acquisition time 1.27 min, delay time 1.0 s and 48 the number of scans. The spectral width was 12019 Hz. For ^{1}H-NMR analysis 2–3 mg of crude extracts were dissolved in 0.6 mL of DMSO-d6, while for ^{13}C-NMR spectra 10–20 mg for each sample were used.

The ^{1}H-NMR analysis of complex matrix such as propolis extracts were complicated for the presence of a high number of similar compounds. A comparison of ^{1}H-NMR spectra of propolis extracts is reported in Figure 3. The most interesting spectral region are between 3.50 and 8.25 ppm, which contains aliphatic and aromatic signals, and between 10.00 and 13.00 ppm which contains chelated phenolic groups and carboxylic proton signals. Figure 3 showed that the samples appear to be very similar to each other; the main signals associated with the secondary metabolites characterizing the extract appear to be present both in plain and hill samples, in some samples a slightly different quantitative ratio. Diagnostic signals related to chelated phenolic groups, typical of flavonoids, or hydroxyl groups of carboxylic functions can be detected at low fields between 12.0 and 13.3 ppm. The signals of the known compounds were determined by comparison of their physical and spectroscopic features with standard compounds and with those reported in the literature [34]. Variable amounts of flavonoid were identified: pinocembrin, chrysin, galangin, pinobanskin 3 O acetate, and pinostrobin. Several phenolic acids, pinobanskin, kaempheride, apigenin, and other compounds were also present as minority components.

The phenyl ester of caffeic acid, known as CAPE, had the signals of the methylene protons resonating at 2.95 and 4.32 ppm; at the examined concentrations these signals were lacking and therefore CAPE was not detectable in our samples. Quercetin was not detectable in both extracts.

Specific resonances attributable to glycerol esters (such as 1,3-di-p-coumaryl-2-acetyl-glycerol and 1,3-diferulyl-p-coumarate-glycerol) were given by the presence of signals in the zone 4.2–5.3 ppm (glycerol moiety), an area crowded with several overlapping signals, and a singlet resonance for methyl groups at 2.05 ppm. The presence of the acetyl groups and the ester groups were also confirmed by ^{13}C-NMR spectrum due to the presence of signals in the area between 160 and 170 ppm, the portion of glycerol, instead, gives signals at 63–71 ppm, the methyl groups of acetyl at 19.7 and 21.9 ppm in agreement with the literature data (Figure 4) [25]. The definition of the type of glycerol ester was given with LC–MS orbitrap.

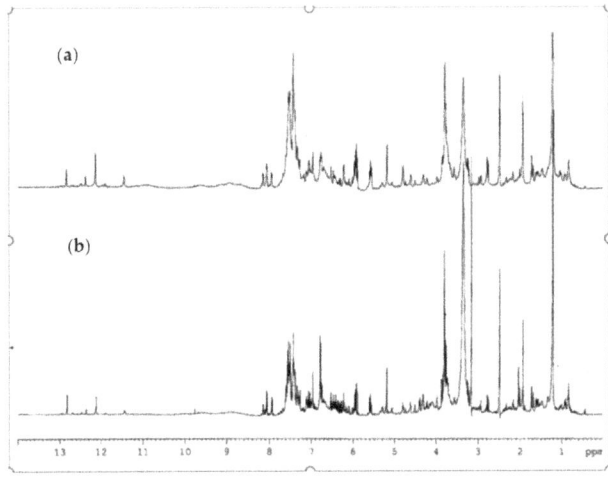

Figure 3. Comparison of ^{1}H-NMR spectra of propolis extract (**a**) plain sample (**b**) hill sample.

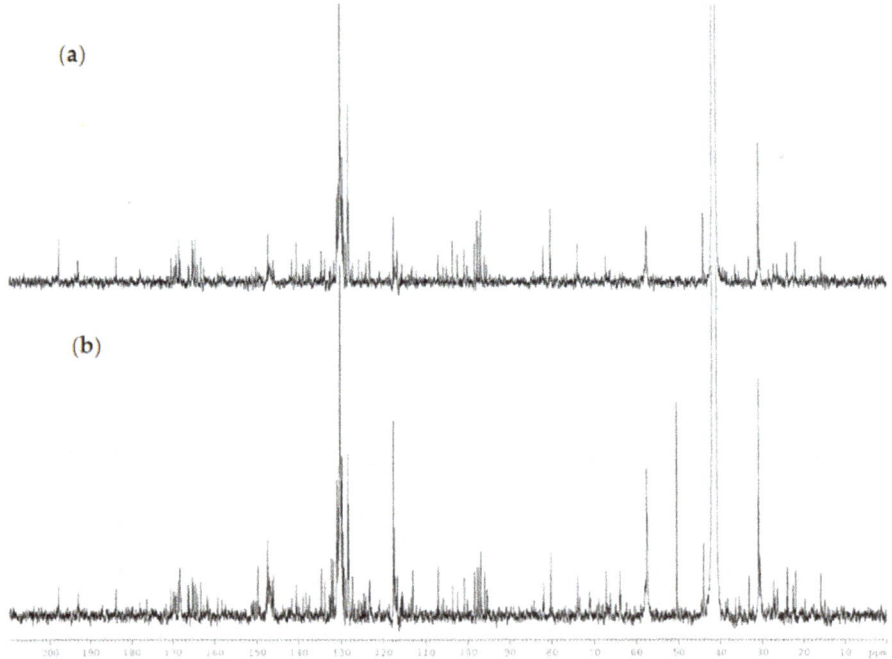

Figure 4. Comparison of ^{13}C-NMR spectra of propolis extract (**a**) plains sample (**b**) hills sample.

2.5. HPLC–Q-Exactive-Orbitrap®–MS Analysis

Crude extracts that were used for HPLC–UV analysis were diluted (1:100) and were subsequently subjected to HPLC-Q-Exactive-Orbitrap®-MS analysis (activated in negative mode) in order to perform untargeted profiling of propolis collected from hills (Ligurian–Piedmont Apennines) and plain (Po Valley) with subsequent data processing performed by Compound Discoverer™ (CD) software. Two type of Q-Exactive-Orbitrap®-acquisition mode were executed:

- First one was full scan (FS) at maximum resolution of 140,000 that involved generation of the lists of compounds that are potentially present in the samples (307 candidates). Using Compound Discoverer platform compounds were identified applied workflow that includes RT alignment, blank subtraction, and molecular formula assignment. Also, in FS acquisition mode the additional detection settings were applied: (1) Selecting the unknown peaks with criteria such as mass tolerance (<2 ppm); (2) minimum peak intensity (100,000), 3) integrating isotope and adduct peaks of the same compound into one group to reduce the incidence of false positives. This phase involved also the differential analysis with Volcano Plot (VP) (Figure 5) and principal component analysis (PCA) (Figure 6). PCA clearly distinguished the hills and plains samples, where VP analysis gave more precise response which signals are the main contributors along with the statistical evaluation presented in Table 4.
- Second type of analysis regards FS-data dependent (FS-DDA) acquisition mode and was performed on the inclusion list of 307 signals extracted from the FS data collection. MS–MS fragmentation performed in FS–DDA modality enabled the putative identification beyond the available standards. This phase comprises molecular formula assignment according to the accurate mass, adduct state, isotopes and fragmentation patterns with selecting best-fit candidates for the non-target peaks after comparison and evaluation with the software-linked MS2 libraries (mzCloud, *m/z*

Valut and ChemSpider). To make the results more reliable especially when the mzCloud did not give any well-defined response the matching results are further filtered and checked with other on-line databases (human Metaboloeme at the first place). In some cases, as we have not found any satisfactory confirmation from existing databases, the tentative deduction of the final structure was performed manually assigning the fragments structure in concordance with available literature [35,36].

Using the above described platform, it was possible to single out and speculate ninety compounds divided in the categories listed in the Table 4. Most phenols have been previously confirmed in the literature for "poplar type" propolis [35,36]. The method described allowed the hypothesis of the presence of some new compounds not previously found in propolis: 4-ethyl-7-hydroxy-3-(p-methoxyphenyl)-coumarin in plain samples and 4-hydroxy-4'-methoxychalcone in hills propolis. As showed in our study, the use of NMR analysis of complex matrix such as propolis extracts is complicated for the presence of a high number of similar compounds but further investigation with this and other analytical instruments should be effectuated to verify the presence of these new compounds. The most important results regard the strongly upregulated phenolic glycerides in hills samples. The main characteristic of hills samples was the unambiguous occurrence of different glycerol esters which HRMS signals were very poor in plains group. For the Po-Vally samples the two isomers of abscisic acid were dominant in samples from this area. Also, plains propolis revealed higher concentration of trans-cinnamic acid which is accompanied with its caffeic esters form. The Volcano Plot (Figure 5) segregated the selected metabolites as those responsible for grouping, while the PCA projection clearly demonstrated that same geographic regions are closely grouped, and the first two components describe 59% variability.

Table 4. Tentative identification of compounds based on Compound Discoverer evaluation.

Compounds	Ret. Time	Formula	Exact Mass	Differential Analysis *
Phenolic acids and their derivatives				
p-hydroxybenzoic acid	5.48	$C_7H_8O_4$	137.0244	ns
cinnamic acid isomer	5.87	$C_9H_8O_2$	147.0452	up-regulated in hills
Salicylic acid	9.92	$C_7H_8O_4$	137.0244	ns
Caffeic acid	10.52	$C_9H_8O_4$	179.035	ns
m-Coumaric acid	11.03	$C_9H_8O_3$	163.0401	up-regulated in hills
p-Coumaric acid	12.61	$C_9H_8O_3$	163.0401	up-regulated in hills
Cinnamic acid	13.12	$C_9H_8O_2$	147.0452	up-regulated in plains
Ferulic acid	13.43	$C_{10}H_{10}O_4$	193.0506	up-regulated in hills
Cinnamic acid isomer	13.81	$C_9H_8O_2$	147.0452	up-regulated in hills
Isoferulic acid	15.22	$C_{10}H_{10}O_4$	193.0506	ns
Cinnamic acid isomer	16.15	$C_9H_8O_2$	147.0452	up-regulated in hills
p-Coumaroylquinic acid	17.12	$C_{16}H_{18}O_8$	337.0922	ns
Chlorogenic acid	21.05	$C_{16}H_{18}O_9$	353.0876	ns
Benzyl-caffeate	21.17	$C_{16}H_{13}O_4$	269.0819	ns
Prenyl-caffeate	21.7	$C_{14}H_{16}O_4$	247.0976	up-regulated in hills
Caffeic acid phenethyl-ester (CAPE)	21.93	$C_{17}H_{16}O_4$	283.0976	ns
p-Coumaric acid prenyl-ester	22.08	$C_{14}H_{16}O_3$	231.1027	up-regulated in hills
Coniferyl-ferulate isomer	22.61	$C_{20}H_{20}O_6$	355.1187	ns
Caffeic acid cinnamyl-ester	22.93	$C_{18}H_{16}O_4$	295.0976	up-regulated in plains
Dupunin (Prenylated phenyl-propanoic acid)	23.01	$C_{14}H_{16}O_3$	231.1027	ns
Coniferyl ferulate	25.51	$C_{20}H_{20}O_6$	355.1187	ns
Capillartemisin A (Prenylated phenyl-propanoic acid)	25.74	$C_{18}H_{24}O_4$	315.16	up-regulated in hills
Phenolic glycerides				
Caffeoyl-glycerol	10.01	$C_{12}H_{14}O_6$	253.0713	highly up-regulated in hills
Dicaffeoyl-acetyl-glycerol	19.14	$C_{23}H_{21}O_{10}$	457.1142	highly up-regulated in hills
Diferuloyl-glycerol	19.33	$C_{23}H_{24}O_9$	443.1349	highly up-regulated in hills
acetyl-caffeoyl-feruloyl-glycerol	20.09	$C_{24}H_{24}O_{10}$	471.1298	highly up-regulated in hills
Coumaroyl-caffeoyl-acetyl-glycerol	20.21	$C_{23}H_{22}O_9$	441.1185	highly up-regulated in hills
Acetyl-coumaroyl-feruloyl-glycerol	21.17	$C_{24}H_{24}O_9$	455.1347	highly up-regulated in hills
Di-p-coumaroyl-acetyl-glycerol	21.28	$C_{26}H_{22}O_8$	425.1242	highly up-regulated in hills
Coumaroyl-acetyl-glycerol	24.22	$C_{18}H_{16}O_3$	279.086	highly up-regulated in hills
Coumarins				
4-Ethyl-7-hydroxy-3-(p-methoxyphenyl)-coumarin	23.05	$C_{18}H_{16}O_4$	295.0976	up-regulated in plains

Table 4. Cont.

Compounds	Ret. Time	Formula	Exact Mass	Differential Analysis *
Flavanones				
Pinostrobin	23.65	C₁₆H₁₄O₄	269.0819	up-regulated in plains
Strobopinin	19.55	C₁₆H₁₄O₄	269.0819	up-regulated in plains
Pinocembrin	21.88	C₁₅H₁₂O₄	255.0663	ns
Sakuranetin	21.46	C₁₆H₁₄O₅	285.0768	ns
3′,5,7-Trihydroxy-4′-methoxyflavanone	19.15	C₁₆H₁₄O₆	301.0718	up-regulated in hills
Hesperetin	20.01	C₁₆H₁₄O₆	301.0718	up-regulated in hills
Chalcones				
Pinostrobin-chalcone	19.62	C₁₆H₁₄O₄	269.0819	up-regulated in hills
4-Hydroxy-4′-methoxychalcone	23.02	C₁₆H₁₄O₃	253.087	highly up-regulated in hills
Flavonols				
Quercetin	18.2	C₅H₁₀O₇	301.0354	ns
Quercetin-3-O-methyl-ether	19.01	C₁₆H₁₂O₇	315.051	up-regulated in hills
Rhamnetin	20.16	C₁₆H₁₂O₇	315.051	up-regulated in hills
Kaempferol	19.86	C₁₅H₁₀O₆	285.0405	up-regulated in hills
Isorhamnetin	21.24	C₁₆H₁₂O₇	315.051	up-regulated in hills
Kaempferide	22.93	C₁₆H₁₂O₆	299.0563	up-regulated in hills
Bis-methylated quercetin	21.79	C₁₇H₁₄O₇	329.0667	ns
Galangin	23.87	C₁₅H₁₀O₅	269.0455	ns
Rhamnocitrin	23.18	C₁₆H₁₂O₆	299.0563	up-regulated in hills
Flavanonols				
Pinobanksin	18.05	C₁₅H₁₂O₅	271.0613	ns
Pinobanksin-5-methyl-ether	17.59	C₁₆H₁₄O₅	285.0768	ns
Pinobanksin-5-methylether-3-O-acetate	20.2	C₁₈H₁₆O₆	327.0869	up-regulated in hills
Pinobanksin-3-O-propionate	22.78	C₁₈H₁₆O₆	327.0869	up-regulated in hills
Pinobanksin-3-O-butyrate	24.02	C₁₉H₁₈O₆	341.103	ns
Pinobanksin-3-O-acetate	21.43	C₁₇H₁₄O₆	313.0712	ns
Aromadendrin	14.85	C₁₅H₁₂O₆	287.0651	ns
Isoflavones				
Genistein	19.08	C₁₅H₁₀O₅	269.0542	ns
Formononetin glucoside	18.28	C₂₂H₂₂O₉	429.1191	highly up-regulated in hills

Table 4. Cont.

Compounds	Ret. Time	Formula	Exact Mass	Differential Analysis *
Formononetin (biochanin B)	20.8	$C_{16}H_{12}O_4$	267.0663	ns
Hispiludin	20.52	$C_{16}H_{12}O_6$	299.0563	up-regulated in hills
Flavones				
Apigetrin (Apigenin-7-O-glucoside)	18.01	$C_{21}H_{20}O_{10}$	431.0983	highly up-regulated in hills
Apigenin	20.44	$C_{15}H_{10}O_5$	269.0542	up-regulated in hills
Dihydroxyflavone	21.43	$C_{15}H_{10}O_4$	253.0506	ns
Chrysin	22.04	$C_{15}H_{10}O_4$	253.0506	ns
Methoxy-chrysin	23.01	$C_{16}H_{12}O_5$	283.0612	ns
Tricin	20.59	$C_{17}H_{14}O_7$	329.0667	ns
Chrysoeriol	20.52	$C_{16}H_{12}O_6$	299.0563	up-regulated in hills
Terpenoids				
Ursolic acid	29.65	$C_{30}H_{48}O_3$	455.3531	highly up-regulated in hills
trans,trans-Abscisic acid	16.58	$C_{15}H_{20}O_4$	263.1289	up-regulated in plains
cis,trans-Abscisic acid	23.66	$C_{15}H_{20}O_4$	263.1289	up-regulated in plains
Unknowns				
Unknown 1 (phenylacetaldehide or isomer)	12.51	C_8H_8O	119.0502	highly up-regulated in hills
Unknown 2 (p-coumaric derivate)	14.97	$C_{19}H_{20}O_6$	359.11	highly up-regulated in hills
Unknown 3 (ferulic acid derivative)	15.01	$C_{20}H_{22}O_8$	389.124	highly up-regulated in hills
Unknown 4	18.28	$C_{27}H_{42}O_4$	429.1191	highly up-regulated in hills
Unknown 5	19.05	$C_{17}H_{16}O_6$	315.0851	highly up-regulated in hills
Unknown 6	19.11	/	597.1007	highly up-regulated in hills
Unknown 8 (p-coumaric acid derivative)	20.05	$C_{28}H_{27}O_7$	475.1762	ns
Unknown 9 (Chrysin derivate)	21.09	/	639.1112	highly up-regulated in hills
Unknown 10 (p-Hydroxybenzoic acid derivative)	21.15	$C_{28}H_{27}O_7$	475.1762	ns
Unknown 11	21.19	/	461.1007	highly up-regulated in hills
Unknown 12	22.29	$C_{20}H_{37}O_9$	421.2441	highly up-regulated in hills
Unknown 13 (flavone ester of caffeic acid)	22.78	/	565.1509	up-regulated in plains
Unknown 14 (Chrysin derivate)	23.44	$C_{25}H_{22}O_6$	417.135	highly up-regulated in hills
Unknown 15 (Chrysin derivate)	24	$C_{25}H_{22}O_6$	417.135	highly up-regulated in hills
Unknown 16	25.17	/	413.1963	highly up-regulated in hills
Unknown 17	26.5	$C_{18}H_{30}O_3$	293.212	highly up-regulated in hills

* Differential analysis performed applying Volcano Plot Model (Figure 5); ns, not significant.

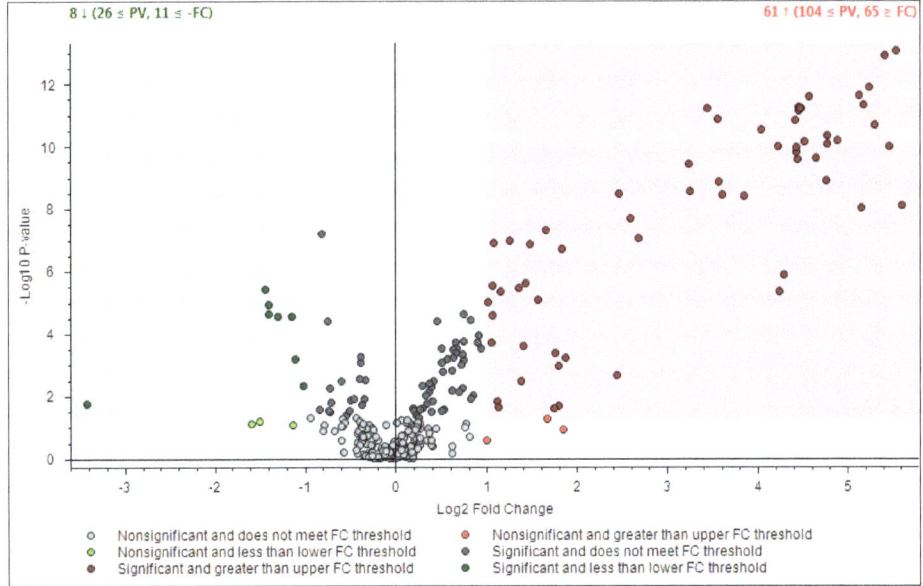

Figure 5. Differential analysis presented by Volcano Plot Model for the comparison between the relative intensity of chromatographic peak from two propolis groups. *P*-value (PV) was set on 0.05. Red region contains up-regulated signal where the quantities from Apennines was significantly higher than those found in Po Valley samples and was greater than the upper fold-change (FC) threshold. The green region comprises up-regulated peaks in Po Valley samples.

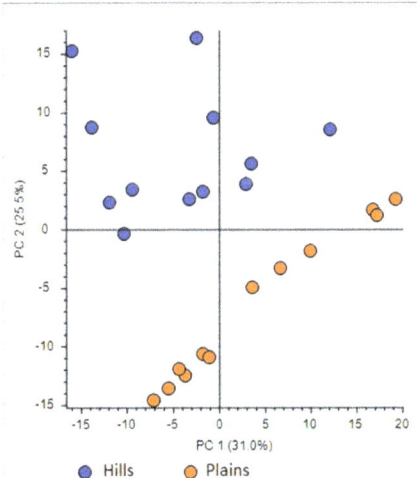

Figure 6. Principal Component Analysis (PCA) projection on the distribution of samples respect to PC-1 with PC-2 when high resolution mass spectrometry (HRMS) spectra were used.

3. Discussion

No significant differences were found between the two sampling sites for what concerns the balsam content. Also, no significative differences were found for total phenols, total flavones and flavonols and scavenging activity. This is probably due to the very low sensitivity of spectrophotometric evaluation, as the other analytical methods showed both quantitative and qualitative differences in the composition of the two propolis samples. As already reported in the literature [28], the difficulty in the measurement of antioxidant activity arises from the different standard reference compounds and there is a wide variety of methods to assess antioxidant capacity, each having advantages and disadvantages. For this aim, it would be important to find a standardized method to evaluate the antioxidant power of each kind of propolis, to valorize quality productions and avoid falsification.

Moisture content was lower than 1% in all samples, as propolis is a material produced by bees to last long time in the beehive.

VOCs determination can be considered an important aspect for propolis characterization [37]. The main influence on the aroma composition of propolis is formed by volatile compounds that may come from the material collected by bees and then it may largely depend on the plant of propolis origin. Other factors had been demonstrated, as the state of propolis maturity and the honeybees as well. Aldehydes and alcohols may also be a consequence of microbiological activity or heat exposure, while linear aldehydes are considered as characteristic compounds associated with certain herbal origin as terpenes and terpenoids. Acetic acid, hexanal, alpha pinene, camphene, benzaldehyde, octanal, nonanal, beta-ciclocitrale, cinnamaldeide, cinnamyl alcohol, alpha copaene, cinnamyc alcohol, acetocinnamone, cinnamic acid, gamma cadinene, guaiol, gamma and beta eudesmols, benzyl benzoate has been previously detected in propolis in [28] and the chemical composition of propolis volatile fraction determined in the present study by means of HS–SPME-GC–MS was found to be in agreement with previous reports [32,38].

Two of the compounds found significantly higher in the plains are α-eudesmol and β-eudesmol. The last is the most abundant compound in resins from leaf buds of black poplar (*Populus nigra* L.) [39], which represents one of the main botanical sources of propolis constituents in temperate regions [40]. Indeed, several volatile compounds identified in propolis volatile fraction have been previously detected from leaf buds of *Populus nigra*. As the *Populus nigra* is a plant that grows in perifluvial environments of the Po Valley [41], we can assume that the apiary in the plains found easier to collect resins from this plant. According to [32], propolis from temperate zones can be classified in two types, based on the presence of representative amounts of β-eudesmol (40%–60%) or benzyl benzoate (20%–40%) in the essential oil. In our research as well, the two considered type of propolis differs for eudesmol content.

Likewise, flavonoid aglycones and esters of substituted cinnamic acids are the major constituents of propolis in the temperate zone where the basic plant source of bee glue are the bud exudates of trees of the genus *Populus*, mainly the black poplar *P. nigra*. [42]. In our study cinnamaldehyde, acetocinnamone, and cinnamyl alcohol were found significantly higher in the plain's samples while the two unknown sesquiterpenes characterized the hills samples in absence of characteristic compounds of poplar exudates. This could be of interest because it has been shown that bees can find in their environment and use as propolis source the best agent to protect their hives against bacterial and fungal infections [43].

The acetic acid, a carboxylic acid, was found in high quantity in our propolis samples and significantly higher in the plains samples. Acetic acid was found in headspace volatiles (dynamic headspace sampling, DHS) of Chinese propolis from 23 regions of China, as one of the main aroma-active components [44]. SPME with GC–MS was used for analysis of volatiles of Chinese propolis from the Beijing and Hebei provinces and again acetic acid and phenethyl acetate were among the main volatile constituents, together with phenethyl alcohol [45]. Their composition was somewhat similar to the volatiles of gum from poplar growing in China [46]. So, also this compound confirms the poplar exudates as the main origin of propolis plant-based component.

In our study we found low quantity of pinene (both in the plains and in the hills), that may be the consequence of collecting the resin of coniferous trees only when the other preferred sources are not sufficiently available [47]. This is probably due to the fact that conifers are scarce in the two sampling areas.

Until now, the propolis from areas without black poplar has been very poorly investigated. Out of 114 propolis samples analyzed [24] only 17 originated from "northern and mountain groups". As it turned out, these samples contained considerably less (approximately 25%) biologically active polyphenols characteristic for "poplar type" propolis, "but did not have significantly lower antibacterial activity". If it can be assumed that in the composition of propolis not deriving mainly from poplar exudates, there must be some active substances of unknown origin and unidentified chemical structure [47].

With chromatographic condition applied herein it was possible to perform the unambiguous detection and subsequent identification of phenolic acids (caffeic, p-and m-coumaric, ferulic, *trans*-cinammic) as well as two flavonoids namely pinocembrin and chrysin. The other compounds that were afterward defined by HRMS and NMR analysis were not quantified due to matrix complexity. As our aim was to use a simple HPLC–UV run for the fast characterization of propolis, any modification of chromatographic/detector condition did not bring any improvement in separation of other propolis components. The HPLC–UV analysis confirmed the presence of a complex mixture of compounds; the chromatographic resolution was slightly improved to obtain the optimal conditions for HPLC–Q-Exactive-Orbitrap®–MS analysis. The HRMS analysis demonstrated the presence of phenolic acids, flavanones, flavones, chalcones and isoflavones in the composition of the plains and hills propolis. Plains propolis, as expected, reveled higher concentration of trans-cinnamic acids which is accompanied with its caffeic esters form, displaying the typical pattern of "poplar" propolis.

HPLC–Q-Exactive-Orbitrap®–MS and NMR showed to be complementary methods for propolis characterization as they confirmed the principal compounds. For for minor compounds, as CAPE, which was not detected by NMR, the technique of HPLC–Q-Exactive-Orbitrap®–MS was fundamental; on the other hand, NMR allowed the detection of of some very descriptive minor molecules as phenolic glycerides. In the hills' samples, in fact, phenolic glycerides (dicoumaroyl acetyl glycerol, diferuloyl acetyl glycerol, feruloyl coumaroyl acetyl glycerol, caffeoyl coumaroyl acetyl glycerol) were upregulated. These compounds have been isolated [48] from North-Russian propolis and the exudate of *Populus tremula* L. (aspen) was found to be their plant source. Phenolic glycerides were previously detected in propolis samples from a mountain region at about 700 m a.s.l. in Switzerland where there are relatively high numbers of young *P. tremula* trees, and relatively few *P. nigra* [49].

P. tremula is present in the hills areas of the Ligurian–Piedmont Apennines. This species grows in abandoned fields in plant communities of *Sambuco-Salicion capreae* phytosociological alliance [50]. This vegetation is expanding both in the Apennines and in the Alps due to the drop out of agricultural practices [51]. Our study confirm that the determination of the "type" of propolis, according to its plant source, has to be the first step in quality control of bee glue and that bees have the ability to find in their environment and use as propolis source the best agent to protect their hives against bacterial and fungal infections [42].

It is also interesting the presence of abscisic acid in the Po Valley propolis samples. Abscisic acid is in fact a plant hormone with many functions, including seed and bud dormancy, the control of organ size and stomatal closure and it is sometimes involved in leaves abscission. The production of this and similar unusual compounds is a common ecological strategy in plants [52,53]. Therefore, the finding of high amounts of abscisic acid in the Po Valley samples is probably due to greater plant exposure to stress (dry, high temperature etc.) compared to those growing in the mountain areas.

So far, there is no evidence that individual propolis components are chemically modified by bee enzymes [54]. Our results support this finding as the agreement between the fingerprints of Apennines and Po-Vally propolis was surprisingly good. This points toward the conclusion that the difference revealed in relative amount of some components between two geographically different propolis sampling are due to botanical (vegetation) surrounding where bees collected the propolis.

The obtained results revealed importance of combined approach in analysis of complex biological matrices which composition can vary significantly depending on environmental conditions where is produced.

Propolis knowledge has registered an important evolution over time, due to exhaustive studies regarding its chemical composition and biological activities. In the 60's, it was thought that, despite its complexity, propolis chemical composition was more or less constant [55]. Spectrophotometric analysis was thought to be enough descriptive of this matrix and also now spectrophotometric analysis of total phenols, flavones and flavonols and antioxidant activity are widely used.

In the last decades analysis of a large number of samples from different geographic origins revealed that chemical composition of propolis is highly variable and also difficult to standardize because it depends on factors such as the vegetation, season, and environmental conditions of the site of collection [55]. Different resin types were proposed: poplar propolis, birch, green, red, "Pacific," and "Canarian" and also classification by propolis color (green, red, brown etc.) [56,57]. We demonstrated anyway that propolis from the same geographic area (Northern Italy) and of the same color (brown) differs significantly for many bioactive compounds. In fact, HPLC–Q-Exactive-Orbitrap®–MS analysis has been crucial in the appropriate recognition of evaluate number of metabolites, but also NMR itself could give more detailed information especially when isomeric compounds should be identified. However, for full metabolomics profiling combining single instrumental technique is indispensable in propolis characterization as identified metabolites belong to different chemical groups, so different analytical techniques are required.

4. Materials and Methods

4.1. Sample Collection and Study Area

The samples of propolis were collected from a professional beekeeper conducting 170 beehives for the harvests and around 100/200 breeding nucleuses for the company comeback and the sale. Propolis is harvested using both mesh and scraping. In our study we consider only propolis harvested by mesh. Samples were collected in a randomized selection of six beehives in the plains (Po Valley) and six beehives on the Ligurian–Piedmont Apennines. Considering that there are many factors which influence the phytochemical composition of propolis, the samples that were subjected to our evaluation originate from the same bees' strain and were harvested by same method. The only variable was the two different apiary geographical locations.

The apiaries are sedentary and placed: (A) in the valley bottom, on the edge of the plain (municipality: Visone—AL; elevation: 100 m a.s.l.; Latitude: 44°35′21″ N; Longitude: 8°27′37″ E) and (B) on the hill (Municipality: Ponzone - AL; elevation: 550 m a.s.l.; Latitude: 44°39′46″N; Longitude: 8°30′06″E) and distant about 25 km (Figure 7). The two sampling areas belong to the "Alpi Marittime" Ecoregional Subsection of Italy (Western Alps Section, Alpine Province) [58] with "Temperate continental submediterranean" bioclimate [59]. The two sampling areas are different as regards their vegetation series (*sigmeta*).

The sampling area A belong to *Physospermo cornubiensis-Querco petraeae sigmetum* where the mature stage of the vegetation series is the forest of *Quercus petraea* (dominant species) with other trees (*Castanea sativa, Sorbus aria, Fraxinus ornus*), shrubs (*Corylus avellana, Erica arborea, Frangula alnus, Juniperus communis*) and herbs (*Physospermum cornubiense, Pteridium aquilinum, Molinia arundinacea, Sesleria cylindrica, Carex montana, Euphorbia flavicoma, Brachypodium rupestre*) [59]. The main plants growing nearby the apiary and suited for honeybees' visit are *Castanea sativa, Erica arborea, Calluna vulgaris, Genista pilosa, Populus tremula* and *Salix caprea*.

The sampling area B belong to the vegetation series of lower Po Valley where the mature stage is the forest of the *Carpinion betuli* phytosociological alliance with *Quercus robur, Carpinus betuli, Fraxinus excelsior, Tilia cordata* and *Robinia pseudoacacia*, this latter species is very common where anthropic disturbance is greater [60–62]. Moreover, the area B is close to the Bormida di Spigno river near which there is riparian vegetation with willows (*Salix alba, Salix eleagnos, Salix purpurea*) and poplars (*Populus nigra* and *Populus alba*) [63]. The main trees growing nearby the apiary and suited for honeybees' visit are *Salix alba, Salix eleagnos, Salix purpurea, Populus nigra* and *Populus alba, Robinia pseudoacacia, Tilia cordata,* and *Ailanthus altissima*.

Figure 7. Localization of the two sampling areas: (**A**) hill (Ponzone); (**B**) plain (Visone).

4.2. Extraction, Balsam and Moisture Content

Propolis was pulverized by freezing it at −80 °C for an hour and pounding it in a mortar.

One gram of pulverized sample was weighed and dissolved in 30 mL of 70% ethanolic solution (70:30 ethanol:water) and stirred constantly for 2.5 h in a dark room at medium strength (200 rpm). The ethanol/water mixture (70/30) is the most commonly used extraction method for propolis as it is non-toxic and efficient in particular for polyphenols and flavonoids, responsible for the properties of the substance [24]. After that, ethanolic extract was separated by a 5 min centrifugation (5000 rpm at 5 °C) and the supernatant was separated from the residue by filtration (Whatman 3), as described in [21]. The supernatant was collected in a volumetric flask and topped up to 100 mL using the same 70% ethanol solvent. The final filtrates represent the balsam (tincture) of propolis and are referred to as PEE (propolis ethanolic extract). The yield was expressed as balsam content (soluble ethanolic fraction) and determined according to [24]. To this end, an aliquot (50.0 mL) of each ethanolic extract was evaporated to dryness on a rotary evaporator under reduced pressure at 40 °C.

The moisture content was determined as percentage weighting 1 g of propolis oven dried at 40 °C for 16 h.

4.3. Total Phenolic Content, Total Flavones and Flavonols, and Free Radical-Scavenging Activity

The ethanolic extract was diluted 1:10 to calculate the total phenolic content. The method used to determine the total phenolic content of the propolis extract was the one described in [21]. One hundred microliters of each extract of propolis plus 1900 µL distilled water were placed in a glass tube and then the solution was oxidized by adding 100 µL of FolinCiocalteau reagent. After exactly 2 min, 800 µL of 5% sodium carbonate (*w/v*) was added. This solution was maintained in a water bath at 40 °C for 20 min, and then the tube was rapidly cooled with crushed ice to stop the reaction. The generated blue color was measured using a spectrophotometer at 760 nm. In order to prepare the stock standard solutions, 25 mg of gallic acid or a were dissolved to a final volume of 25 mL methanol and stored at −20 °C. The calibration curve was carried out at the beginning of the working day and was prepared by appropriate dilution of each stock standard solution with 70% ethanol (y = 2.3454x + 0.0047; R^2 = 0.9998). The ethanolic solution was used as a blank.

The total flavones and flavonols (TFF) were estimated according to an aluminum chloride method following [63]. For the calibration curve, four standard solutions of quercetin in 80% ethanol (25, 50, 100, and 200 µg/mL) were prepared (y = 0.0099x − 0.055; R^2 = 0.9999). A 0.5 mL portion of standard solutions was separately mixed with 1.5 mL of 95% ethanol, 0.1 mL of 10% AlCl$_3$ in water (w/v), 0.1 mL of 1 M potassium-acetate, and 2.8 mL of 80% ethanol. After incubation at 20 °C for 30 min, the absorbance was measured at 425 nm. The 10% AlCl$_3$ was substituted by the same quantity of distilled water in the blank sample. Similarly, 0.5 mL of each extract diluted to 1:50 (v/v) in 80% ethanol was analyzed as described above. The results are expressed as TFF% w/w.

DPPH radical scavenging activity measured with the method described in [64]. Fifty µL of various concentrations of propolis samples were added to 2 mL of 60 µM methanolic solution of 1,1-diphenyl-2-picrylhydrazyl (DPPH). Absorbance measurements were read at 517 nm, after 20 min of incubation time at room temperature (A1). Absorption of a blank sample containing the same amount of methanol and DPPH solution acted as the negative control (A0). The results are expressed as % inhibition of the free radical with DPPH, as described in [65].

4.4. HPLC Analysis

The standards used: kaempferol, caffeic acid, p-coumaric acid, ferulic acid, *m*-coumaric acid, quercetin, trans-cinnamic acid, apigenin, genistein, chrysin, pinocembrin, formic acid, acetonitrile, and ethanol were purchased from Sigma-Aldrich (St. Louis, MO, USA). Gallic acid, Folin-Ciocalteau reagent and 1,1-diphenyl-2-picrylhydrazyl (DPPH) were also purchased from Sigma–Aldrich (St. Louis, MO, USA). All reagents and standards used were HPLC grade, and purified water from a Milli Q system was used throughout the experiments.

Individual stock solutions of each standard were prepared using ethanol for kaempeferol, pinocembrin, ferulic acid and m coumaric acid, DMSO and methanol (1:9) for quercetin, apigenin and chrysin, methanol for genistein, *p*-coumaric acid, caffeic acid and trans-cinnamic at 10 mg/mL, and stored at −20 °C. The working standard mixture solutions were made by diluting the appropriate amount of each stock standard solution to obtain 5 calibration levels (final concentrations of 31, 25, 62, 5, 125, 250, and 500 µg/mL).

The HPLC system used to determine the quantity of the most present phenols was a LC Agilent series 1200 (Waldbronn, Germany) consisting of a degasser, a quaternary gradient pump, an auto-sampler and a UV-Vis detector (Waldbronn, Germany). A Phenomenex Lichrospher C18, 4.6 × 250 mm, 5 µm column (Torrance, CA, USA) was used for this analysis with a column flow of 1 mL min^{-1}. Sample injections were made at 10 µL for all samples and standards. The run time was 35 min, with 1 min post run time. Details about the method are as follows: column oven (20 °C); mobile phase A (0.05% formic acid); mobile phase B (acetonitrile); flowrate (1 mL/min); needle wash (100% acetonitrile); injection volume (10 µL); detection at 295 nm, 325 nm, and 375 nm. The gradient applied was: 0 min (15% B); 5 min (40% B); 25 min (50% B); 30 min (90% B); a low gradient between 40% and 50% B was used to separate the acid compounds. A blank injection was performed in all the trials to check chromatographic interference in the resolution. The retention times of all the standards were confirmed by individual standard injections. A fortification of random samples was used to check further the retention factors. A standard mixture to check the retention times was injected each working day. The samples were filtered through a 0.2 µm pore size membrane filter prior to chromatographic analysis. LOD (0.5 µg/mL) and LOQ (1 µg/mL) was calculated according S/N ratio 3 and 10, respectively.

4.5. Solid Phase Microextraction (SPME) and Gas Chromatography Mass Spectrometry (GC–MS) Procedure

HS–SPME and GC–MS analysis were performed following the method in [28] opportunely modified. A 2 g amount of finely powdered raw propolis was weighed and put into 20 mL glass vials along with 100 µL of the IS (4-nonylphenol, 2000 µg/mL in 2-propanol). Each vial was fitted with a cap equipped with a silicon/PTFE septum (Supelco, Bellefonte, PA, USA). At the end of the sample equilibration time a conditioned SPME fiber was exposed to the headspace of the sample for 120 min using a CombiPAL system injector autosampler (CTC Analytics, Zwingen, Switzerland).

After sampling, the SPME fiber was immediately inserted into the GC injector and thermally desorbed. A desorption time of 1 min at 230 °C was used in the splitless mode. Before sampling, each fiber was reconditioned for 5 min in the GC injector port at 230 °C.

Analyses were performed with a Trace GC Ultra coupled to a Trace DSQII quadrupole mass spectrometer (MS) (Thermo-Fisher Scientific, Waltham, MA, USA) equipped with an Rtx-Wax column (30 m × 0.25 mm i.d. × 0.25 µm film thickness) (Restek, Bellefonte, PA, USA).

The identification was accomplished using computer searches on a NIST98 MS data library. In some cases, when identical spectra have not been found, only the structural type of the corresponding component was proposed on the basis of its mass-spectral fragmentation. If available, reference compounds were co-chromatographed to confirm GC retention times. The components of ethanol extracts of propolis were determined by considering their areas as percentage of the total ion current. Some components remained unidentified because of the lack of authentic samples and library spectra of the corresponding compounds.

4.6. NMR

The NMR spectra of samples were recorded on a Bruker Avance (Santa Barbara, CA, USA) spectrometer with proton operating frequency 600.13 MHz with a 5mm TBI probe. The spectra were performed at 300 K using 16K of TD (time domain), acquisition time 1.27 min, delay time 1.0 s and the number of scans 48. Was used a spectral width of 12019 Hz. For ^{1}H-NMR analysis 2–3 mg of crude extracts were dissolved in 0.6 mL of DMSO-d6, while for ^{13}C-NMR spectra 10–20 mg for each sample were used.

4.7. HPLC–Q-Exactive-Orbitrap®–MS Analysis: Untargeted Metabolomics Approach

In order to perform HPLC–Q-Exactive-Orbitrap®-MS analysis, samples that were subjected to HPLC–UV analysis were diluted (1:100) in starting mobile phase. Chromatography was accomplished on an HPLC Surveyor MS quaternary pump, a Surveyor AS autosampler with a column oven and a Rheodyne valve with a 20 µL loop system (Thermo Fisher Scientific, San Jose, CA, USA). Analytical separation was carried out using a reverse-phase HPLC column 150 × 2 mm i.d., 4 µm, Synergi Hydro RP, with a 4 × 3 mm i.d. C18 guard column (Phenomenex, Torrance, CA, USA). The mobile phase was run as a gradient that consisted of water and methanol both acidified with 0.1% formic acid. The gradient (flow rate 0.3 mL/min) was initiated with 80% eluent 0.1% aqueous formic acid with a linear decrease up to 5% in 30 min. The mobile phase was returned to initial conditions at 36 min, followed by a 9-min re-equilibration period. The The column and sample temperatures were 30 °C and 5 °C, respectively. The mass spectrometer Thermo Q-Exactive Plus (Thermo Scientific, San Jose, CA, USA) was equipped with a heated electrospray ionisation (HESI) source. Capillary temperature and vaporizer temperature were set at 330 and 380 °C, respectively, while the electrospray voltage operating in positive was adjusted at 3.30 kV. Sheath and auxiliary gas were 35 and 15 arbitrary units, with S lens RF level of 60. The mass spectrometer was controlled by Xcalibur 3.0 software (Thermo Fisher Scientific, San Jose, CA, USA). The exact mass of the compounds was calculated using Qualbrowser in Xcalibur 3.0 software. The full scan (FS) with resolving power 140,000 in negative mode was used for the screening and statistical evaluation of obtained chromaptografic profiles. FS-dd-MS^2 (full scan data-dependent acquisition) was used for confirmation. Resolving power of FS adjusted on 70,000 FWHM at m/z 200, with scan range of m/z 100–900. Automatic gain control (AGC) was set at 3e^6, with an injection time of 200 ms. The AGC target was set to 2e^5, with the maximum injection time of 100 ms. Fragmentation of precursors was optimised as three-stepped normalized collision energy (NCE) (20, 40, and 40 eV). Detection was based on retention time and on calculated exact mass of the protonated molecular ions, with at least one corresponding fragment of target compounds). Good peak shape of extracted ion chromatograms (EICs) for targeted compounds was ensured by manual inspection, as well.

Raw data from Xcalibur 3.0 software were processed with Compound Discoverer™ (Thermo Scientific, Waltham, MA, USA). In particular, this platform enables peak detection, retention time adjustment, profile assignment, and isotope annotation. A list of potential compounds was suggested for each chromatographic peak depending on the mass fragmentation of the parent pseudomolecular ion. Accurate mass determination generating elemental composition within a narrow mass tolerance window for identification based on accurate precursor mass. For some signals, the putative identification was confirmed by analysis performed on authentic standard. Compounds identification was based on accurate mass and mass fragmentation pattern spectra against MS–MS spectra of compounds available on mzCloud database (HighChem LLC, Bratislava, Slovakia). The ChemSpider Web services platform and Human metabolome [66] were used as additional confirmation tool. If mass fragmentation pattern did not correspond to any of databases annotated by Compound Discoverer™ software, manual confirmation using program ChemDrow of their fragments was performed.

4.8. Statistical Analysis

The relative intensity of chromatographic peak from two propolis types were processed by Compound Discoverer platform that enabled differential analysis applying Volcano Plot Model and setting p-value (PV) on 0.05. In addition, the propolis samples were ordered by Principal Component Analysis (PCA) using HRMS spectra.

Data of moisture content, balsam content, total phenols, total flavones and flavonols, DPPH radical scavenging activity were analyzed using Student's t- test at 95% confidence level in order to compare the two propolis types. The same statistical analysis was done for VOCs and single compounds quantified with HPLC.

VOCs that resulted significant at the t-test were employed in the PCA to highlight the most important differences between the two batches of propolis type. T-test and PCA were performed using R 3.5.2 software [67].

Author Contributions: A.G. and V.L. conceptualized the research activities; V.L. realized the GC–MS methodology and formal analysis; G.B. and A.B. realized the NMR methodology and formal analysis; R.P. performed LC Orbitrap analysis; V.L. realized the methodology and formal physical, spectrophotometric, and HPLC analysis and the most of literature investigation; L.G. performed statistical analysis and botanical evaluations; V.L., L.G., G.B., R.P., G.C., and S.S. analyzed the data and wrote the article. The first three authors have equally contributed to the realization of the research. A.G. is responsible of the funding acquisition and of the team coordination. All authors have read and agreed to the published version of the manuscript.

Funding: The present paper is funded and realized within the "Italian mountain Lab" project and by DARA-CRC Ge.S.Di.Mont. agreement.

Acknowledgments: We thank the professional beekeeper Kovac Paolo of Terra di Mezzo Agribusiness for the input and the support

Conflicts of Interest: The authors declare no conflict of interest. The funders had no role in the design of the study; in the collection, analyses, or interpretation of data; in the writing of the manuscript, or in the decision to publish the results

References

1. Kuropatnicki, A.K.; Szliszka, E.; Krol, W. Historical Aspects of Propolis Research in Modern Times. *Evid. Complementary Altern. Med.* **2013**, 11. [CrossRef] [PubMed]
2. Righi, A.A.; Alves, T.R.; Negri, G.; Marques, L.M.; Breyer, H.; Salatino, A. Brazilian red propolis: Unreported substances, antioxidant and antimicrobial activities. *J. Sci. Food. Agri.* **2010**, *91*, 2363–2370. [CrossRef] [PubMed]
3. Ghisalberti, E.L. Propolis—Review. *BeeWorld* **1979**, *60*, 59–84. [CrossRef]
4. Silva, R.O.; Andrade, V.M.; Bullé Rêgo, E.S.; Azevedo Dòria, G.A.; Santos Lima, B.; Silva, F.A.; Araújo, A.A.S.; de Albuquerque, R.L.C., Jr.; Cardoso, J.C.; Gomes, M.Z. Acute and sub-acute oral toxicity of Brazilian red propolis in rats. *J. Ethnopharmacol.* **2015**, *170*, 66–71. [CrossRef] [PubMed]

5. Houghton, P.J. Propolis as a Medicine. Are there Scientific Reasons for its Reputation? In *Beeswax and Propolis for Pleasure and Profit*; Munn, P., Ed.; International Bee Research Association: Cardiff, UK, 1998; p. 10.
6. Koo, H.; Rosalen, P.L.; Cury, J.A.; Park, Y.K.; Bowen, W.H. Effects of Compounds Found in Propolis on Streptococcus mutans growth and on glucosyltransferase activity. *Antimicrob. Agents Chemother.* **2002**, *46*, 1302–1309. [CrossRef] [PubMed]
7. Silva Cunha, I.B.; Salomão, K.; Shimizu, M.; Bankova, V.S.; Custódio, A.R.; Castro, S.L.; Marcucci, M.C. Antitrypanosomal activity of Brazilian propolis from Apis mellifera. *Chem. Pharm. Bull.* **2004**, *52*, 602–604. [CrossRef] [PubMed]
8. Borrelli, F.; Maffia, P.; Pinto, L.; Ianaro, A.; Russo, A.; Capasso, F.; Ialenti, A. Phytochemical compounds involved in the anti-inflammatory effect of propolis extract. *Fitoterapia* **2002**, *73*, S53–S63. [CrossRef]
9. Ahn, M.; Kumazawa, S.; Usui, Y.; Nakamura, J.; Matsuka, M.; Zhu, F.; Nakayama, T. Antioxidant activity and constituents of propolis collected in various areas of China. *Food Chem.* **2007**, *101*, 1383–1392. [CrossRef]
10. Oršolić, N.; Knežević, A.H.; Šver, L.; Terzić, S.; Bašić, I. Immunomodulatory and antimetastatic action of propolis and related polyphenolic compounds. *J. Ethnopharmacol.* **2004**, *94*, 307–315. [CrossRef]
11. Amoros, M.; Lurton, E.; Boustie, J.; Girre, L.; Sauvager, F.; Cormier, M. Comparison of the anti-herpes simplex virus activities of propolis and 3-methyl-but-2-enyl caffeate. *J. Nat. Prod.* **1994**, *57*, 644–647. [CrossRef]
12. Seo, K.W.; Park, M.; Song, Y.J.; Kim, S.J.; Yoon, K.R. The protective effects of propolis on hepatic injury and its mechanism. *Phytother. Res.* **2003**, *17*, 250–253. [PubMed]
13. Ota, C.; Unterkircher, C.; Fantinato, V.; Shimizu, M.T. Antifungal activity of propolis on different species of Candida arten. *Mycoses* **2001**, *44*, 375–378. [CrossRef] [PubMed]
14. Silici, S.; Kutluca, S. Chemical composition and antibacterial activity of propolis collected by three different races of honeybees in the same region. *J. Ethnopharmacol.* **2005**, *99*, 69–73. [CrossRef]
15. Castaldo, S.; Capasso, F. Propolis, an old remedy used in modern medicine. *Fitoterapia* **2002**, *73*, S1–S6. [CrossRef]
16. Stan, L.; Liviu, A.; Dezmirean, D. Quality Criteria for Propolis Standardization. *Sci. Pap. Anim. Sci. Biotechnol.* **2011**, *44*, 137–140.
17. Sforcin, J.M.; Bankova, V. Propolis: Is there a potential for development of new drugs? *J. Ethnopharmacol.* **2011**, *133*, 253–260. [CrossRef] [PubMed]
18. Katekhaye, S.; Fearnley, H.; Fearnley, J.; Paradkar, A. Gaps in propolis research: Challenges posed to commercialization and the need for a holistic approach. *J. Apic. Res.* **2019**, *58*, 604–616. [CrossRef]
19. Giupponi, L.; Pentimalli, D.; Manzo, A.; Panseri, S.; Giorgi, A. Effectiveness of fine root fingerprinting as a tool to identify plants of the Alps: Results of preliminary study. *Plant Biosyst.* **2017**, *152*, 464–473. [CrossRef]
20. Cottica, S.M.; Sabik, H.; Antoine, C.; Fortin, J.; Graveline, N.; Visentainer, J.V.; Britten, M. Characterization of Canadian propolis fractions obtained from two-step sequential extraction. *LWT Food Scie Technol.* **2015**, *60*, 609–614. [CrossRef]
21. Escriche, I.; Juan-Borrás, M. Standardizing the analysis of phenolic profile in propolis. *Food Res. Int.* **2018**, *106*, 834–841. [CrossRef]
22. Cicco, N.; Lanorte, M.T.; Paraggio, M.; Viggiano, M.; Lattanzio, V. A reproducible, rapid and inexpensive Folin–Ciocalteu micro-method in determining phenolics of plant methanol extracts. *Microchem. J.* **2009**, *91*, 107–110. [CrossRef]
23. Castro, C.; Mura, F.; Valenzuela, G.; Figueroa, C.; Salinas, R.; Zuñiga, M.C.; Torres, J.L.; Fuguet, E.; Delporte, C. Identification of phenolic compounds by HPLC-ESI-MS/MS and antioxidant activity from Chilean propolis. *Food Res. Int.* **2014**, *64*, 873–879. [CrossRef]
24. Popova, M.; Bankova, V.; Bogdanov, S.; Tsvetkova, I.; Naydenski, C.; Marcazzan, G.L.; Sabatini, A.G. Chemical characteristics of poplar type propolis of different geographic origin. *Apidologie* **2007**, *38*, 306–311. [CrossRef]
25. Anđelković, B.; Vujisić, L.; Vučković, I.; Tešević, V.; Vajs, V.; Gođevac, D. Metabolomics study of Populus type propolis. *J. Pharm. Biomed.* **2017**, *135*, 217–226. [CrossRef] [PubMed]
26. Xu, Y.; Luo, L.; Chen, B.; Fu, Y.; Chaz Du, P. Recent development of chemical components in propolis. *Front. Biol. China* **2009**, *4*, 385. [CrossRef]
27. Graça, M.M.; Dulce, A.M. Is propolis safe as an alternative medicine? *J. Pharm. Bioallied Sci.* **2011**, *3*, 479–495.
28. Pellati, F.; Prencipe, F.P.; Benvenuti, S. Headspace solid-phase microextraction-gas chromatography–mass spectrometry characterization of propolis volatile compounds. *J Pharm. Biomed.* **2013**, *84*, 103–111. [CrossRef] [PubMed]

29. Jelen', H.H.; Majchera, M.; Dziadas, M. Microextraction techniques in the analysis of food flavor compounds: A review. *Anal. Chim. Acta* **2012**, *738*, 13–26. [CrossRef]
30. Plutowska, B.; Chmiel, T.; Dymerski, T.; Wardencki, W. A headspace solid phase microextraction method development and its application in the determination of volatiles in honeys by gas chromatography. *Food Chem.* **2011**, *126*, 1288–1298. [CrossRef]
31. Belliardo, F.; Bicchi, C.; Cordero, C.; Liberto, E.; Rubiolo, P.; Sgorbini, B. Headspace solid-phase microextraction in the analysis of the volatile fraction of aromatic and medicinal plants. *J. Chromatogr. Sci.* **2006**, *44*, 416–429. [CrossRef]
32. Petri, G.; Lemberkovics, E.; Foldvari, F. Examination of Differences Between Propolis (Bee Glue) Produced from Different Flora Environment. In *Flavors and Fragrances: A World Perspective*; Lawrence, B.M., Mookherjee, B.D., Willis, B.J., Eds.; Elsevier Science Publishers B.V.: Amsterdam, The Netherlands, 1988; pp. 439–446.
33. Miguel, M.M.; Nunes, S.; Cruz, C.; Duarte, J.; Antunes, M.D.; Cavaco, A.M.; Mendes, M.D.; Lima, A.S.; Pedro, L.G.; Barroso, J.G.; et al. Propolis volatiles characterization from acaricide-treated and -untreated beehives maintained at Algarve (Portugal). *Nat. Prod. Res. Ifirst* **2012**, 1–7.
34. Papotti, G.; Bertelli, D.; Plessi, M.; Rossi, M.C. Use of HR-NMR to classify propolis obtained using different harvesting methods. *Int. J. Food Sci. Tech.* **2010**, 1610–1618. [CrossRef]
35. Ristivojević, P.; Trifković, J.; Gašić, U.; Andrić, F.; Nedić, N.; Tešić, Ž.; Milojković-Opsenica, D. Ultrahigh-performance Liquid Chromatography and Mass Spectrometry (UHPLC–LTQ/Orbitrap/MS/MS) Study of Phenolic Profile of Serbian Poplar Type Propolis. *Phytochem. Anal.* **2015**, *26*, 127–136. [CrossRef] [PubMed]
36. Saftić, L.; Peršurić, Ž.; Fornal, E.; Pavlešić, T.; Pavelić, S.K. Targeted and untargeted LC-MS polyphenolic profiling and chemometric analysis of propolis from different regions of Croatia. *J. Pharm. Biomed.* **2019**, *165*, 162–172. [CrossRef]
37. Papotti, G.; Bertelli, D.; Bortolotti, L.; Plessi, M. Chemical and Functional Characterization of Italian Propolis Obtained by Different Harvesting Methods. *J. Agric. Food Chem.* **2012**, *60*, 2852–2862. [CrossRef]
38. Borčić, I.; Radonić, A.; Grzunov, K. Comparison of the volatile constituents of propolis gathered in different regions of Croatia. *lavour Fragr. J.* **1996**, *11*, 311–313.
39. Jerković, I.; Mastelić, J. Volatile compounds from leaf-buds of Popolus nigra L. (Salicaceae). *Phytochem* **2003**, *63*, 109–113. [CrossRef]
40. Salatino, A.; Fernandes-Silva, C.C.; Righi, A.A.; Salatino, M.L.F. Propolis research and the chemistry of plant products. *Nat. Prod. Rep.* **2011**, *28*, 925–936. [CrossRef]
41. Blasi, C. *La Vegetazione d'Italia*; Palombi & Partner: Roma, Italy, 2010.
42. Bankova, V.; de Castro, S.L.; Marcucci, M.C. Propolis: Recent advances in chemistry and plant origin. *Apidologie* **2000**, *31*, 3–15. [CrossRef]
43. Bankova, V.; Popova, M.; Trusheva, B. Propolis volatile compounds: Chemical diversity and biological activity: A review. *Chem. Cent. J.* **2014**, *8*, 8–28. [CrossRef]
44. Yang, C.; Luo, L.; Zhang, H.; Yang, X.; Lv, Y.; Song, H. Common aroma-active components of propolis from 23 regions of China. *J. Sci. Food. Agric.* **2010**, *90*, 1268–1282. [CrossRef] [PubMed]
45. Xu, X.; Dong, J.; Li, J. Analysis of volatile compounds of propolis by solidphase microextraction combined with GC-MS. *Sci. Technol. Food Ind.* **2008**, *5*, 57–59.
46. Cheng, H.; Qin, Z.H.; Hu, X.S.; Wu, J.H. Analysis of volatile compounds of propolis and poplar tree gum by SPME/DHS-GC-MS. *J. Food Saf. Qual.* **2012**, *3*, 1–9.
47. Isidorov, V.A.; Szczepaniak, L.; Bakier, S. Rapid GC/MS determination of botanical precursors of Eurasian propolis. *Food Chem.* **2014**, *142*, 101–106. [CrossRef] [PubMed]
48. Popravko, S.A.; Sokolov, I.V.; Torgov, I.V. New natural phenolic triglycerides. *Chem. Nat. Compd.* **1982**, *18*, 153–157. [CrossRef]
49. Bankova, V.; Popova, M.; Bogdanov, S.; Sabatini, A.G. Chemical Composition of European Propolis: Expected and Unexpected Results. *Zeitschrift für Naturforschung C* **2014**, *57*, 530–533. [CrossRef]
50. Landolt, E.; Bäumler, B.; Erhardt, A.; Klötzli, F.; Lämmler, W.; Urmi, E. *Flora Indicativa. Ecological Indicator Values and Biological Attributes of the Flora of Switzerland and the Alps*; Haupt Verlag: Bern, Switzerland; Stuttgart, Germany; Vienna, Austria, 2010.

51. Cislaghi, A.; Giupponi, L.; Tamburini, A.; Giorgi, A.; Bischetti, G.B. The effects of mountain grazing abandonment on plant community, forage value and soil properties: Observations and field measurements in an alpine area. *Catena* **2019**, *181*, 104086. [CrossRef]
52. Radmila, P.; Panseri, S.; Giupponi, L.; Leoni, V.; Citti, C.; Cattaneo, C.; Cavaletto, M.; Giorgi, A. Phytochemical and Ecological Analysis of Two Varieties of Hemp (*Cannabis sativa* L.) Grown in a Mountain Environment of Italian Alps. *Front Plant Sci.* **2019**, *10*, 1265.
53. Giorgi, A.; Bononi, M.; Tateo, F.; Cocucci, M. Yarrow (*Achillea millefolium* L.) growth at different altitudes in central italian alps: Biomass yield, oil content and quality. *J. Herbs Spices Med. Plants* **2005**, *11*, 47–58. [CrossRef]
54. Bankova, V.; Popova, M.; Trusheva, B. Plant Sources of Propolis: An Update from a Chemist's Point of View. *Nat. Prod. Commun.* **2006**. [CrossRef]
55. Silva-Carvalho, R.; Baltazar, F.; Almeida-Aguiar, C. Propolis: A Complex Natural Product with a Plethora of Biological Activities That Can Be Explored for Drug Development. *Evid. Based Complementary Altern. Med.* **2015**, *2015*, 29. [CrossRef] [PubMed]
56. Bankova, V. Chemical diversity of propolis and the problem of standardization. *J. Ethnopharmacol.* **2005**, *100*, 114–117. [CrossRef] [PubMed]
57. Fokt, H.; Pereira, A.; Ferreira, A.M.; Cunha, A.; Aguiar, C. How do Bees Prevent Hive Infections? The Antimicrobial Properties of Propolis. In *Current Research, Technology and Education Topics in Applied Microbiology and Microbial Biotechnology*; Mendez-Vilas, A., Ed.; Microbiology Book Series—Number 2; Researchgate: Berlin, Germany, 2010; Volume 1, pp. 481–493.
58. Blasi, C.; Capotorti, G.; Copiz, R.; Guida, D.; Mollo, B.; Smiraglia, D.; Zavattero, L. Classification and mapping of the ecoregions of Italy. *Plant Biosyst.* **2014**, *148*, 1255–1345. [CrossRef]
59. Rivas-Martínez, S.; Penas, A.; Díaz, T.E. Biogeographic Map of Europe. 2004. Available online: www.globalbioclimatics.org/form/maps.htm (accessed on 13 March 2019).
60. Giupponi, L.; Corti, C.; Manfredi, P. Onopordum acanthium subsp. acanthium in una ex-discarica della Pianura Padana (Piacenza). *Ital. Bot.* **2013**, *45*, 213–219.
61. Giupponi, L.; Corti, C.; Manfredi, P.; Cassinari, C. Application of the floristic-vegetational indexes system for the evaluation of the environmental quality of a semi-natural area of the Po Valley (Piacenza, Italy). *Plant Sociol.* **2013**, *50*, 47–56.
62. Giupponi, L.; Corti, C.; Manfredi, P. The vegetation of the Borgotrebbia landfill (Piacenza, Italy): Phytosociological and ecological characteristics. *Plant Biosyst.* **2015**, *149*, 865–874. [CrossRef]
63. Siniscalco, C.; Bouvet, L. Le Serie di Vegetazione della Regione Piemonte. In *La vegetazione d'Italia*; Blasi, C., Ed.; Palombi & Partner: Roma, Italy, 2010; pp. 17–37.
64. Miguel, M.G.; Nunes, S.; Dandlen, S.A.; Cavaco, A.M.; Antunes, M.D. Phenols and antioxidant activity of hydro-alcoholic extracts of propolis from Algarve, South of Portugal. *Food Chem. Toxicol.* **2010**, *48*, 3418–3420. [CrossRef] [PubMed]
65. Marinova, G.V.; Batcharov, V. Evaluation of the methods for determination of the free radical scavenging activity by DPPH. *Bulg. J. Agric. Sci.* **2011**, *17*, 11–24.
66. Wishart, D.S.; Feunang, Y.D.; Marcu, A.; Guo, A.C.; Liang, K.; Vázquez-Fresno, R.; Sayeeda, Z. HMDB 4.0—The Human Metabolome Database for 2018. *Nucleic Acids Res.* **2018**, *46*, D608–D617. [CrossRef]
67. R Development Core Team. R: A Language and Environment for Statistical Computing. Wien, R Foundation for Statistical Computing. 2018. Available online: http://www.r-project.org (accessed on 1 October 2019).

© 2020 by the authors. Licensee MDPI, Basel, Switzerland. This article is an open access article distributed under the terms and conditions of the Creative Commons Attribution (CC BY) license (http://creativecommons.org/licenses/by/4.0/).

Article

Grouping, Spectrum–Effect Relationship and Antioxidant Compounds of Chinese Propolis from Different Regions Using Multivariate Analyses and Off-Line Anti-DPPH Assay

Xiasen Jiang [1,2], Linchen Tao [1], Chunguang Li [2], Mengmeng You [1], George Q. Li [2], Cuiping Zhang [1] and Fuliang Hu [1,*]

1 College of Animal Science, Zhejiang University, Hangzhou 310058, China; jxsen@zju.edu.cn (X.J.); 11817025@zju.edu.cn (L.T.); mmyou@zju.edu.cn (M.Y.); lgzcplyx@aliyun.com (C.Z.)
2 NICM Health Research Institute, Western Sydney University, Westmead, NSW 2145, Australia; c.li@westernsydney.edu.au (C.L.); g.li@westernsydney.edu.au (G.Q.L.)
* Correspondence: flhu@zju.edu.cn; Tel.: +86-27-8898-2952

Academic Editors: Gavino Sanna and Wilfried Rozhon
Received: 2 June 2020; Accepted: 14 July 2020; Published: 16 July 2020

Abstract: 49 samples of propolis from different regions in China were collected and analyzed for their chemical compositions, contents of total flavonoids (TFC), total phenolic acid (TPC) and antioxidant activity. High-performance liquid chromatography (HPLC) analysis identified 15 common components, including key marker compounds pinocembrin, 3-O-acetylpinobanksin, galangin, chrysin, benzyl p-coumarate, pinobanksin and caffeic acid phenethyl ester (CAPE). Cluster analysis (CA) and correlation coefficients (CC) analysis showed that these propolis could be divided into three distinct groups. Principal component analysis (PCA) and multiple linear regression analysis (MLRA) revealed that the contents of isoferulic acid, caffeic acid, CAPE, 3,4-dimethoxycinnamic acid, chrysin and apigenin are closely related to the antioxidant properties of propolis. In addition, eight peak areas decreased after reacting with 1,1-Diphenyl-2-picrylhydrazyl (DPPH) radicals, indicating that these compounds have antioxidant activity. The results indicate that the grouping and spectrum–effect relationship of Chinese propolis are related to their chemical compositions, and several compounds may serve as a better marker for the antioxidant activity of Chinese propolis than TFC and TPC. The findings may help to develop better methods to evaluate the quality of propolis from different geographic origins.

Keywords: propolis; antioxidant activity; spectrum–effect relationships; cluster analysis; principal component analysis; multiple linear regression analysis

1. Introduction

Propolis is a biologically active natural product produced by honeybees collecting substances from parts of plants, buds and exudates [1]. Bees use propolis to build and repair their hives, such as for controlling the size of the hive door and repairing any cracks [2].

Propolis has a complex composition, with more than 500 compounds having been identified within it [3,4]. Many factors such as plant origin, geographic location and seasonality can influence the chemical composition of propolis by affecting plant bud exudates [5–7]. It is known that bees collect resin from more than 16 plant families, particularly the *Populus* family, with at least seven different *Populus* species having proven to be plant sources of propolis [6]. Previous studies indicate that the main plant sources of Chinese propolis are *Populus* species [8,9]. However, there are few studies on the classification of Chinese propolis to find different propolis types in China.

Propolis has demonstrated various bioactivities, and been used as a health supplement and food additive [10–12]. Among them, the antioxidant activity of propolis may play a key role in protection against the damage caused by free radicals in some chronic diseases [13]. The antioxidant activity of Chinese propolis is largely attributed to the high levels of various phenolic compounds, such as flavonoids and phenolic acids [14]. In this regard, analysis of the chemical composition and its relationship to biological activity, or the spectrum–effect relationship, is important to evaluate the quality of natural products [15–17].

Thus, we set out to investigate the spectrum–effect relationship of Chinese propolis by using high-performance liquid chromatography (HPLC) to separate and identify the chemical composition in 49 Chinese propolis samples collected from different regions. Furthermore, the total phenolic acid content (TPC), total flavonoid content (TFC) and antioxidant capacity of these propolis samples were determined. Chromatographic data were processed by multivariate analyses such as cluster analysis (CA), principal component analysis (PCA) and multi-linear regression analysis (MLRA), in order to classify samples and obtain the relationship between spectral and antioxidant capacity. The off-line anti-1,1-Diphenyl-2-picrylhydrazyl (DPPH) assay was performed to identify the antioxidant compounds in Chinese propolis.

2. Results and Discussion

2.1. HPLC Analysis of 49 Chinese Propolis

The chromatographic profiles of 49 Chinese propolis samples (as detailed in Table 1) were analyzed using the previously established method [8,18,19]. The results of precision showed that the relative standard deviation (RSD) of the intraday and interday for retention times were less than 0.38% and 0.44%, respectively, and for peak areas were less than 2.54% and 2.69%. The RSD for the repeatability of retention time and peak areas were less than 0.31% and 5.71% (Table S1), respectively. The HPLC fingerprints of representative samples are shown in Figure 1, and the fingerprints of all samples are shown in Figure S1. 15 common peaks were identified by comparison with standard compound retention times, and the content of those compounds was quantified by the regression equation of standard compounds (Table S2). The contents of those compounds varied significantly with geographic origins (Table S3). The compounds with relative higher contents include pinocembrin (ranging from 20.14 to 104.90 mg/g, mean value 41.93 mg/g), 3-O-acetylpinobanksin (ranging from 3.26 to 73.08 mg/g, mean value 35.69 mg/g), galangin (ranging from 10.59 to 52.58 mg/g, mean value 33.45 mg/g), chrysin (ranging from 5.26 to 52.91 mg/g, mean value 33.12 mg/g), benzyl p-coumarate (ranging from 5.04 to 121.39 mg/g, mean value 27.98 mg/g), pinobanksin (ranging from 2.27 to 51.27 mg/g, mean value 23.51 mg/g) and caffeic acid phenethyl ester (CAPE) (ranging from 0 to 49.58 mg/g, mean value 12.26 mg/g). The chemical composition of propolis in all regions showed a similar characteristic as the poplar-type propolis [20]. This may be related to the fact that *Populus* is widespread throughout China [21,22], and research shows that *Apis mellifera* prefer *Populus* as plant sources [6]. The content of common compounds varies between different samples, as the chemical compositions of propolis could be influenced by botanical origin, collecting season or other factors [23].

2.2. Similarity of HPLC Fingerprints among 49 Chinese Propolis

The similarity of 49 propolis fingerprints was analyzed using the software "Similarity Evaluation System for Chromatographic Fingerprint of Tradition Chinese Medicine (TCM)" developed by the Chinese Pharmacopoeia Commission. The results showed that the correlation coefficients (CC) values of most Chinese propolis fingerprints were higher than 0.6, indicating that their compositions were very similar. However, the CC values of five samples (S21, 0.512; S23, 0.306; S25, 0.356; S27, 0.407; and S29, 0.499) collected from Northeast China were significantly lower than that of other propolis, indicating the difference of these samples with other Chinese propolis (Table 1).

The CA identified three distinctive groups: Group 1 contains 42 propolis samples; Group 2 contains S19 and S26; and Group 3 contains S21, S25, S29, S23 and S27 (Figure 2). All samples in Groups 2 and 3 were collected in Northeast China, which contained rich *p*-coumaric acid and benzyl *p*-coumarate. This is consistent with the results of the CC values; the CC values of all samples in Group 3 are below 0.5.

Table 1. Collection site correlation coefficients of 49 Chinese propolis.

Samples No.	Collected Site (city, province)	Correlation Coefficients	Samples No.	Collected Site (City, Province)	Correlation Coefficients
S1	Lujiang, Anhui	0.891	S26	Yichun, Heilongjiang	0.822
S2	Tongcheng, Anhui	0.861	S27	Yilan, Heilongjiang	0.407
S3	Shouxian, Anhui	0.675	S28	Baoqing, Heilongjiang	0.892
S4	Mingcong, Anhui	0.896	S29	Muling, Heilongjiang	0.499
S5	Huaibei, Anhui	0.887	S30	Xinye, Henan	0.924
S6	Fuyang, Anhui	0.914	S31	Shangqiu, Henan	0.881
S7	Mengcheng, Guizhou	0.533	S32	Fangcheng, Henan	0.915
S8	Meishan, Sichuan	0.926	S33	Shangshui, Henan	0.861
S9	Xicuan, Sichuan	0.890	S34	Nanyang, Henan	0.889
S10	Qixia, Shandong	0.880	S35	Huaibin, Henan	0.875
S11	Dezhou, Shandong	0.825	S36	Liutai, Jiangsu	0.894
S12	Heze, Shandong	0.878	S37	Jianhu, Jiangsu	0.901
S13	Longkou, Shandong	0.921	S38	Yicheng, Hubei	0.818
S14	Donge, Shandong	0.890	S39	Gucheng, Hubei	0.895
S15	Penglai, Shandong	0.879	S40	Dongkou, Hubei 1	0.796
S16	Yanbailou, Shaanxi	0.811	S41	Dongkou, Hubei 2	0.789
S17	Mizhi, Shaanxi	0.869	S42	Xianju, Zhejiang	0.767
S18	Fusong, Jilin	0.906	S43	Jinyun, Zhejiang	0.829
S19	Baishan, Jilin	0.846	S44	Penghu, Ningxia	0.889
S20	Jian, Jilin	0.848	S45	Gongliu, Xinjiang	0.836
S21	Dashiqiao, Liaoning	0.512	S46	Beijing	0.826
S22	Faku, Liaoning	0.930	S47	Baotou, Inner Mongolia	0.926
S23	Suizhong, Liaoning	0.306	S48	Huailai, Hebei	0.687
S24	Shuangyashan, Heilongjiang	0.921	S49	Qiuxian, Hebei	0.910
S25	Gannan, Heilongjiang	0.356			

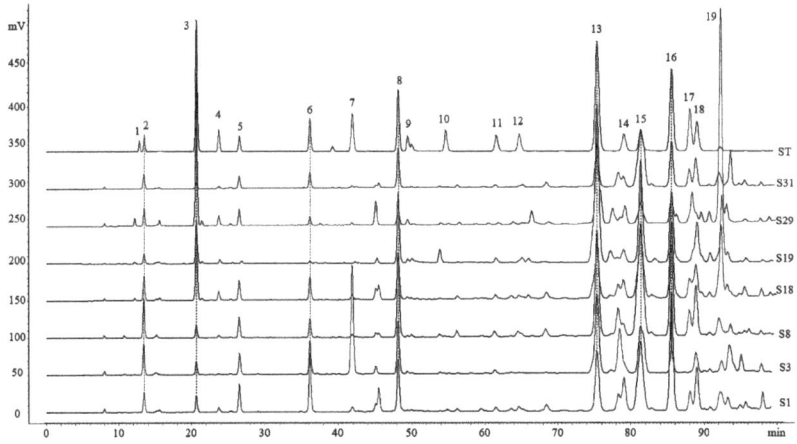

Figure 1. High-performance liquid chromatography (HPLC) chromatograms of the standard solution (ST) and Chinese propolis (S1,S3, S8, S18, S19, S29, S31): 1. Vanillic; 2. Caffeic acid; 3. *p*-Coumaric acid; 4. Ferulic acid; 5. Isoferulic acid; 6. 3,4-Dimethoxycinnamic acid; 7. Cinnamic acid; 8. Pinobanksin; 9. Naringenin; 10. Quercetin; 11. Kaempferol; 12. Apigenin; 13. Pinocembrin; 14. Benzyl caffeate; 15. 3-*O*-acetylpinobanksin; 16. Chrysin; 17. Caffeic acid phenethyl ester (CAPE); 18. Galangin; and 19. Benzyl *p*-coumarate.

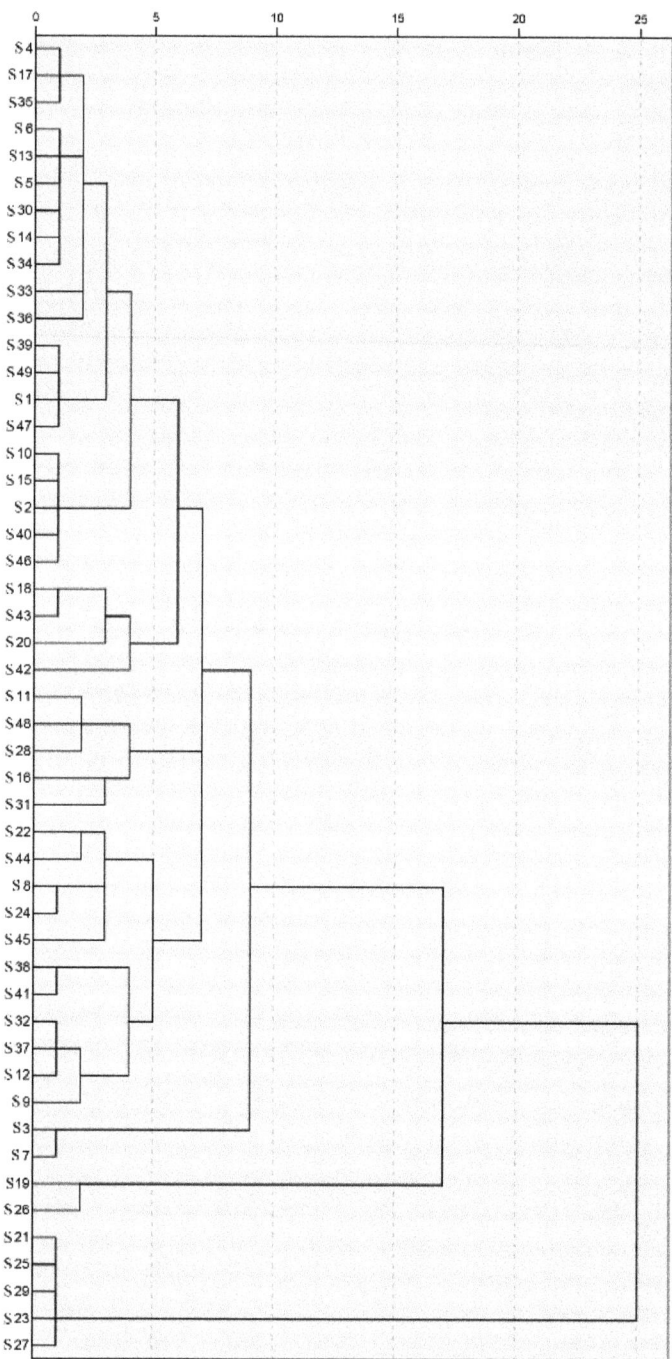

Figure 2. Dendrogram of cluster analysis of 49 Chinese propolis samples (S1–S49).

These results indicate that some propolis from Northeast China was special, and could be classified as a new type of propolis. We have previously shown that the propolis collected from the

Changbai Mountain area (CBM) in Northeast China had a higher content of p-coumaric acid and benzyl p-coumarate [18]. Accordingly, samples in Group 1 were ordinary Chinese propolis, samples in Group 2 were mixed propolis and samples in Group 3 were CBM propolis. Samples in the same group may also be subdivided into different subgroups, but further research is needed to establish this. As mentioned earlier, many factors can affect the chemical composition of propolis [5,6]. Chinese propolis could be divided into different groups, which may be caused by differences in plant sources, climate and other factors in different regions.

2.3. Contents of Flavonoids and Phenolics of 49 Chinese Propolis

It has been well established that flavonoids and phenolic acids, as the secondary metabolites in plants with broad biological activities [24,25], are the main active substances in poplar-type propolis [26]. TFC and TPC have been widely used as indicators for evaluating the quality of propolis. Table 2 shows the results of TFC and TPC of 49 propolis samples, determined by the methods described previously [27]. There is a significant variation of TFC and TPC among these samples, ranging from 63.75 ± 1.92 mg/g to 454.92 ± 32.67 mg/g for TFC and 150.83 ± 2.75 mg/g to 556.3 ± 5.55 mg/g for TPC, respectively. These results are consistent with the previous report [28]. The variations of TFC and TPC in these samples are in line with the chemical variations as shown above, indicating that the differences in these propolis samples are related to their geographic origins.

Table 2. Total phenolic acid (TPC) contents, contents of total flavonoids (TFC) and 1,1-Diphenyl-2-picrylhydrazyl (DPPH) scavenging activity of 49 Chinese propolis.

Samples No.	TPC (mg/g, GAE)	TFC (mg/g, QE)	DPPH Scavenging Activity (IC$_{50}$, μg/mL)	Samples No.	TPC (mg/g, GAE)	TFC (mg/g, QE)	DPPH Scavenging Activity (IC$_{50}$, μg/mL)
S1	404.47 ± 10.72	361.59 ± 8.54	90.51 ± 0.89	S26	232.59 ± 3.18	146.63 ± 3.33	294.04 ± 10.26
S2	249.95 ± 4.52	162.39 ± 13.93	113.19 ± 5.93	S27	199.70 ± 3.83	75.43 ± 4.19	216.06 ± 3.30
S3	223.12 ± 1.69	110.56 ± 1.24	101.21 ± 1.87	S28	275.54 ± 2.91	167.16 ± 6.34	83.47 ± 3.55
S4	248.62 ± 2.22	135.56 ± 4.49	122.88 ± 1.73	S29	208.31 ± 2.45	77.76 ± 4.67	171.61 ± 8.54
S5	529.49 ± 4.10	427.56 ± 17.03	124.35 ± 1.36	S30	231.11 ± 2.03	141.17 ± 2.84	112.14 ± 5.71
S6	556.3 ± 5.55	454.92 ± 32.67	113.10 ± 4.67	S31	170.67 ± 0.94	106.06 ± 1.34	171.52 ± 8.21
S7	236.4 ± 3.17	102.50 ± 1.17	92.45 ± 8.31	S32	219.01 ± 2.30	204.76 ± 4.05	233.6 ± 10.59
S8	257.15 ± 3.76	191.19 ± 1.55	104.17 ± 6.69	S33	232.35 ± 5.43	210.53 ± 4.05	153.93 ± 3.81
S9	236.92 ± 6.38	205.32 ± 6.12	164.27 ± 7.47	S34	242.33 ± 1.11	149.68 ± 2.55	119.5 ± 4.51
S10	276.21 ± 4.99	190.72 ± 12.24	108.17 ± 5.66	S35	228.94 ± 2.76	192.23 ± 6.62	165.23 ± 4.23
S11	272.91 ± 3.51	186.96 ± 5.35	73.56 ± 2.61	S36	223.90 ± 2.22	216.31 ± 4.29	241.85 ± 11.63
S12	210.23 ± 4.14	205.95 ± 1.68	308.11 ± 6.36	S37	228.39 ± 2.45	223.83 ± 1.15	269.08 ± 1.94
S13	242.81 ± 2.60	197.64 ± 2.02	124.54 ± 5.94	S38	197.97 ± 1.84	203.46 ± 6.06	303.87 ± 6.00
S14	237.32 ± 3.06	179.05 ± 7.48	124.16 ± 9.21	S39	252.31 ± 5.44	167.96 ± 4.77	100.20 ± 1.46
S15	246.23 ± 5.42	159.25 ± 1.50	109.00 ± 3.77	S40	231.12 ± 4.63	143.61 ± 5.02	132.98 ± 3.55
S16	254.92 ± 5.02	143.97 ± 3.06	76.06 ± 3.15	S41	205.82 ± 5.02	208.93 ± 3.23	404.56 ± 11.9
S17	216.69 ± 2.12	158.80 ± 3.48	133.82 ± 2.93	S42	150.83 ± 2.75	70.08 ± 4.15	354.31 ± 5.12
S18	256.96 ± 3.37	163.24 ± 5.62	122.64 ± 9.30	S43	234.55 ± 5.82	126.88 ± 7.89	146.45 ± 6.61
S19	217.37 ± 3.40	133.04 ± 4.02	352.75 ± 9.75	S44	268.44 ± 2.77	194.49 ± 4.98	124.45 ± 6.19
S20	139.92 ± 1.90	63.75 ± 1.92	432.08 ± 6.42	S45	312.01 ± 4.34	207.96 ± 9.33	132.46 ± 4.53
S21	223.74 ± 1.30	100.90 ± 0.42	197.34 ± 6.78	S46	252.51 ± 1.40	149.41 ± 9.75	91.01 ± 9.85
S22	302.55 ± 6.12	219.2 ± 2.56	87.14 ± 7.38	S47	251.63 ± 0.80	172.76 ± 9.53	125.20 ± 5.37
S23	178.37 ± 0.89	53.45 ± 3.41	277.77 ± 10.99	S48	274.21 ± 2.41	134.62 ± 3.64	71.19 ± 5.31
S24	240.99 ± 4.03	171.38 ± 6.69	124.92 ± 6.32	S49	247.34 ± 4.43	173.99 ± 7.50	113.08 ± 6.15
S25	188.86 ± 1.80	73.51 ± 9.97	309.97 ± 10.23				

Data are shown as the mean ± SD ($n = 3$). GAE, gallic acid equivalent; QE, quercetin equivalent.

2.4. DPPH Scavenging Activity of 49 Chinese Propolis

DPPH assay has been widely used as a sensitive method to assess the antioxidant capacity of various samples [29]. We determined the DPPH scavenging activity (IC$_{50}$) of these Chinese propolis samples. As shown in Table 2, all propolis samples showed strong antioxidant activity. The IC$_{50}$ values of these propolis samples varied widely, ranging from 71.19 ± 5.31 µg/mL to 432.08 ± 6.42 µg/mL, indicating that the antioxidant activity of Chinese propolis is also region-dependent. In addition, there was a significant negative correlation between DPPH scavenging activity (IC_{50}) and TPC (R = −0.469, $p < 0.01$), but not TFC (R = −0.260, $p > 0.05$). These results indicate that the phenolic acids have a greater influence on the propolis antioxidant capacity than the flavonoids, which was consistent with previous research [27,30]. The antioxidative activity of propolis is the most appreciated property, and a variety of biological activities of propolis largely results from their antioxidative effects [31,32]. Therefore, the antioxidant capacity is an important indicator of the quality of propolis.

2.5. Spectrum–Effect Relationship of 49 Chinese Propolis

MLRA is a useful method to quantify the relationship between spectrum and bioactivities [33]. However, when independent variables have collinear relationships, the MLRA model is unreliable. PCA could reduce the dimensionality of data and convert correlated data into a few integrated variables without collinearity [34,35]. In this study, we firstly conducted PCA and identified four principal components (PC) which contained 79.99% information of the original data (PC1, 35.44%; PC2, 25.81%; PC3, 10.10%; and PC4, 8.64%). As shown in Table 3 and Figure 3, the PC1 was highly positively correlated to the contents of isoferulic acid, caffeic acid, CAPE and 3,4-dimethoxycinnamic acid. The PC2 was positively correlated to the contents of kaempferol, but negatively correlated to the contents of ferulic acid benzyl p-coumarate and p-coumaric acid. The PC3 was positively correlated to the contents of galangin, and negatively correlated to benzyl caffeate content. The PC4 was positively correlated to the contents of chrysin and apigenin. Based on PCA results, we further performed MLRA and found that PC1 and PC4 had a greater impact on the antioxidant capacity, while the impacts of PC2 and PC3 were less significant. The equations were: IC_{50} = 169.795-68.899PC1-23.475PC4. These findings indicate that isoferulic acid, caffeic acid, CAPE and 3,4-dimethoxycinnamic acid in PC1 and chrysin and apigenin in PC4 contribute more to the antioxidant capacity of Chinese propolis. The total content of these compounds showed a superior negative correlation with the DPPH scavenging activity (IC_{50}) (R = −0.716, $p < 0.001$). Thus, the content of these compounds could be used as an indicator to predict the antioxidant capacity of propolis. However, this does not mean that all these compounds have antioxidant activity. Conversely, the antioxidant compounds in propolis may not only be these compounds. In addition to MLRA and PCA, there are many other types of statistical analysis methods on spectrum–efficacy studies, and different methods have different emphases [36–38].

Table 3. The loadings of the first four rotated principal components.

Peak No.	Compound	The Loading			
		PC1 (35.44%)	PC2 (25.81%)	PC3 (10.10%)	PC4 (8.64%)
5	Isoferulic acid	**0.911**	0.083	−0.211	0.064
2	Caffeic acid	**0.869**	0.115	0.032	0.015
17	CAPE	**0.825**	−0.069	−0.009	0.312
6	3,4-Dimethoxycinnamic acid	**0.745**	0.277	−0.292	0.236
4	Ferulic acid	−0.034	**−0.914**	−0.119	0.087
19	Benzyl p-coumarate	−0.238	**−0.850**	0.090	−0.357
3	p-Coumaric acid	−0.249	**−0.849**	0.015	−0.354
11	Kaempferol	−0.041	**0.659**	0.587	0.169
13	Pinobanksin	−0.386	0.599	0.398	0.427
14	Benzyl caffeate	0.063	0.092	**−0.769**	−0.024
18	Galangin	−0.396	0.242	**0.709**	0.370
15	3-O-acetylpinobanksin	0.496	0.486	0.589	0.015
8	Pinocembrin	−0.218	0.127	0.541	−0.495
16	Chrysin	0.103	0.356	0.023	**0.829**
12	Apigenin	0.309	0.170	0.198	**0.757**

Bold fonts indicate high score (absolute value greater than 0.6).

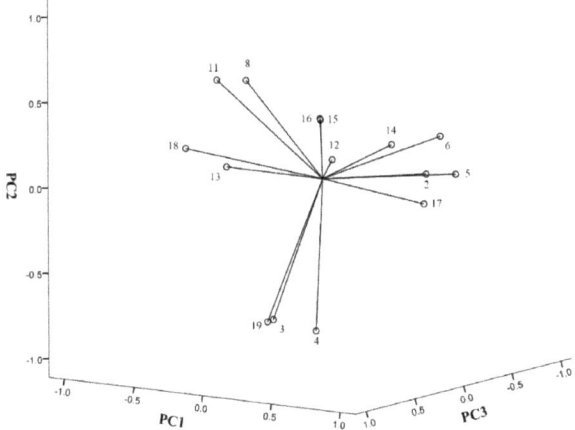

Figure 3. The loadings plots of the principal components. See Table 3 for compound names.

2.6. Determination of Antioxidant Compounds in Propolis by Off-Line Anti-DPPH Assay

To directly determine the nature of antioxidant compounds in Chinese propolis, an off-line anti-DPPH assay was performed. Eight compounds (caffeic acid, ferulic acid, kaempferol, unknown compound 1, benzyl caffeate, 3-O-acetylpinobanksin, CAPE and galangin) were found with a decrease in peak area after the reaction, indicating that these compounds have antioxidant activity (Figure 4). Other studies on the antioxidant capacity of Chinese propolis have also confirmed that these compounds have the DPPH scavenging activity [14,39,40]. Furthermore, in the related research about Brazil green propolis, nine compounds, including caffeic acid and kaempferol, show a decrease in peak area [27]. Unlike the ambiguity and integrity of the spectrum–effect relationship, this method could detect antioxidant compounds in propolis, as well as in other natural products. In addition, there is an on-line anti-DPPH assay similar to this method, which can monitor the reaction in real time, but requires an additional post-column system [41].

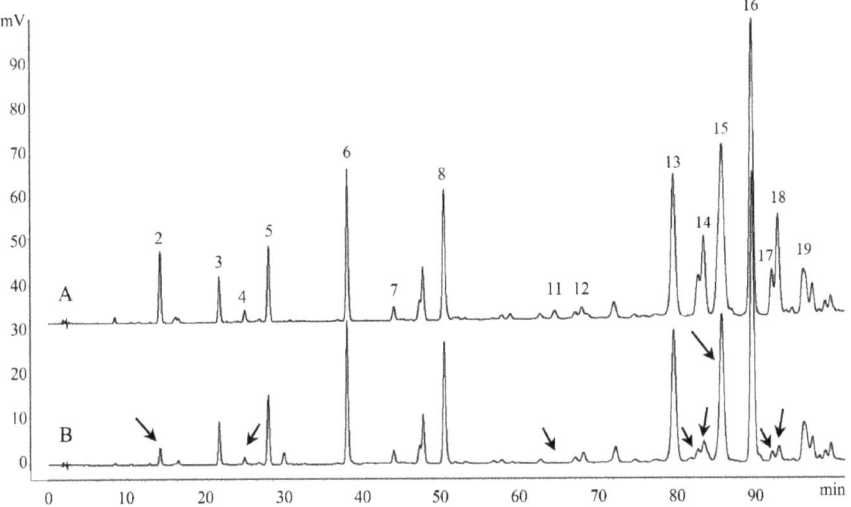

Figure 4. HPLC chromatograms of the Chinese propolis before (**A**) and after (**B**) reacting with DPPH radical. Arrows indicate peaks with a decreasing peak area.

3. Materials and Methods

3.1. Chemicals and Reagents

HPLC-grade methanol was purchased from Merck (Merck & Co., Inc., Billerica, MA, USA), and analytical grade acetic acid and absolute ethanol were purchased from Chemical Reagent Factory of Zhejiang University (Hangzhou, Zhejiang, China). Absolute alcohol and acetic acid were purchased from Shanghai Chemical Reagent Company of Chinese Medical Group (Shanghai, China). Ultra-Pure water was purified by the Yjd-upws Ultra-Pure water system (Hangzhou, Zhejiang, China).

DPPH, vanillic acid, caffeic acid, ferulic acid, isoferulic acid, p-coumaric acid, cinnamic acid, 3,4-dimethoxycinnamic acid, CAPE, myricetin, apigenin, galangin, chrysin, pinocembrin, quercetin, kaempferol, luteolin and naringenin were purchased from Sigma–Aldrich (St. Louis, MI, USA), pinobanksin and 3-O-acetylpinobanksin were purchased from Ningbo Haishu Apexocean Biochemicals Co., Ltd (Ningbo, Zhejiang, China), and benzyl p-coumarate was purchased from Kunming BioBioPha Co., Ltd (Kunming, Yunnan, China).

3.2. Samples Collection and Preparation

49 propolis samples (S01–S49) used in this study were harvested by scratching from beehives in 49 cities of 16 provinces (Table 1). The propolis samples sites cover the main Apis mellifera breeding areas and propolis production areas in China.

The frozen propolis samples were extracted, as reported previously [8]. The raw propolis samples (3.0 g) were extracted with 50 mL of a 95% hydro-alcoholic solution in an ultrasonic water bath for 45 min. The mixture was then centrifuged, and the sediment was re-extracted twice under the same conditions. The supernatant was kept in a refrigerator overnight and filtered to remove impurities. After that, the filtered solution was evaporated to dryness. The dry residue powder of propolis (0.2 g) was then redissolved in 10 mL ethanol (20 mg/mL).

3.3. HPLC Procedures

Chromatographic analysis was performed with Agilent 1200 Series (Agilent Technologies Inc., Santa Clara, CA, USA) equipment. Separation was achieved on a Sepax HP-C18 column (150 × 4.6 mm, 5 μm; Sepax Technologies Inc., Newark, DE, USA) and maintained at 33 °C. The mobile phase was maintained at a constant flow rate of 1 mL/min. The gradient elution, which consisted of aqueous phase A (1% acetic acid) and organic phase B (anhydrous methanol), was adjusted as we previously reported, in detail: 15% to 35% (B) 0 to 30 min; 35% to 44% (B) 30 to 46 min; 44% to 50% (B) 46 to 70 min; 50% to 52% (B) 70 to 77 min; 52% to 60% (B) 77 to 92 min; 60% to 75% (B) 92 to 115 min; 75% to 100% (B) 115 to 125 min; and 100% to 15% (B) 125 to 135 min [18]. Each sample (5 μL) was purified with 0.45 μm filters, and then injected through an automatic sampler system and monitored by a UV detector at 280 nm.

The methodology was validated through intraday precision, intraday precision and repeatability tests. The contents of identified compounds in propolis were quantified using the respective regression equation of standard substances. The peaks-area was quantitated by external calibration, the standards compounds were dissolved in methanol, the mixed standard solution was prepared and a series of working standard solutions were prepared according to the level of these reference standards expected in samples.

3.4. Determinations of Total Flavonoids and Total Phenolics

TFC was determined using the method reported previously, with minor modifications [27,42]. Briefly, 60 μL of propolis ethanol solution (0.2 mg/mL) was mixed with 40 μL (100 g/L) aluminum nitrate and 40 μL (9.8 g/L) potassium acetate, and adjusted to 200 μL with distilled water. The mixed solution was kept in a dark room for 1 h at room temperature, and then measured the absorbance at 415 nm using a microplate reader (Bio-Rad, Hercules, CA, USA).

TPC was measured by the Folin–Ciocalteu method [27,43]. Briefly, a 100 µL Folin–Ciocalteu reagent was added to 100 µL of propolis ethanol solution (10 mg/mL), and then mixed with 500 µL (1 mol/L) sodium carbonate and adjusted to 1 mL with distilled water. The mixed solution was kept in a dark room for 1 h at room temperature, and then measured the absorbance at 760 nm using the microplate reader.

3.5. Antioxidant Capacity

Fresh DPPH stock solution was prepared by dissolving 30 mg DPPH in 10 mL ethanol (3 mg/mL). The DPPH scavenging activity was determined according to the method reported previously, with modifications [27,44]. In brief, 100 µL DPPH working solution was mixed with 100 µL propolis extract in a 96-well plate, and incubated for 30 min in the dark. The absorbance of the reaction solutions was then measured at 517 nm using the microplate reader. The results were expressed as IC_{50} (µg/mL, the concentration of scavenging 50% DPPH radical).

3.6. Off-line Anti-DPPH Assay

The off-line anti-DHHP assay was based on the method reported previously, with some modifications [27,45]. Briefly, 10 mg/mL DPPH ethanol solution was mixed with an equal volume of 5 mg/mL Chinese propolis ethanol solution, and then placed in a dark room for 30 min. After filtration, the mixed solution was determined by HPLC, using the procedures as described above.

3.7. Statistical Analysis and Chemometric Application

The software "Similarity Evaluation System for Chromatographic Fingerprint of TCM", published by the Chinese Pharmacopoeia Commission (Chinese Pharmacopoeia Commission, Version 2012.130723), was used to synchronize and conduct qualitative and quantitative comparisons for all propolis samples. The reference fingerprint was formed by the system using the median method from the chromatograms of 49 propolis, and the similarity values of each propolis extract and reference fingerprint were also determined. The means and standard deviations (SD) were calculated using the Microsoft Excel 2016 software (Microsoft Inc., Redmond, WA, USA). The CA, HCA, MLRA and IC_{50} of 49 propolis were performed using SPSS 22 statistics software (SPSS Inc., Armonk, NY, USA).

4. Conclusion

In this study, 49 propolis samples collected from different regions in China were studied for their chemical profiles, antioxidant activity and spectrum–effect relationship. The results showed that the Chinese propolis could be divided into three different types according to the similarity of their HPLC fingerprints, with propolis collected from Changbai Mountain in Northeast China, which contained a higher content of p-coumaric acid and benzyl p-coumarate as a distinct type. The spectrum–effect relationship showed that the contents of isoferulic acid, caffeic acid, CAPE, 3,4-dimethoxycinnamic acid, chrysin and apigenin in Chinese propolis could be related to the antioxidant activity of propolis samples. Furthermore, eight active compounds were identified with anti-DPPH activities in Chinese propolis. The results indicate that the grouping and spectrum–effect relationship of Chinese propolis are related to their chemical compositions, and several compounds may serve as a better marker for the antioxidant activity of Chinese propolis than TFC and TPC. The findings may help to develop better methods to evaluate the quality of propolis from different geographic origins.

Supplementary Materials: The following are available online at http://www.mdpi.com/1420-3049/25/14/3243/s1, Table S1. Precision and repeatability data of 15 compounds in Chinese propolis. Table S2: The regression equation of standard compounds in Chinese propolis. Table S3: The content of common compounds in different Chinese propolis. Figure S1: HPLC chromatograms of the standard solution and Chinese propolis.

Author Contributions: Conceptualization, X.J. and F.H.; methodology, X.J., C.L. and C.Z.; software, X.J. and G.Q.L.; validation, X.J., L.T. and M.Y.; data curation, X.J. and L.T.; writing—original draft preparation, X.J.; writing—review and editing, C.L., G.Q.L. and C.Z.; supervision, C.Z. and F.H.; project administration, F.H.; and funding acquisition, C.Z. and F.H. All authors have read and agreed to the published version of the manuscript.

Funding: This research was funded by the earmarked fund for the National Natural Science Foundation of China (Project number: 31972627), Modern Agro-Industry Technology Research System from the Ministry of Agriculture of China (CARS-44) and the Public Welfare Research Program of Zhejiang Province, China (No. LGN18C170001).

Acknowledgments: We appreciated the Experimental Teaching Center of Animal Sciences College of Zhejiang University for the help of the multimode reader and Yuan-qiang Hu for helping to collect propolis samples.

Conflicts of Interest: The authors declare no competing financial interest.

References

1. Burdock, G. Review of the biological properties and toxicity of bee propolis (propolis). *Food Chem. Toxicol.* **1998**, *36*, 347–363. [CrossRef]
2. Wagh, V.D. Propolis: A wonder bees product and its pharmacological potentials. *Adv. Pharm. Sci.* **2013**, *2013*, 308249. [CrossRef]
3. Marcucci, M. Propolis: Chemical composition, biological properties and therapeutic activity. *Apidologie* **1995**, *26*, 83–99. [CrossRef]
4. Huang, S.; Zhang, C.P.; Wang, K.; Li, G.Q.; Hu, F.L. Recent advances in the chemical composition of propolis. *Molecules* **2014**, *19*, 19610–19632. [CrossRef] [PubMed]
5. Valencia, D.; Alday, E.; Robles-Zepeda, R.; Garibay-Escobar, A.; Galvez-Ruiz, J.C.; Salas-Reyes, M.; Jiménez-Estrada, M.; Velazquez-Contreras, E.; Hernandez, J.; Velazquez, C. Seasonal effect on chemical composition and biological activities of Sonoran propolis. *Food Chem.* **2012**, *131*, 645–651. [CrossRef]
6. Bankova, V.; Popova, M.; Trusheva, B. The phytochemistry of the honeybee. *Phytochemistry* **2018**, *155*, 1–11. [CrossRef] [PubMed]
7. Bankova, V.S.; de Castro, S.L.; Marcucci, M.C. Propolis: Recent advances in chemistry and plant origin. *Apidologie* **2000**, *31*, 3–15. [CrossRef]
8. Zhang, C.; Huang, S.; Wei, W.; Ping, S.; Shen, X.; Li, Y.; Hu, F. Development of high-performance liquid chromatographic for quality and authenticity control of Chinese propolis. *J. Food Sci.* **2014**, *79*, 1315–1322. [CrossRef]
9. Ristivojević, P.; Trifković, J.; Andrić, F.; Milojković-Opsenica, D. Poplar-type propolis: Chemical composition, botanical origin and biological activity. *Nat. Prod. Commun.* **2015**, *10*, 1934578X1501001117. [CrossRef]
10. Sawicka, D.; Car, H.; Borawska, M.H.; Nikliński, J. The anticancer activity of propolis. *Folia Histochem. Cytobiol.* **2012**, *50*, 25–37. [CrossRef]
11. Demir, S.; Aliyazicioglu, Y.; Turan, I.; Misir, S.; Mentese, A.; Yaman, S.O.; Akbulut, K.; Kilinc, K.; Deger, O. Antiproliferative and proapoptotic activity of Turkish propolis on human lung cancer cell line. *Nutr. Cancer* **2016**, *68*, 165–172. [CrossRef] [PubMed]
12. Xuan, H.; Yuan, W.; Chang, H.; Liu, M.; Hu, F. Anti-inflammatory effects of Chinese propolis in lipopolysaccharide-stimulated human umbilical vein endothelial cells by suppressing autophagy and MAPK/NF-κB signaling pathway. *Inflammopharmacology* **2019**, *27*, 561–571. [CrossRef] [PubMed]
13. Lobo, V.; Patil, A.; Phatak, A.; Chandra, N. Free radicals, antioxidants and functional foods: Impact on human health. *Pharm. Rev.* **2010**, *4*, 118–126. [CrossRef] [PubMed]
14. Kumazawa, S.; Hamasaka, T.; Nakayama, T. Antioxidant activity of propolis of various geographic origins. *Food Chem.* **2004**, *84*, 329–339. [CrossRef]
15. Zhu, C.-S.; Lin, Z.-J.; Xiao, M.-L.; Niu, H.-J.; Zhang, B. The spectrum–effect relationship—A rational approach to screening effective compounds, reflecting the internal quality of Chinese herbal medicine. *Chin. J. Nat. Med.* **2016**, *14*, 177–184. [CrossRef]
16. Chen, Y.; Yu, H.; Wu, H.; Pan, Y.; Wang, K.; Liu, L.; Jin, Y.; Zhang, C. A Novel Reduplicate Strategy for Tracing Hemostatic Compounds from Heating Products of the Flavonoid Extract in Platycladi cacumen by Spectrum–effect Relationships and Column Chromatography. *Molecules* **2015**, *20*, 16970–16986. [CrossRef]
17. Zhu, Y.; Zheng, Q.; Sun, Z.; Chen, Z.; Zhao, Y.; Wang, Z.; Yang, H.; Li, J.; Li, Y.; Xiao, X. Fingerprint–efficacy study of Radix Aconiti Lateralis Preparata (Fuzi) in quality control of Chinese herbal medicine. *J. Therm. Anal. Calorim.* **2014**, *118*, 1763–1772. [CrossRef]

18. Jiang, X.; Tian, J.; Zheng, Y.; Zhang, Y.; Wu, Y.; Zhang, C.; Zheng, H.; Hu, F. A New Propolis Type from Changbai Mountains in North-east China: Chemical Composition, Botanical Origin and Biological Activity. *Molecules* **2019**, *24*, 1369. [CrossRef]
19. Zhang, C.-P.; Zheng, H.-Q.; Liu, G.; Hu, F.-L. Development and validation of HPLC method for determination of salicin in poplar buds: Application for screening of counterfeit propolis. *Food Chem.* **2011**, *127*, 345–350. [CrossRef]
20. Popova, M.P.; Bankova, V.S.; Bogdanov, S.; Tsvetkova, I.; Naydenski, C.; Marcazzan, G.L.; Sabatini, A.-G. Chemical characteristics of poplar type propolis of different geographic origin. *Apidologie* **2007**, *38*, 306–311. [CrossRef]
21. Wang, Y.-L.; Lu, Q.; Decock, C.; Li, Y.-X.; Zhang, X.-Y. Cytospora species from Populus and Salix in China with C. davidiana sp. nov. *Fungal Biol.* **2015**, *119*, 420–432. [CrossRef]
22. Xu, X.; Yang, F.; Xiao, X.; Zhang, S.; Korpelainen, H.; Li, C. Sex-specific responses of Populus cathayana to drought and elevated temperatures. *Plant Cell Environ.* **2008**, *31*, 850–860. [CrossRef]
23. Teixeira, E.W.; Negri, G.; Meira, R.M.; Message, D.; Salatino, A. Plant Origin of Green Propolis: Bee Behavior, Plant Anatomy and Chemistry. *Evid. Based Complement Altern. Med.* **2005**, *2*, 85–92. [CrossRef]
24. Tapas, A.R.; Sakarkar, D.; Kakde, R. Flavonoids as nutraceuticals: A review. *Trop. J. Pharm. Res.* **2008**, *7*, 1089–1099. [CrossRef]
25. Lafay, S.; Gil-Izquierdo, A. Bioavailability of phenolic acids. *Phytochem. Rev.* **2007**, *7*, 301–311. [CrossRef]
26. Popova, M.; Bankova, V.; Butovska, D.; Petkov, V.; Nikolova-Damyanova, B.; Sabatini, A.G.; Marcazzan, G.L.; Bogdanov, S. Validated methods for the quantification of biologically active constituents of poplar-type propolis. *Phytochem. Anal. Int. J. Plant Chem. Biochem. Tech.* **2004**, *15*, 235–240. [CrossRef] [PubMed]
27. Zhang, C.; Shen, X.; Chen, J.; Jiang, X.; Hu, F. Identification of Free Radical Scavengers from Brazilian Green Propolis Using Off-Line HPLC-DPPH Assay and LC-MS. *J. Food Sci.* **2017**, *82*, 1602–1607. [CrossRef] [PubMed]
28. Ahn, M.; Kumazawa, S.; Usui, Y.; Nakamura, J.; Matsuka, M.; Zhu, F.; Nakayama, T. Antioxidant activity and constituents of propolis collected in various areas of China. *Food Chem.* **2007**, *101*, 1383–1392. [CrossRef]
29. Foti, M.C. Use and Abuse of the DPPH* Radical. *J. Agric. Food Chem.* **2015**, *63*, 8765–8776. [CrossRef] [PubMed]
30. Shi, H.; Yang, H.; Zhang, X.; Yu, L. Identification and quantification of phytochemical composition and anti-inflammatory and radical scavenging properties of methanolic extracts of Chinese propolis. *J. Agric. Food Chem.* **2012**, *60*, 12403–12410. [CrossRef]
31. Kurek-Górecka, A.; Rzepecka-Stojko, A.; Górecki, M.; Stojko, J.; Sosada, M.; Świerczek-Zięba, G. Structure and antioxidant activity of polyphenols derived from propolis. *Molecules* **2014**, *19*, 78–101.
32. Gülcin, I. Antioxidant activity of food constituents: An overview. *Arch. Toxicol.* **2012**, *86*, 345–391.
33. Wang, J.; Tong, X.; Li, P.; Liu, M.; Peng, W.; Cao, H.; Su, W. Bioactive components on immuno-enhancement effects in the traditional Chinese medicine Shenqi Fuzheng Injection based on relevance analysis between chemical HPLC fingerprints and in vivo biological effects. *J. Ethnopharmacol.* **2014**, *155*, 405–415. [CrossRef] [PubMed]
34. Kaski, S. Dimensionality reduction by random mapping: Fast similarity computation for clustering. In *1998 IEEE International Joint Conference on Neural Networks Proceedings. IEEE World Congress on Computational Intelligence (Cat. No. 98CH36227)*; IEEE: Anchorage, AK, USA, 1998; pp. 413–418.
35. Schölkopf, B.; Smola, A.; Müller, K.-R. *Kernel principal component analysis*, In International Conference on Artificial Neural Networks; Springer: Berlin, Germany, 1997; pp. 583–588.
36. Li, Q.; Guan, H.; Wang, X.; He, Y.; Sun, H.; Tan, W.; Luo, X.; Su, M.; Shi, Y. Fingerprint–efficacy study of the quaternary alkaloids in Corydalis yanhusuo. *J. Ethnopharmacol.* **2017**, *207*, 108–117. [CrossRef] [PubMed]
37. Wang, L.; Yan, T.; Yang, L.; Zhang, K.; Jia, J. Monitoring the consistency quality and antioxidant activity of Da Hong Pao teas by HPLC fingerprinting. *J. Chromatogr. Sci.* **2017**, *55*, 528–535. [CrossRef]
38. Kong, W.; Wang, J.; Zang, Q.; Xing, X.; Zhao, Y.; Liu, W.; Jin, C.; Li, Z.; Xiao, X. Fingerprint–efficacy study of artificial Calculus bovis in quality control of Chinese materia medica. *Food Chem.* **2011**, *127*, 1342–1347. [CrossRef] [PubMed]
39. Yang, H.; Dong, Y.; Du, H.; Shi, H.; Peng, Y.; Li, X. Antioxidant compounds from propolis collected in Anhui, China. *Molecules* **2011**, *16*, 3444–3455. [CrossRef] [PubMed]

40. Russo, A.; Longo, R.; Vanella, A. Antioxidant activity of propolis: Role of caffeic acid phenethyl ester and galangin. *Fitoterapia* **2002**, *73*, S21–S29. [CrossRef]
41. Niederlander, H.A.; van Beek, T.A.; Bartasiute, A.; Koleva, I.I. Antioxidant activity assays on-line with liquid chromatography. *J. Chromatogr. A* **2008**, *1210*, 121–134. [CrossRef]
42. Simonetti, P.; Pietta, P.; Testolin, G. Polyphenol content and total antioxidant potential of selected Italian wines. *J. Agric. Food Chem.* **1997**, *45*, 1152–1155. [CrossRef]
43. Woisky, R.G.; Salatino, A. Analysis of propolis: Some parameters and procedures for chemical quality control. *J. Apic. Res.* **2015**, *37*, 99–105. [CrossRef]
44. Bondet, V.; Brand-Williams, W.; Berset, C. Kinetics and mechanisms of antioxidant activity using the DPPH. free radical method. *Lwt-Food Sci. Technol.* **1997**, *30*, 609–615. [CrossRef]
45. Matsuda, A.H.; de Almeida-Muradian, L.B. Validated method for the quantification of artepillin-C in Brazilian propolis. *Phytochem. Anal. Int. J. Plant Chem. Biochem. Tech.* **2008**, *19*, 179–183. [CrossRef] [PubMed]

Sample Availability: The propolis samples are available from the authors.

© 2020 by the authors. Licensee MDPI, Basel, Switzerland. This article is an open access article distributed under the terms and conditions of the Creative Commons Attribution (CC BY) license (http://creativecommons.org/licenses/by/4.0/).

Article

Determination of Ascorbic Acid, Total Ascorbic Acid, and Dehydroascorbic Acid in Bee Pollen Using Hydrophilic Interaction Liquid Chromatography-Ultraviolet Detection

Meifei Zhu [1,2,†], Jian Tang [1,2,†], Xijuan Tu [1,2,*] and Wenbin Chen [1,2,*]

1. College of Bee Science, Fujian Agriculture and Forestry University, Fuzhou 350002, China; zmf563563@163.com (M.Z.); t1074643655@163.com (J.T.)
2. College of Animal Science, Fujian Agriculture and Forestry University, Fuzhou 350002, China
* Correspondence: xjtu@fafu.edu.cn (X.T.); wbchen@fafu.edu.cn (W.C.)
† These authors contributed equally to this work.

Academic Editors: Gavino Sanna, Marco Ciulu, Yolanda Picò, Nadia Spano and Carlo I.G. Tuberoso
Received: 17 November 2020; Accepted: 30 November 2020; Published: 3 December 2020

Abstract: Ascorbic acid (AA) is one of the essential nutrients in bee pollen, however, it is unstable and likely to be oxidized. Generally, the oxidation form (dehydroascorbic acid (DHA)) is considered to have equivalent biological activity as the reduction form. Thus, determination of the total content of AA and DHA would be more accurate for the nutritional analysis of bee pollen. Here we present a simple, sensitive, and reliable method for the determination of AA, total ascorbic acids (TAA), and DHA in rape (*Brassica campestris*), lotus (*Nelumbo nucifera*), and camellia (*Camellia japonica*) bee pollen, which is based on ultrasonic extraction in metaphosphoric acid solution, and analysis using hydrophilic interaction liquid chromatography (HILIC)-ultraviolet detection. Analytical performance of the method was evaluated and validated, then the proposed method was successfully applied in twenty-one bee pollen samples. Results indicated that contents of AA were in the range of 17.54 to 94.01 µg/g, 66.01 to 111.66 µg/g, and 90.04 to 313.02 µg/g for rape, lotus, and camellia bee pollen, respectively. In addition, percentages of DHA in TAA showed good intra-species consistency, with values of 13.7%, 16.5%, and 7.6% in rape, lotus, and camellia bee pollen, respectively. This is the first report on the discriminative determination between AA and DHA in bee pollen matrices. The proposed method would be valuable for the nutritional analysis of bee pollen.

Keywords: bee pollen; ascorbic acid; total ascorbic acids; dehydroascorbic acid; HILIC

1. Introduction

Bee pollen is one of the main sources of food for honeybees. In the process of collecting, honeybees moisten pollen grains with secretions and packs them into the pollen basket. Bee pollen is obtained by pollen trap, which is generally made of a grid and placed on the entrance of hive to remove pollen pellets from honeybee's legs. As the important nutrient source of honeybee colony, bee pollen is rich in proteins, lipids, sterols, vitamins, and minerals [1,2]. It has been regarded as a natural health food with balanced nutrients, and its potential bioactive and therapeutic properties have made bee pollen a promising therapeutic and nutritional food supplement [3–5].

Ascorbic acid (AA) is a water-soluble vitamin, which plays a significant role in human health [6,7]. Since human body does not synthesize AA, it is supplied by exogenous intake. AA can be oxidized to dehydroascorbic acid (DHA) in the condition of high temperature, high pH, light, and in the presence of oxygen [8]. AA has been found to be lost at the food processing and storage stage [9,10]. This loss may be resulted from the oxidation of AA to DHA, and the further degradation to diketogulonic acid [11]. As the processing of bee pollen normally includes a hot-air-drying procedure [12], AA in bee pollen is potentially oxidized to DHA. This oxidation reaction is reversible and DHA can be reduced back to AA in the presence of suitable reductants such as mercapto [13,14] and phosphorus compounds [10,15–17]. Studies have shown that the membrane transport of DHA is more efficient than AA, and after crossing membranes DHA can be rapidly reduced to AA by reductase [18]. Thus, the biological activity of DHA has been considered to be equivalent to that of AA, and the determination of total ascorbic acids (TAA) which defined as the sum of both AA and DHA contents would be more accurate for the evaluation of nutritional quality in food analysis [19].

HPLC, which is a powerful technique for the selective and sensitive quantification of natural compounds in complex matrices [20], is the preferred analysis method for determination of AA and DHA [19]. Based on the reversible reaction between AA and DHA, HPLC methods for the analysis of TAA can be divided into two groups. The first one is the oxidation method, in which AA is oxidized to DHA before HPLC separation [21]. Because DHA has little absorbance above 220 nm, the most used ultraviolet (UV) detection technique cannot be utilized directly. Furthermore, DHA is unstable and irreversible hydrolyzed to diketogulonic acid [8,11], thus the oxidized condition must be estimated carefully. Another possibility is the reduction method, in which DHA is reduced to AA by reductants [10,13,15,16]. Since AA is stable in the presence of excessive reductants and shows UV absorption, the reduction method is commonly used [19].

AA content in bee pollen was reported by using 2,6-dichlorophenol-indophenol (DCIP) assays [22,23]. In this assay, the blue dye DCIP is reduced to colorless product by AA, then the content of AA is measured by titrimetric or colorimetric methods. However, this reaction lacks specificity, and many substances in the matrix which are capable of reducing the dye may cause serious interferences [24]. In this paper, we reported the specific determination of AA, TAA, and DHA in three major commercial bee pollen in China, rape (*Brassica campestris*), lotus (*Nelumbo nucifera*), and camellia (*Camellia japonica*), using the hydrophilic interaction liquid chromatography-ultraviolet (HILIC-UV) method to provide high specificity and sensitivity for the analysis of AA. To the best of our knowledge, this is the first report on discriminative determination of AA and DHA in bee pollen by using the proposed methodology.

2. Results and Discussion

The AA in bee pollen samples were extracted by metaphosphoric acid (MPA) solution under ultrasonication. It has been demonstrated that MPA is beneficial for the inhibition of AA oxidation [25]. This acidic condition could prevent the formation of ascorbate, and provide a better recovery of AA. Effect of ultrasonic extraction time on the extracted AA is shown in Figure 1. Results indicated that for all three species of bee pollen, one minute of ultrasonication was sufficient for the extraction of AA into MPA solution. This fast extraction may be attributed to the excellent solubility of AA in MPA solution, and consequently makes the sample preparation very simple and rapid.

Following the extraction procedure, extract was separated in HILIC with UV detector for the analysis of AA. The column with amide-bonded silica as the stationary phase was used for the HILIC separation due to the compatible with high concentration of MPA solution [26]. The mobile phase consisted of acetonitrile (ACN) and 0.1% formic acid aqueous solution was examined to obtain a general condition for separation of AA in all the three species of bee pollen. Different ratios (v/v) of ACN ranged from 88% to 94% with interval of 2% were investigated. Under concentration of 88%, the AA peak could not be isolated from interferences in all the three matrices. With the increase of ACN concentration, resolution of AA peak was improved. For rape bee pollen, ACN concentrations with 90% and 92% were both suitable for the separation. While for lotus and camellia bee pollen, only larger than 92% was acceptable for the separation. Meanwhile, as the ACN concentration further increased to 94%, the retention time and peak width of AA were significantly increased. In addition, the peak symmetry was found to be decreased. Therefore, mobile phase composed of 92% ACN was finally proposed for the HILIC separation. Figure 2 represented the chromatograms of AA standard and bee pollen samples under the optimized mobile phase condition with flow rate at 0.8 mL/min. As could be seen in Figure 2a, chromatogram of AA standard showed a well-resolved peak with a retention time (RT) at 10.68 min. Chromatograms of extract from rape (Figure 2b), lotus (Figure 2c), and camellia bee pollen (Figure 2d) all showed the clearly defined AA peaks.

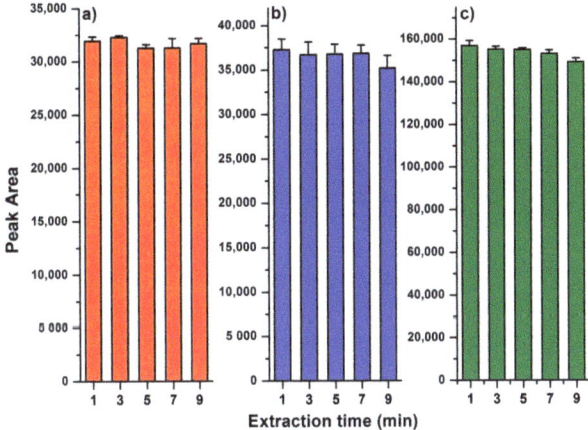

Figure 1. Effect of ultrasonic extraction time on extracted ascorbic acid from different bee pollen samples, (**a**) rape (*Brassica campestris*), (**b**) lotus (*Nelumbo nucifera*), and (**c**) camellia (*Camellia japonica*). The extract under different extraction times were analyzed by hydrophilic interaction liquid chromatography-ultraviolet (HILIC-UV), and the peak area of ascorbic acid (AA) were presented.

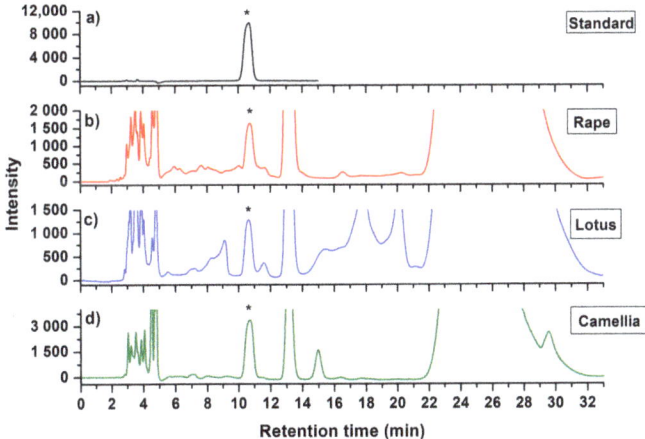

Figure 2. Representative chromatograms of (**a**) ascorbic acid standard, and the final extract of (**b**) rape (*Brassica campestris*), (**c**) lotus (*Nelumbo nucifera*), and (**d**) camellia (*Camellia japonica*) bee pollen. Asterisks indicate the peak of ascorbic acid.

Quantification of AA was performed by external calibration. Good linearity ($R^2 > 0.999$) was achieved in the range from 0.2 to 16 µg/mL. According to the literature [27], the instrumental limits of detection (LOD) and quantification (LOQ) were estimated to be 0.06 µg/mL (S/N = 3) and 0.2 µg/mL (S/N = 10) in MPA solution, respectively. Then the LOD and LOQ in bee pollen were calculated by the reported method [15]. Considering that 0.2 g of bee pollen were diluted in 10 mL, this LOD corresponded to a concentration of 3 μg/g bee pollen, and the LOQ corresponded to 10 μg/g bee pollen. Recovery was tested at two fortification levels. Mean recovery percentages were in the range of 90.07~94.82%, 96.98~104.91%, and 90.45~95.41% for rape, lotus, and camellia bee pollen, respectively (Table 1). The coefficient variation of the repeatability (intra-day precision) was 2.25~4.88%, 1.54~4.84%, and 1.69~4.49% for rape, lotus, and camellia bee pollen, respectively. The intermediate precision (inter-day precision) was ranged from 1.85% to 3.87% in these three species of bee pollen. All recovery and precision values were in the acceptable range [28].

For the determination of TAA, reducing reagent was introduced to transform DHA into AA. The reported reagents typically include mercapto based compounds [13,14] and phosphorus-based tris-[2-carboxyethyl] phosphine (TCEP) [10,15–17]. Mercapto compounds are efficient only at mildly acidic and neutral conditions, while TCEP shows high reduction efficiency in the MPA solution [29]. Thus, TCEP was selected in the present work. Wechtersbach et al. reported that in MPA solution, DHA was fully reduced to AA by TCEP in less than 20 min [29]. To confirm the conversion of DHA into AA in bee pollen matrices, the recovery of DHA reduced to AA was studied. Bee pollen samples were fortified with DHA at the level of 50 μg/g, and the obtained recoveries under different concentration of TCEP were shown in Figure 3. Results indicated that the recoveries were increased as the concentration of TCEP varied from 0.2 to 1.0 mg/mL, then reached a plateau when the concentration was larger than 1.5 mg/mL. Eventually, TCEP with concentration of 2.0 mg/mL was applied to ensure the accomplishment of DHA reduction. It was important to notice that under this concentration, the calculated recovery values were from 65.6% to 75.4%, which was lower than the recovery of spiked AA as illustrated in Table 1. This lower calculated recovery value may be resulted from the lower purity of commercially available DHA, which was reported to be 80% [29]. After correcting the purity, the accurate recovery should be between 82.0% and 95.4%. These above results indicated that the DHA could be efficiently reduced to AA by the proposed reduction protocol and the present sample preparation method could be applicable for the determination of AA and DHA in these three bee pollen matrices.

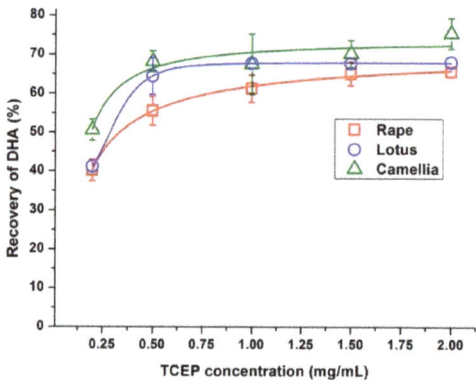

Figure 3. The trends of dehydroascorbic acid (DHA) recovery with increasing the concentration of tris-[2-carboxyethyl] phosphine (TCEP) in rape (*Brassica campestris*), lotus (*Nelumbo nucifera*), and camellia (*Camellia japonica*) bee pollen. The spiked content of DHA in bee pollen samples were 50 μg/g, and the concentration of TCEP in the 10 mL final extract were presented.

Table 1. The recovery and precision for AA at spiked bee pollen samples.

Bee Pollen	AA, µg/g (Mean ± SD, n = 6)	Fortified AA Contents, µg/g	Intra-Day Recovery and Precision						Inter-Day Recovery and Precision	
			Day 1		Day 2		Day 3			
			Mean ± SD (%, n = 6)	RSD (%, n = 6)	Mean ± SD (%, n = 6)	RSD (%, n = 6)	Mean ± SD (%, n = 6)	RSD (%, n = 6)	Mean ± SD (%, n = 18)	RSD (%, n = 18)
Rape (Brassica campestris)	46.55 ± 0.66	20	90.38 ± 2.21	2.44	90.21 ± 3.80	4.21	94.82 ± 4.63	4.88	91.80 ± 3.55	3.87
		40	92.43 ± 2.63	2.84	94.14 ± 2.11	2.25	90.07 ± 2.82	3.13	92.21 ± 2.52	2.73
Lotus (Nelumbo nucifera)	71.05 ± 0.97	35	104.91 ± 2.86	2.73	96.98 ± 4.69	4.84	101.53 ± 3.20	3.15	101.14 ± 3.58	3.54
		70	98.75 ± 1.52	1.54	98.35 ± 1.97	2.00	97.47 ± 1.96	2.01	98.19 ± 1.82	1.85
Camellia (Camellia japonica)	311.22 ± 5.70	160	95.41 ± 4.28	4.49	90.45 ± 2.03	2.25	90.75 ± 2.30	2.53	92.20 ± 2.87	3.11
		320	90.71 ± 1.53	1.69	95.06 ± 2.21	2.32	92.32 ± 2.14	2.31	92.70 ± 1.96	2.11

Twenty-one bee pollen samples were analyzed using the proposed method. Results of the detected AA, TAA, and DHA contents in each sample were summarized in Table 2. The distribution of AA contents in different species of bee pollen were compared in Figure 4. It could be observed that the mean AA content was increased as the order of rape (46.63 µg/g) < lotus (89.19 µg/g) < camellia (199.06 µg/g). Additionally, AA contents were distributed with intra-species differences. For instance, camellia bee pollen showed the largest distribution range between 90.04 µg/g and 313.02 µg/g, rape bee pollen was in the range from 17.54 µg/g to 94.01 µg/g, while lotus bee pollen exhibited a relative concentrated range from 66.01 µg/g to 111.66 µg/g. These intra-species differences might be caused by environment conditions such as temperature, light, and storage time, which have significant effects on the oxidization of AA and the degradation of DHA. Furthermore, it was interesting that although the contents of AA were distributed in a wide range, the percentages of AA and DHA in the TAA showed good intra-species consistency. As shown in Figure 5, when the contents of AA were plotted with the contents of TAA, good linearity with the R^2 value larger than 0.996 were achieved in all the three species of bee pollen. According to the fitting results, percentages of AA in TAA were 86.3%, 83.5%, and 92.4% in rape, lotus, and camellia bee pollen, respectively. This meant that the percentages of DHA in TAA would be about 13.7%, 16.5%, and 7.6% in rape, lotus, and camellia bee pollen, respectively. This stable value of DHA percentage suggests that the irreversible degradation of DHA could be happened in bee pollen samples, and the oxidation of AA into DHA and the further degradation of DHA might be in different equilibrium in these three species of bee pollen.

Table 2. The AA, total ascorbic acids (TAA), and DHA contents in the investigated bee pollen samples.

Bee Pollen	Sample ID	AA, µg/g (Mean ± SD, $n = 3$)	TAA, µg/g (Mean ± SD, $n = 3$)	DHA, µg/g
Rape (Brassica campestris)	1	64.77 ± 0.35	75.53 ± 3.43	10.76
	2	94.01 ± 0.58	108.10 ± 0.69	14.09
	3	26.90 ± 0.84	32.99 ± 0.49	6.09
	4	17.54 ± 0.66	22.98 ± 0.57	5.44
	5	47.59 ± 0.70	57.41 ± 1.65	9.82
	6	66.99 ± 0.70	71.15 ± 1.24	4.16
	7	25.55 ± 0.32	36.21 ± 0.38	10.66
	8	29.75 ± 1.26	36.11 ± 1.76	6.36
Lotus (Nelumbo nucifera)	1	71.69 ± 0.67	83.13 ± 1.50	11.44
	2	85.31 ± 3.03	103.61 ± 0.46	18.30
	3	110.29 ± 1.51	138.90 ± 2.53	28.61
	4	111.02 ± 1.12	137.65 ± 1.22	26.63
	5	68.37 ± 2.75	71.08 ± 1.18	2.71
	6	111.66 ± 0.54	127.26 ± 1.52	15.6
	7	66.01 ± 0.49	79.67 ± 0.42	13.66
Camellia (Camellia japonica)	1	289.05 ± 7.30	307.32 ± 7.85	18.27
	2	195.77 ± 3.10	207.15 ± 2.12	11.38
	3	147.53 ± 0.88	170.69 ± 1.74	23.16
	4	90.04 ± 2.89	91.83 ± 0.54	1.79
	5	158.95 ± 3.29	172.71 ± 1.20	13.76
	6	313.02 ± 3.56	341.65 ± 0.71	28.63

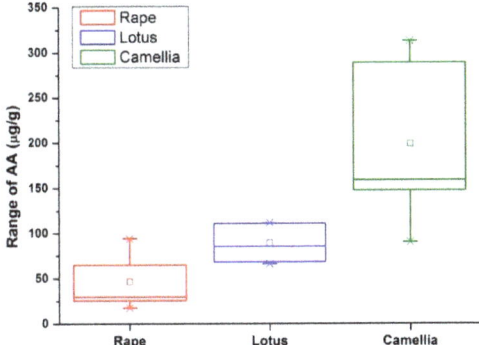

Figure 4. Distribution of the contents of AA in the investigated bee pollen samples.

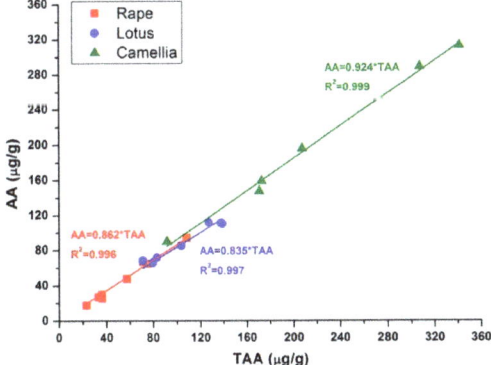

Figure 5. Plots of AA vs. TAA in the investigated bee pollen samples. Linear equation and correlation coefficient (R^2) were presented.

3. Materials and Methods

3.1. Chemicals

HPLC grade acetonitrile and methanol were obtained from Merck (Darmstadt, Germany). Formic acid and metaphosphoric acid were obtained from Sinopharm Chemical Reagent Co., Ltd. (Shanghai, China), AA analytical standard and DHA were purchased from Sigma-Aldrich (Shanghai, China), and TCEP was purchased from Macklin Biochemical Co., Ltd. (Shanghai, China). Commercial bee pollen samples were collected from local markets and stored at −18 °C. Stock solution of AA standard was weekly prepared in MPA solution (5%, *w/v*) at the concentration of 0.1 mg/mL. Working solutions of AA standard were prepared daily by further dilution with the MPA solution. TCEP was also prepared in MPA solution at the concentration of 10 mg/mL. DHA was dissolved in methanol with concentration of 0.1 mg/mL. All these solutions were stored at 4 °C until used. Ultrapure water (18.2 MΩ) was used throughout the experiments.

3.2. Sample Preparation

Bee pollen samples were homogenized in pulverizer (LINDA-DFY300, Wenling, China) to break up the pollen pellet into powder. For the determination of AA, 0.2 g of homogenized bee pollen powder were placed in a 15 mL tube, then 8 mL of MPA (5%) was added, and the mixture was extracted in an ultrasonic bath (40 Hz, KQ2200B, Kunshan Ultrasonic Instruments Co., Ltd., Kunshan, China) for 1 min. Then the solution was transferred to a 10 mL amber volumetric flask and made up to 10 mL with the MPA solution. Aliquot of the final solution were centrifuged at 10,000 rpm for 10 min, and the supernatant was filtered through a 0.22 μm syringe filters prior to HPLC analysis.

For the determination of TAA, 0.2 g of homogenized bee pollen powder were placed in a 15 mL tube, 2.0 mL of TCEP (10 mg/mL prepared in the MPA solution) and additional MPA solution were added to make the final volume 8 mL, and the mixture was extracted in the ultrasonic bath for 1 min. Then the solution was transferred to a 10 mL amber volumetric flask and made up to 10 mL with the MPA solution. This solution was allowing stay for 20 min at room temperature to accomplish the reduction of DHA. After centrifugation at 10,000 rpm for 10 min, the supernatant was filtered through a 0.22 μm syringe filters for HPLC analysis.

3.3. Effect of Ultrasonic Extraction Time

Methods were the same as the determination of AA in Section 3.2, with ultrasonic extraction in different times (1, 3, 5, 7, 9 min).

3.4. Effect of TCEP Concentration on the Recovery of DHA

Bee pollen samples were fortified with DHA at the level of 50 μg/g, then the determination of TAA were performed as described in Section 3.2 with different volumes of TCEP solutions (0.2, 0.5, 1.0, 1.5, and 2.0 mL). Control experiments were carried out as above without the fortification of DHA. Then the recovery of DHA was calculated as the percentage of the measured spike relative to the amount of spike added to the sample.

3.5. Liquid Chromatography

AA was analyzed using Shimadzu chromatographic system with LC-20AT pump, SIL-20A auto-injector, CBM-20A controller, CTO-20AS oven, and SPD-20A detector. Separation of AA was performed in an Inertsil Amide (GL Sciences) column (5 μm, 4.0 mm × 250 mm). Mobile phase consisted of 0.1% formic acid (solvent A) and ACN (solvent B). Chromatographic conditions were as follows: Elution with 92% solvent B for 15 min for the separation of AA, then post run with 92% to 60% for 1 min and maintained at 60% for 5 min to wash out the matrix compounds, and back to 92% for 2 min and maintained at 92% for 10 min. Flow rate was fixed at 0.8 mL/min, column was maintained at 25 °C, the injection volume was 10 μL, and the UV detector was used at 254 nm.

3.6. Method Validation

Quantification of AA was based on calibrations of external standards. The linearity on a seven-point calibration curve was checked ranging from 0.2 to 16 μg/mL, by constructing peak area vs. the concentration of AA. The LOD and LOQ were defined as 3 times and 10 times the signal-noise ratio, respectively [27]. They were determined by serial dilution of the standard solution using the described HPLC conditions and calculated by using the reported method [15]. To evaluate the accuracy and precision of the method, recovery experiment was performed by analyzing samples spiked at two concentration levels. Precision was determined as relative standard deviation (RSD) to the mean recovery in repeatability (intra-day precision, $n = 6$) and intermediate precision (inter-day precision, three days, $n = 18$) analysis.

4. Conclusions

We demonstrated that the proposed HILIC-UV method could be used for the determination of AA, TAA, and DHA in rape, lotus, and camellia bee pollen. The sample preparation procedure was simple, rapid, and effective by using ultrasonic extraction in MPA solution. In addition, TCEP conferred the highly efficient reduction of DHA into AA in bee pollen matrices. Finally, the proposed method was fully validated and successfully applied in twenty-one bee pollen samples. The present method would be a suitable approach for the analysis of AA in bee pollen and be valuable for the better understand of bee pollen nutrition.

Author Contributions: Conceptualization, X.T. and W.C.; Data curation, M.Z. and J.T.; Funding acquisition, X.T. and W.C.; Investigation, M.Z. and J.T.; Methodology, X.T. and W.C.; Project administration, X.T. and W.C.; Resources, X.T.; Supervision, X.T. and W.C.; Validation, M.Z. and J.T.; Writing—original draft, X.T. and W.C.; Writing—review & editing, X.T. and W.C. All authors have read and agreed to the published version of the manuscript.

Funding: This research was funded by Natural Science Foundation of Fujian Province, grant number 2020J01535, and 2019J01409.

Acknowledgments: Meifei Zhu and Jian Tang would like to acknowledge Fujian Agriculture and Forestry University Training Program of Innovation and Entrepreneurship for Undergraduates (Project number 202010389170).

Conflicts of Interest: The authors declare no conflict of interest.

References

1. Campos, M.G.R.; Bogdanov, S.; Almeida-Muradian, L.B.; Szczesna, T.; Mancebo, Y.; Frigerio, C.; Ferreira, F. Pollen composition and standardisation of analytical methods. *J. Apicult. Res.* **2008**, *47*, 156–163. [CrossRef]
2. Ares, A.M.; Valverde, S.; Bernal, J.L.; Nozal, M.J.; Bernal, J. Extraction and determination of bioactive compounds from bee pollen. *J. Pharm. Biomed. Anal.* **2018**, *147*, 110–124. [CrossRef]
3. Thakur, M.; Nanda, V. Composition and functionality of bee pollen: A review. *Trends Food Sci. Technol.* **2020**, *98*, 82–106. [CrossRef]
4. Kroyer, G.; Hegedus, N. Evaluation of bioactive properties of pollen extracts as functional dietary food supplement. *Innov. Food Sci. Emerg. Technol.* **2001**, *2*, 171–174. [CrossRef]
5. Denisow, B.; Denisow-Pietrzyk, M. Biological and therapeutic properties of bee pollen: A review. *J. Sci. Food Agric.* **2016**, *96*, 4303–4309. [CrossRef]
6. Padayatty, S.J.; Katz, A.; Wang, Y.; Eck, P.; Kwon, O.; Lee, J.H.; Chen, S.; Corpe, C.; Dutta, A.; Dutta, S.K.; et al. Vitamin C as an antioxidant: Evaluation of its role in disease prevention. *J. Am. Coll. Nutr.* **2003**, *22*, 18–35. [CrossRef]
7. Arrigoni, O.; De Tullio, M.C. Ascorbic acid: Much more than just an antioxidant. *Biochim. Biophys. Acta* **2002**, *1569*, 1–9. [CrossRef]
8. Deutsch, J.C. Dehydroascorbic acid. *J. Chromatogr. A* **2000**, *881*, 299–307. [CrossRef]
9. Asami, D.K.; Hong, Y.J.; Barrett, D.M.; Mitchell, A.E. Comparison of the total phenolic and ascorbic acid content of freeze-dried and air-dried marionberry, strawberry, and corn grown using conventional, organic, and sustainable agricultural practices. *J. Agric. Food Chem.* **2003**, *51*, 1237–1241. [CrossRef]
10. Phillips, K.M.; Tarragó-Trani, M.T.; Gebhardt, S.E.; Exler, J.; Patterson, K.Y.; Haytowitz, D.B.; Pehrsson, P.R.; Holden, J.M. Stability of vitamin C in frozen raw fruit and vegetable homogenates. *J. Food Compos. Anal.* **2010**, *23*, 253–259. [CrossRef]
11. Serpen, A.; Gökmen, V. Reversible degradation kinetics of ascorbic acid under reducing and oxidizing conditions. *Food Chem.* **2007**, *104*, 721–725. [CrossRef]
12. Domínguez-Valhondo, D.; Gil, D.B.; Hernández, M.T.; González-Gómez, D. Influence of the commercial processing and floral origin on bioactive and nutritional properties of honeybee-collected pollen. *Int. J. Food Sci. Technol.* **2011**, *46*, 2204–2211. [CrossRef]
13. Odriozola-Serrano, I.; Hernández-Jover, T.; Martín-Belloso, O. Comparative evaluation of UV-HPLC methods and reducing agents to determine vitamin C in fruits. *Food Chem.* **2007**, *105*, 1151–1158. [CrossRef]
14. Brause, A.R.; Woollard, D.C.; Indyk, H.E. Determination of total vitamin C in fruit juices and related products by liquid chromatography: Interlaboratory study. *J. AOAC Int.* **2003**, *86*, 367–374. [CrossRef]
15. Fontannaz, P.; Kilinc, T.; Heudi, O. HPLC-UV determination of total vitamin C in a wide range of fortified food products. *Food Chem.* **2006**, *94*, 626–631. [CrossRef]
16. Lykkesfeldt, J. Determination of ascorbic acid and dehydroascorbic acid in biological samples by high-performance liquid chromatography using subtraction methods: Reliable reduction with tris [2-carboxyethyl]phosphine hydrochloride. *Anal. Biochem.* **2000**, *282*, 89–93. [CrossRef]
17. Tarrago-Trani, M.T.; Phillips, K.M.; Cotty, M. Matrix-specific method validation for quantitative analysis of vitamin C in diverse foods. *J. Food Compos. Anal.* **2012**, *26*, 12–25. [CrossRef]

18. Laggner, H.; Besau, V.; Goldenberg, H. Preferential uptake and accumulation of oxidized vitamin C by THP-1monocytic cells. *Eur. J. Biochem.* **1999**, *262*, 659–665. [CrossRef]
19. Nováková, L.; Solich, P.; Solichová, D. HPLC methods for simultaneous determination of ascorbic and dehydroascorbic acids. *Trends Anal. Chem.* **2008**, *27*, 942–958. [CrossRef]
20. Wolfender, J. HPLC in natural product analysis: The detection issue. *Planta Med.* **2009**, *75*, 719–734. [CrossRef]
21. Burini, G. Development of a quantitative method for the analysis of total l-ascorbic acid in foods by high-performance liquid chromatography. *J. Chromatogr. A* **2007**, *1154*, 97–102. [CrossRef]
22. Oliveira, K.C.L.S.; Moriya, M.; Azedo, R.A.B.; de Almeida-Muradian, L.B.; Teixeira, E.W.; Alves, M.L.T.M.F.; Moreti, A.C.d.C.C. Relationship between botanical origin and antioxidants vitamins of bee-collected pollen. *Quim. Nova* **2009**, *32*, 1099–1102. [CrossRef]
23. Melo, I.L.P.d.; Almeida-Muradian, L.B.d. Stability of antioxidants vitamins in bee pollen samples. *Quim. Nova* **2010**, *33*, 514–518. [CrossRef]
24. Arya, S.P.; Mahajan, M.; Jain, P. Photometric methods for the determination of Vitamin C. *Anal. Sci.* **1998**, *14*, 889–895. [CrossRef]
25. Kim, S.R.; Abdel-Ghany, M.; Stitt, K.R. Influence of extraction on l-ascorbic acid recovery from selected foods and beverages. *J. Food Quality* **1987**, *10*, 1–7. [CrossRef]
26. Barros, A.I.R.N.A.; Silva, A.P.; Gonçalves, B.; Nunes, F.M. A fast, simple, and reliable hydrophilic interaction liquid chromatography method for the determination of ascorbic and isoascorbic acids. *Anal. Bioanal. Chem.* **2010**, *396*, 1863–1875. [CrossRef]
27. ICH Q2(R1). Validation of Analytical Procedures: Text and Methodology. 2005. Available online: https://database.ich.org/sites/default/files/Q2%28R1%29%20Guideline.pdf (accessed on 26 November 2020).
28. Taverniers, I.; De Loose, M.; Van Bockstaele, E. Trends in quality in the analytical laboratory. II. Analytical method validation and quality assurance. *Trends Anal. Chem.* **2004**, *23*, 535–552. [CrossRef]
29. Wechtersbach, L.; Cigic, B. Reduction of dehydroascorbic acid at low pH. *J. Biochem. Biophys. Methods* **2007**, *70*, 767–772. [CrossRef]

Sample Availability: Not available.

Publisher's Note: MDPI stays neutral with regard to jurisdictional claims in published maps and institutional affiliations.

© 2020 by the authors. Licensee MDPI, Basel, Switzerland. This article is an open access article distributed under the terms and conditions of the Creative Commons Attribution (CC BY) license (http://creativecommons.org/licenses/by/4.0/).

Article

Mining the Royal Jelly Proteins: Combinatorial Hexapeptide Ligand Library Significantly Improves the MS-Based Proteomic Identification in Complex Biological Samples

Eliza Matuszewska [1], Joanna Matysiak [2], Grzegorz Rosiński [3], Elżbieta Kędzia [4], Weronika Ząbek [5], Jarosław Zawadziński [6] and Jan Matysiak [1,*]

1. Department of Inorganic and Analytical Chemistry, Poznan University of Medical Sciences, Grunwaldzka 6 Street, 60-780 Poznań, Poland; eliza.matuszewska@ump.edu.pl
2. Faculty of Health Sciences, Calisia University, Kaszubska 13 Street, 62-800 Kalisz, Poland; jkamatysiak@gmail.com
3. Department of Animal Physiology and Development, Faculty of Biology, Adam Mickiewicz University in Poznan, Uniwersytetu Poznańskiego 6 Street, 61-614 Poznań, Poland; rosin@amu.edu.pl
4. Department of Innovative Biomaterials and Nanotechnologies, Institute of Natural Fibres and Medicinal Plants—National Research Institute, Wojska Polskiego 71B Street, 60-630 Poznań, Poland; elzbieta.kedzia@iwnirz.pl
5. Department of Pediatric Gastroenterology and Metabolic Diseases, Poznan University of Medical Sciences, Szpitalna 27/33 Street, 60-572 Poznań, Poland; weronikazabek@wp.eu
6. Department of Physiotherapy, Poznan University of Medical Sciences, 28 Czerwca 1956 r. 135/147 Street, 61-545 Poznań, Poland; jaroslaw.zawadzinski@gmail.com
* Correspondence: jmatysiak@ump.edu.pl

Abstract: Royal jelly (RJ) is a complex, creamy secretion produced by the glands of worker bees. Due to its health-promoting properties, it is used by humans as a dietary supplement. However, RJ compounds are not fully characterized yet. Hence, in this research, we aimed to broaden the knowledge of the proteomic composition of fresh RJ. Water extracts of the samples were pre-treated using combinatorial hexapeptide ligand libraries (ProteoMiner^{TM} kit), trypsin-digested, and analyzed by a nanoLC-MALDI-TOF/TOF MS system. To check the ProteoMiner^{TM} performance in the MS-based protein identification, we also examined RJ extracts that were not prepared with the ProteoMiner^{TM} kit. We identified a total of 86 proteins taxonomically classified to *Apis* spp. (bees). Among them, 74 proteins were detected in RJ extracts pre-treated with ProteoMiner^{TM} kit, and only 50 proteins were found in extracts non-enriched with this technique. Ten of the identified features were hypothetical proteins whose existence has been predicted, but any experimental evidence proves their in vivo expression. Additionally, we detected four uncharacterized proteins of unknown functions. The results of this research indicate that the ProteoMiner^{TM} strategy improves proteomic identification in complex biological samples. Broadening the knowledge of RJ composition may contribute to the development of standards and regulations, enhancing the quality of RJ, and consequently, the safety of its supplementation.

Keywords: royal jelly; proteins; ProteoMiner^{TM}; MALDI-TOF-MS; proteomics; beehive product

1. Introduction

Bee products are unique, complex mixtures of biologically active compounds produced by honeybees (*Apis mellifera*). They exhibit strong health-promoting properties, appreciated since ancient times and used for medical purposes. Until now, they are recommended as drugs within the branch of alternative medicine called apitherapy [1]. One of the most sterling bee products, along with honey, bee pollen, and propolis, is royal jelly (RJ). It is a semi-liquid, milky white or yellowish secretion of worker bees' salivary glands. Its sour-bitter-sweet taste (resulting from its acidic pH) and characteristic phenolic smell make it mostly recognizable among the so-called "superfoods" [2]. The term "superfoods" refers to foods that have beneficial effects on human health due to their rich nutrient content [3], and RJ fits this description well.

Chemically, fresh RJ consists of about 60–70% water, 11–23% carbohydrates, 9–18% proteins, 3–8% lipids, and small amounts of vitamins, minerals, free amino acids, and other constituents [2,4,5]. According to the literature reports, RJ is nonhomogeneous, displaying high variability between its main components [5–10]. Factors responsible for that variability are the metabolic and physiologic condition of worker bees and larval age [10–12], honeybee strain [13], place of origin, and year and season of the collection [11]. Since the fluctuations in RJ composition may significantly influence its biological activity, standardization, and quality control of RJ and other bee products seem to be an essential issue.

In a beehive, RJ serves as superior nourishment for all bee larvae until the prepupal stage. However, the honeybee queen is fed exclusively with RJ for its lifetime [14]. Since only the queen is able to reproduce, the epigenetic influence of RJ in the sexual maturation of the female bee larvae is unquestionable. Although it is not thoroughly investigated yet, how RJ provokes the queen's development, monomeric major royal jelly protein 1 (MRJP1; royalactin), belonging to a major royal jelly proteins (MRJP) family, has been suggested to influence the larva differentiation into a bee queen [15]. Nevertheless, the mechanism of RJ's action is probably more complicated, dependent not on the single compound but on the unique molecular blend found exclusively in RJ [16]. Hence, taking into account that RJ is the exceptional food of the bee queen, but it is also used in the treatment and prevention of some diseases, there is an urgent need to characterize the complete RJ composition, including macro- and micro-molecules, along with their biological functions. This will contribute to explaining the effects of royal jelly on the human body and also its beneficial and adverse properties in relation to both humans and animals.

Benefits arising from RJ consumption attract people to include it in their diet. However, the use of not thoroughly tested products poses a risk of side effects, such as inflammatory and allergic reactions, dermatitis, asthma, respiratory stress and bronchospasm, gastrointestinal problems, intestinal bleeding, and even death. Most of the factors responsible for both beneficial and harmful effects caused by RJ intake remain uncharacterized. Hence, taking into account that proteins are mostly accountable for the biological activity of the product, in this research, we aimed to broaden the knowledge of the proteomic composition of fresh RJ.

According to the available literature, several analytical methods have been proposed for the proteomic analysis of RJ. Since MRJPs are easily soluble in water, most of the researchers have extracted proteins contained in RJ using ultrapure water or buffer solutions [17–22]. Hence, in this study, we chose water as an optimal solvent. RJ is a complex product eminently rich in proteins [23]. Therefore, in other studies conducted worldwide, before identification, protein extracts were subjected to tryptic digestion and fractionated, using liquid chromatography or isoelectric fractionation approach. The identification analyses were performed mainly by electrophoretic techniques or mass spectrometry [17–21,24–26]. In this study, we decided to utilize proteolytic digestion, nanoLC (nano-liquid chromatography) and tandem MALDI-TOF/TOF (matrix-assisted laser desorption/ionization—time of flight/time of flight) mass spectrometer.

In the proteomic investigation of natural products such as RJ, a major challenge is analysis of low-abundant proteins. Because there are significant differences in the concentrations of different proteins contained in RJ with a high content of MRJPs fraction, it is important to focus on proper samples preparation for proteomic analyses. However, according to the literature data, any previous study devised an efficient method for subproteome isolation. Therefore, we proposed the ProteoMiner™ (Bio-Rad, Hercules, CA, USA) kit for proteins purifying and concentration. ProteoMiner™ uses combinatorial hexapeptide ligand libraries, and it is a protein enrichment strategy allowing capturing low- and very low-abundance proteins. Such prepared RJ protein extracts were analyzed with advanced nanoLC-MALDI-TOF/TOF MS/MS (nano-liquid chromatography—matrix-assisted laser desorption/ionization—time of flight/time of flight mass spectrometry/mass spectrometry) technique. To the best of our knowledge, according to the available literature, it is the first attempt to analyse the proteomic fraction of RJ samples using ProteoMiner™ protein enrichment approach.

2. Results

The methodology proposed for the analysis of RJ samples allowed us to identify a total of 86 proteins taxonomically classified to *Apis* spp. (Table 1). Among them, 50 proteins were detected in "crude" RJ water extracts, which were not pre-treated with ProteoMiner™ kit, and 74 proteins were found in extracts enriched with this commercial technique (total in all fractions). The numbers of proteins identified in "crude" extracts, extracts pre-treated with ProteoMiner™, and both in "crude" and ProteoMiner™ pre-treated extracts are presented in Figure 1.

Table 1. List of proteins identified in royal jelly found in: CRJ—"crude" (non-enriched with ProteoMiner™ technique) water extracts of royal jelly; F1, F2, F3, F4—separate fractions obtained from ProteoMiner™ elution with 1 M sodium chloride, 20 mM HEPES, pH 7.4 (F1), 200 mM glycine, pH 2.4 (F2), 60% ethylene glycol in water (F3), and 33.3% 2-propanol, 16.7% acetonitrile (ACN), 0.1% trifluoroacetic acid (TFA) (F4); MW—molecular weight; pI—isoelectric point. Fractions, in which proteins were detected, marked with "x".

Accession	Protein Name	Function	MW [kDa]	pI	CRJ	F1	F2	F3	F4
gi|110749126	Predicted: glucose dehydrogenase [acceptor] isoform 3 [Apis mellifera]	Enzyme	70.1	6.7		x			
gi|110751029	Predicted: e3 ubiquitin-protein ligase IAP-3-like [Apis mellifera]	Enzyme	43.2	6.4					x
gi|110756431	Predicted: hypothetical protein LOC725074 [Apis mellifera]		8.3	10.2	x				
gi|110758964	Predicted: regucalcin-like [Apis mellifera]	Binding	10.2	9.4	x				x
gi|110763647	Predicted: hypothetical protein LOC726323 [Apis mellifera]		18.5	7.9					x
gi|13184963	defensin [Apis mellifera]	Defense	6.3	6.5					x
gi|166795901	apolipophorin-III-like protein precursor [Apis mellifera]	Binding	21.3	5.4	x	x	x	x	x
gi|202078658	defensin [Apis cerana cerana]	Defense	10.6	7.8	x				
gi|202078660	defensin [Apis cerana cerana]	Defense	10.7	7.8			x		
gi|254548151	defensin precursor [Apis cerana]	Defense	8.7	6.4	x	x		x	x
gi|254548155	defensin precursor [Apis mellifera]	Defense	8.8	5.9	x	x		x	x
gi|254910938	defensin-1 preproprotein [Apis mellifera]	Defense	10.7	6.4			x		
gi|258678306	MRJP9 [Apis cerana]	Honeybee nutrition and development	19.4	9.2			x		
gi|258678310	MRJP5 [Apis dorsata]	Honeybee nutrition and development	21.3	4.7	x				
gi|258678314	MRJP6 [Apis florea]	Honeybee nutrition and development	21.5	4.4		x	x		
gi|258678316	MRJP9 [Apis florea]	Honeybee nutrition and development	21.0	5.5	x	x	x	x	x
gi|283105164	alpha-glucosidase III [Apis dorsata]	Enzyme	65.5	5.0	x				x
gi|284182838	major royal jelly protein 4 [Apis mellifera]	Honeybee nutrition and development	53.0	5.9	x	x	x	x	
gi|284812514	MRJP5 [Apis mellifera]	Honeybee nutrition and development	70.1	6.1	x	x	x	x	x
gi|288872651	major royal jelly protein [Apis mellifera]	Honeybee nutrition and development	61.6	6.7	x	x	x	x	x
gi|28972896	major royal jelly protein-like protein [Apis dorsata]	Honeybee nutrition and development	9.2	9.8	x	x	x	x	x
gi|328775853	Predicted: DNA replication licensing factor MCM4-like [Apis mellifera]	Binding	80.7	6.9		x			
gi|328777366	Predicted: hypothetical protein LOC100577348 [Apis mellifera]		61.1	10.5			x		
gi|328779534	Predicted: hypothetical protein LOC552041 [Apis mellifera]		79.3	4.2					x
gi|328782858	Predicted: hypothetical protein LOC410515 [Apis mellifera]		190.2	6.4	x				

Table 1. Cont.

Accession	Protein Name	Function	MW [kDa]	pI	CRJ	F1	F2	F3	F4
gi|328783362	Predicted: hypothetical protein LOC725249 [Apis mellifera]		6.3	4.9	x				
gi|328783471	Predicted: hypothetical protein LOC725114 isoform 1 [Apis mellifera]		10.1	5.7	x				
gi|328784821	Predicted: hypothetical protein LOC100577210 [Apis mellifera]		11.6	5.9	x	x		x	x
gi|328789019	Predicted: protein SERAC1-like [Apis mellifera]	Remodeling	87.5	9.2		x			
gi|328790726	Predicted: venom acid phosphatase Acph-1-like [Apis mellifera]	Enzyme	42.6	8.5					x
gi|328792767	Predicted: hypothetical protein LOC724993 [Apis mellifera]		43.8	5.4			x		
gi|328793775	Predicted: s-adenosylmethionine decarboxylase proenzyme-like, partial [Apis mellifera]	Enzyme	32.1	4.6	x				
gi|328794347	Predicted: major royal jelly protein 3-like, partial [Apis mellifera]	Honeybee nutrition and development	42.7	6.5	x		x		
gi|33358394	major royal jelly protein MRJP1 [Apis cerana cerana]	Honeybee nutrition and development	49.0	5.4	x	x	x	x	x
gi|380011960	Predicted: slit homolog 2 protein-like [Apis florea]	Binding	155.3	6.0		x			
gi|380012917	Predicted: uncharacterized protein LOC100870850 [Apis florea]		18.9	7.2			x		
gi|380013532	Predicted: Low Quality Protein: clathrin heavy chain-like [Apis florea]	Binding	187.7	5.6		x			
gi|380016522	Predicted: probable bifunctional methylenetetrahydrofolate dehydrogenase/cyclohydrolase 2-like [Apis florea]	Enzyme	39.9	9.6		x			
gi|380017034	Predicted: glucosylceramidase-like [Apis florea]	Enzyme	60.8	5.5				x	x
gi|380019073	Predicted: lysozyme 1-like isoform 2 [Apis florea]	Enzyme	13.8	4.6	x			x	x
gi|380020436	Predicted: regucalcin-like [Apis florea]	Binding	37.7	5.3	x	x		x	
gi|380022658	Predicted: major royal jelly protein 3-like [Apis florea]	Honeybee nutrition and development	62.7	7.8	x	x			
gi|380022660	Predicted: major royal jelly protein 4-like [Apis florea]	Honeybee nutrition and development	56.1	6.1			x		
gi|380022665	Predicted: major royal jelly protein 1-like [Apis florea]	Honeybee nutrition and development	43.9	5.3	x	x	x	x	x
gi|380022667	Predicted: major royal jelly protein 2-like [Apis florea]	Honeybee nutrition and development	49.1	5.7	x	x			
gi|380022669	Predicted: major royal jelly protein 2-like [Apis florea]	Honeybee nutrition and development	49.3	6.0	x	x	x	x	x
gi|380022673	Predicted: major royal jelly protein 5-like [Apis florea]	Honeybee nutrition and development	34.2	4.7			x		
gi|380022681	Predicted: major royal jelly protein 5-like isoform 2 [Apis florea]	Honeybee nutrition and development	47.5	9.2		x			
gi|380023404	Predicted: uncharacterized protein LOC100869599 [Apis florea]		18.7	9.5	x			x	
gi|380024584	Predicted: chymotrypsin inhibitor-like [Apis florea]	Enzyme inhibitor	5.6	10.0	x				
gi|380024588	Predicted: chymotrypsin inhibitor-like [Apis florea]	Enzyme inhibitor	8.0	5.0	x				
gi|380025248	Predicted: alkylated DNA repair protein alkB homolog 8-like [Apis florea]	Enzyme	68.2	9.3	x	x			
gi|380025500	Predicted: venom acid phosphatase Acph-1-like [Apis florea]	Enzyme	39.9	4.9					x
gi|380025661	Predicted: glucose dehydrogenase [acceptor]-like [Apis florea]	Enzyme	67.9	6.4	x	x		x	x
gi|380026601	Predicted: uncharacterized protein LOC100863702 [Apis florea]		9.9	5.0				x	x
gi|380027252	Predicted: uncharacterized protein LOC100864410 [Apis florea]		33.4	9.7			x		
gi|380028593	Predicted: Low Quality Protein: delta-1-pyrroline-5-carboxylate synthase-like [Apis florea]	Enzyme	84.9	9.1			x		

Table 1. Cont.

Accession	Protein Name	Function	MW [kDa]	pI	CRJ	F1	F2	F3	F4
gi	40218299	major royal jelly protein MRJP5 [Apis cerana cerana]	Honeybee nutrition and development	70.5	8.8		x			
gi	40218301	major royal jelly protein MRJP2 [Apis cerana cerana]	Honeybee nutrition and development	53.0	9.1	x		x		x
gi	40557703	major royal jelly protein MRJP1 precursor [Apis cerana]	Honeybee nutrition and development	48.9	5.4	x	x	x	x	x
gi	40557705	major royal jelly protein MRJP2 precursor [Apis cerana]	Honeybee nutrition and development	52.5	8.9	x	x	x	x	x
gi	42601246	major royal jelly protein MRJP5 precursor [Apis cerana]	Honeybee nutrition and development	68.2	9.3		x			x
gi	46358503	major royal jelly protein 2 [Apis cerana]	Honeybee nutrition and development	52.4	8.9	x	x	x	x	x
gi	48094573	Predicted: hypothetical protein LOC408608 [Apis mellifera]		19.4	7.5				x	x
gi	48101366	Predicted: venom serine protease 34 [Apis mellifera]	Enzyme	44.6	5.9				x	
gi	562090	defensin precursor [Apis mellifera]	Defense	10.7	6.4			x		x
gi	56422035	major royal jelly protein 3 [Apis mellifera carnica]	Honeybee nutrition and development	65.7	7.1	x	x	x	x	x
gi	56422037	major royal jelly protein 3 [Apis cerana]	Honeybee nutrition and development	69.2	9.3	x	x		x	
gi	56422041	major royal jelly protein 3 [Apis florea]	Honeybee nutrition and development	59.3	6.4	x				
gi	57546160	major rojal jelly protein 7 [Apis cerana]	Honeybee nutrition and development	24.9	5.5	x	x	x		x
gi	58585090	glucose oxidase [Apis mellifera]	Enzyme	67.9	6.5				x	x
gi	58585098	major royal jelly protein 1 precursor [Apis mellifera]	Honeybee nutrition and development	48.9	5.0	x	x	x	x	x
gi	58585108	major royal jelly protein 2 precursor [Apis mellifera]	Honeybee nutrition and development	51.0	7.0	x	x	x	x	x
gi	58585138	major royal jelly protein 5 precursor [Apis mellifera]	Honeybee nutrition and development	70.2	5.9	x				
gi	58585142	major royal jelly protein 3 precursor [Apis mellifera]	Honeybee nutrition and development	61.6	6.5	x		x		x
gi	58585164	alpha-glucosidase precursor [Apis mellifera]	Enzyme	65.5	4.9					x
gi	58585170	major royal jelly protein 4 precursor [Apis mellifera]	Honeybee nutrition and development	52.9	5.9	x	x	x	x	x
gi	58585188	major royal jelly protein 6 precursor [Apis mellifera]	Honeybee nutrition and development	49.8	5.9	x	x	x	x	x
gi	60115688	icarapin-like precursor [Apis mellifera]	Venom carbohydrate-rich protein	24.8	4.4	x	x		x	x
gi	62198227	major royal jelly protein 7 precursor [Apis mellifera]	Honeybee nutrition and development	50.5	4.8	x	x	x	x	x
gi	66504790	Predicted: integrator complex subunit 7-like isoform 1 [Apis mellifera]	snRNA processing	105.9	9.5	x				
gi	66511554	Predicted: glucosylceramidase-like isoform 1 [Apis mellifera]	Enzyme	59.3	5.2	x	x	x	x	x
gi	66564326	Predicted: plasma glutamate carboxypeptidase-like isoform 1 [Apis mellifera]	Enzyme	52.9	5.1					x
gi	66565246	Predicted: lysozyme isoform 1 [Apis mellifera]	Enzyme	17.1	4.6				x	x
gi	76496465	major royal jelly protein 3 [Apis dorsata]	Honeybee nutrition and development	66.9	6.6	x	x	x	x	x
gi	94471624	icarapin variant 2 precursor [Apis mellifera]	Venom carbohydrate-rich protein	19.6	4.2		x			

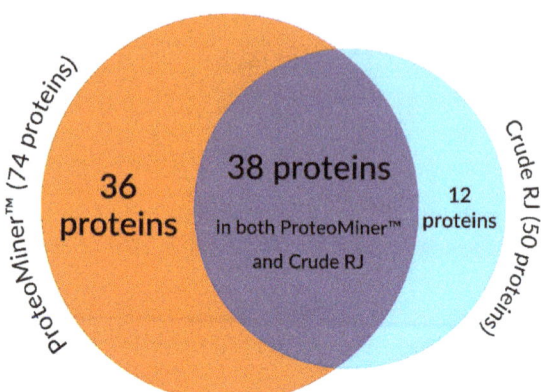

Figure 1. The number of proteins identified in unpurified samples, samples pre-treated with ProteoMiner^{TM}, and both in unpurified and purified samples.

In total, 35 proteins, accounting for 41% of all proteins identified in this study, belonged to the MRJPs family, responsible for honeybee nutrition and development (Figure 2A). Among them, fragments of MRJP1—MRJP7 and MRJP9 have been detected. The next important functional class of proteins were enzymes (20 proteins; 23% of all the identified proteins), with glucose dehydrogenase, alpha-glucosidase, venom acid phosphatase, lysozyme, and chymotrypsin inhibitor, among others. Into the most abundant protein classes, we also ranked defensins (7 proteins; representing 8% of all identified proteins), and binding proteins (6 proteins, accounting for 7% of all identified proteins), including regucalcin-like proteins. Additionally, four of all 86 proteins (5%) were classified into three other functional groups: remodeling proteins, snRNA processing proteins, and venom carbohydrate-rich proteins.

Moreover, ten identified proteins (11% of all identified proteins) were hypothetical proteins whose existence has been predicted, but there is not any experimental evidence proving their in vivo expression. Additionally, four uncharacterized proteins (5%) of unknown functions have been detected.

Proteins identified separately in extracts not pre-treated with ProteoMiner^{TM} enrichment technology were mainly MRJP (28 proteins out of 50, which represents 56% of proteins identified only in non-pretreated samples). The next functional classes of proteins included enzymes (6 identified proteins; 12%), defensins (3 proteins; 6%), binding proteins (3 proteins; 6%), and others (4 proteins; 8%). In these not-enriched samples, five hypothetical proteins (accounting for 10% of proteins identified in non-enriched samples) and one uncharacterized protein (2%) have been detected (Figure 2B).

Considering only extracts pre-treated with ProteoMiner^{TM} kit, the protein functional class distribution was comparable to overall results obtained from all compiled samples. MRJP (32 fragments) constituted 43% of all proteins identified in ProteoMiner^{TM} fractions. Besides, the most numerous classes were enzymes (17 proteins; 23% of proteins identified only in ProteoMiner^{TM} pre-treated samples), followed by defensins (6 proteins; 8%), binding proteins (6 proteins; 8%) and others (3 proteins; 4%). In those enriched samples, we also identified 6 (8% of all proteins identified in pre-treated fractions) hypothetical proteins and 4 (6%) uncharacterized proteins (Figure 2C).

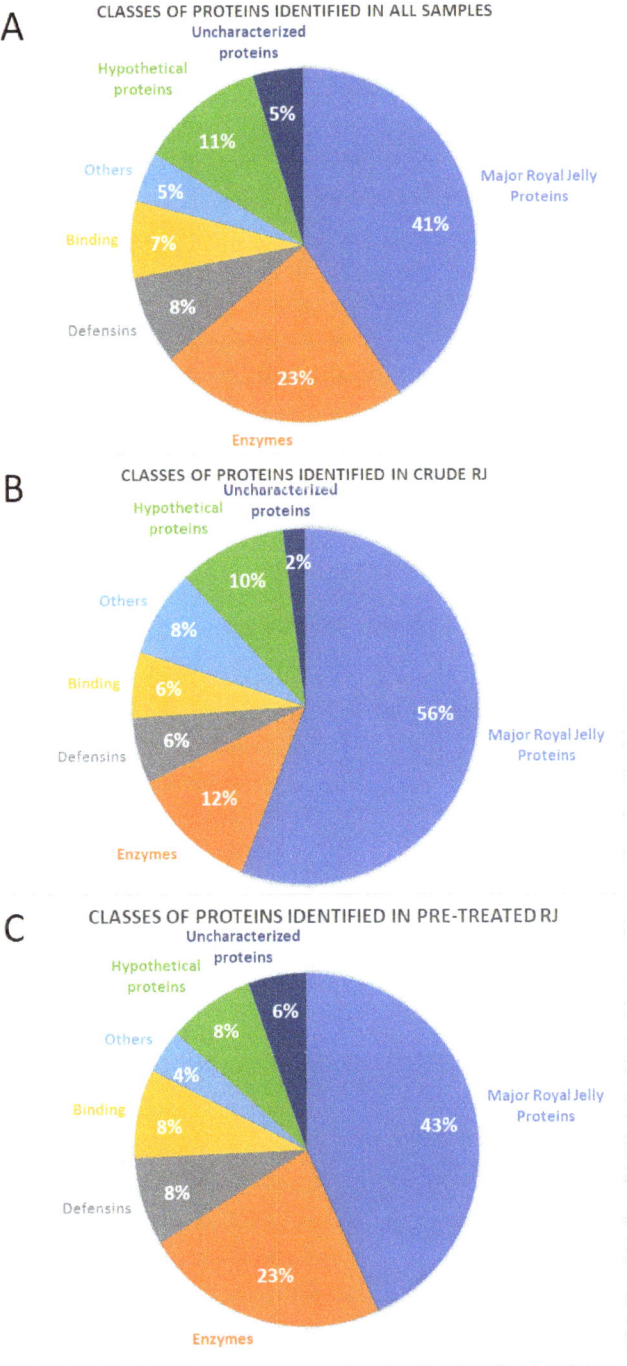

Figure 2. Classes of proteins identified (**A**) in total in all RJ "crude" and pre-treated; (**B**) only in crude (non-ProteoMiner[TM] pre-treated) RJ samples; (**C**) only in ProteoMiner[TM] pre-treated RJ samples.

Application of the ProteoMiner™ technique resulted in expanded identification of proteins that belonged to almost all functional groups. Within the MRJPs group, 9 out of 35 (25.7%) MRJPs identified in total in this study were detected only in ProteoMiner™ fractions. Moreover, 8 out of 19 (42.1%) enzymes identified in total were only detected in pre-treated samples. Regarding defensins, as many as 4 additional proteins belonging to this functional group were identified through the use of ProteoMiner™, whereas only three defensins were identified in the unpurified samples (meaning that 57.1% of all identified defensins were detected using the ProteoMiner™ technique). Similarly, out of 6 identified in total binding proteins, 3 of them (50%) were detected only in ProteoMiner™ fractions. These results clearly indicate that the ProteoMiner™ strategy significantly improves proteomic identification in complex biological samples, increasing the number of some identified functional proteins even more than twice.

Moreover, compared to "crude" extracts, application of the ProteoMiner™ kit enabled the detection of 5 more hypothetical proteins (out of 10 hypothetical proteins identified in total in this study) and 3 more uncharacterized proteins (out of 4 uncharacterized proteins identified in total in this study). These improvements in identification rate are especially beneficial when trying to find new proteomic features in bio-matrices represented by RJ.

3. Discussion

The goal of this study was to characterize the protein-peptide composition of fresh RJ using advanced nanoLC-MALDI-TOF/TOF MS/MS (nano-liquid chromatography—matrix-assisted laser desorption/ionization—time of flight mass spectrometry/mass spectrometry) technique. Despite its many unquestionable advantages, including versatility, ease of use, cost-effectiveness, and possibility to detect biomolecules in extremely low concentrations (even close to sub-femtomoles), the MALDI-TOF MS technique also has some limitations [27]. One of the most challenging issues is the detection of low-abundance proteins and peptides in very complex biological matrices. The presence of high concentration proteins and also other constituents, like lipids and salts, prevent direct analysis of all molecules contained in the sample [28]. Thus, proteins and peptides in low concentration ranges usually remain undetected. To deal with this problem, before MS analyses, biological samples must be purified, and their complexity and dynamic range need to be reduced. In this study, we applied a novel strategy based on a combinatorial hexapeptide ligand library (ProteoMiner™). ProteoMiner™ technique, by binding proteins to hexapeptides (through protein affinity interactions), resulted in reducing high-abundance proteins and enriching low-abundance proteins in RJ samples [29–31]. In this research, using the ProteoMiner™ technique significantly improved the number of identified proteins, from 50 in non-enriched RJ water extracts to 74 in ProteoMiner™ pre-treated extracts.

Application of the proposed methodology resulted in the identification of a total number of 86 proteins (from all samples and ProteoMiner™ fractions) taxonomically restricted to *Apis* spp. (*Apis mellifera*, *Apis cerana*, *Apis dorsata*, and *Apis florea*). As expected on the basis of available literature, the major royal jelly proteins family constituted the largest group. We identified 35 MRJPs or their precursors, which stands for 41% of the number of all identified proteins. Our results reflect previous independent proteomic studies, which reported that MRJPs account for up to 82–90% (w/w) of all proteins found in RJ [32–34]. It is also worth noting that in our study of only crude RJ, the percentage of MRJPs in relation to all identified proteins was as high as 56%. These results prove the utility of the ProteoMiner™ enrichment technique in enhancing the protein identification ability by MS approaches.

MRJPs are a family of nine proteins, of which MRJP1, called apalbumin or royalactin, is the most abundant and also the first MRJP to be identified in 1992 [35]. It has been suggested that MRJPs have mainly nutritional and developmental functions. Moreover, MRJP1 has been reported to exhibit antibacterial and antiproliferative activity [14,36]. Interestingly, the effect of MRJPs and whole RJ on honeybee queen development is not only due to the nutritional properties. MRJP1, along with protein named apisimin, by specific complex formation, provide the optimal RJ viscosity. The viscosity is crucial for holding queen larvae on the surface of the comb cells, which are orientated vertically and open downwards [37,38]. The formation of MRJP1-apisimin polymer is strongly pH-dependent. Long structures are formed at pH about 4.0, measured in fresh RJ [37,39]. The specific acidity is sustained by the fatty acids content produced in the bees' mandibular glands.

The second most-abundant functional group of RJ-identified proteins in this study was enzymes. This group includes some important enzymes crucial for carbohydrate metabolisms, like glucose dehydrogenase, alpha-glucosidase and glucose oxidase. Those enzymes are suggested to play an essential role in facilitating the larvae's digestion of sugars, which are found in RJ in large quantities [19]. The next relevant enzyme identified in our study was lysozyme isoform 1 and 2. Lysozyme is a protein commonly found in animals. It is a natural component of various secretions such as tears, saliva, mucus and milk. Lysozyme exhibits a strong antimicrobial activity, which in beehive prevents bacterial infection and disease [40,41]. Furthermore, because of their human benefits, the antibacterial activities of RJ are worthy of notice. These properties were reported as far back as 1939 [42]. According to the literature, besides the lysozyme, a 5.5 kDa defensin peptide named royalisin, and 10-hydroxy-2-decenoic acid (10-HDA) are mainly responsible for the antibacterial activity of RJ [43–45]. Antimicrobial properties, along with anti-inflammatory, immunomodulatory, antioxidant, moisturising, toning, and anti-ageing activities, make RJ a perfect dermatological product. Hence, RJ is used for treating various skin lesions and wounds [46,47]. Antibacterial defensins, including royalisin preproprotein (named defensin-1 preproprotein), were also identified in this study.

However, among the enzymes included in RJ, we found venom acid phosphatase Acph-1-like and venom serine protease 34. These proteins, along with MRJP8, MRJP9 and icarapin variant 2 precursor (also found in this study), are classified as bee venom allergens (based on Allergen Nomenclature Database, www.allergen.org, accessed on 8 March 2021). Allergens are responsible for the occurrence of allergic reactions in humans and animals after exposition to bee products [48–50]. The highest risk of an allergic reaction, both local or systemic, is associated with exposure to *Hymenoptera* venom, especially after the insect sting. Since allergens are present also in other bee products, like RJ (what was proven in this study), bee products should be consumed with caution and in moderation.

According to literature data, allergy to royal jelly is most common IgE-mediated, but contact allergic reactions (contact dermatitis) are also possible. IgE allergic reactions to royal jelly cause symptoms such as urticaria, angioedema, eczema, allergy rhinitis, conjunctivitis, bronchospasm or even anaphylactic shock [51–53]. Patients with bronchial asthma and atopic dermatitis are particularly predisposed to severe reactions to RJ. Even 17% adult patient with asthma [54] and one third of patient with atopic dermatitis [55] have specific IgE to RJ. Reported cases of allergy to royal jelly concern both people who have consumed this product for the first time or after long use. Allergy reaction to RJ after the first use is possible because of cross-reactivity between RJ and other allergens. Cross-reactions between RJ and bee venom can be associated with proteins found in our study—venom acid phosphatase Acph-1-like, venom serine protease 34, MRJP8, MRJP9 and icarapin variant 2 precursor. Moreover, cross-reactions are also possible with very typical allergens like house dust mite (*Dermatophagoides farinae*, *Dermatophagoides pteronyssinus*), German cockroach or crab [55]. In conclusion, RJ can cause an allergic reaction because it contains allergenic proteins and, further, because of cross-reactions with typical allergens. Patients with venom allergy, asthma, atopic dermatitis, and allergic rhinitis should be careful before consuming RJ.

Additionally, to minimalize the risk of dangerous adverse effects, bee products should be thoroughly tested for allergens, both qualitatively and quantitatively. However, despite its undeniable beneficial pro-health properties, the use of RJ in people with local and/or systemic *Hymenoptera* venom allergic reactions should be avoided.

Although in this study we focused mainly on nutritive MRJPs, and enzymes, defensins, and allergens, we also identified other functional groups of proteins included binding proteins, chymotrypsin inhibitors, protein SERAC1-like, and integrator complex subunit 7-like isoform 1. Those proteins show mainly regulatory activities.

The most interesting group of the identified biomolecules, regarding the possibility of learning about new properties and functions of royal jelly, are hypothetical and uncharacterized proteins. In this study, we detected 10 hypothetical proteins whose existence has been predicted based on the genetic information, but there is any experimental evidence proving their in vivo expression. The functions of these proteins, as in the case of uncharacterized proteins, have not been investigated yet. For these proteins, we performed BLAST analysis (https://blast.ncbi.nlm.nih.gov/, accessed on 11 January 2021) to find similar regions in various biological sequences. This approach helps to assess functional relationships between proteins and identify members of protein families. Results of our analysis are presented in Table 2.

Table 2. Results of BLAST analysis.

Accession	Protein Name	Significant Alignment Sequence	Query Coverage	Percent Identity	E-Value
gi|110756431	Predicted: hypothetical protein LOC725074 [Apis mellifera]	omega-conotoxin-like protein 1 [Apis mellifera]	100%	100.00%	2.00×10^{-47}
gi|110763647	Predicted: hypothetical protein LOC726323 [Apis mellifera]	uncharacterized protein LOC726323 isoform X1 [Apis mellifera]	100%	99.39%	1.00×10^{-116}
gi|328777366	Predicted: hypothetical protein LOC100577348 [Apis mellifera]	uncharacterized protein LOC100577348 isoform X2 [Apis mellifera]	98%	98.37%	0.0
gi|328779534	Predicted: hypothetical protein LOC552041 [Apis mellifera]	uncharacterized protein LOC102655185 isoform X1 [Apis mellifera]	100%	71.55%	5.00×10^{-92}
gi|328782858	Predicted: hypothetical protein LOC410515 [Apis mellifera]	uncharacterized protein LOC410515 [Apis mellifera]	93%	100.00%	0.0
		chorion peroxidase [Habropoda laboriosa]	93%	62.86%	0.0
gi|328783362	Predicted: hypothetical protein LOC725249 [Apis mellifera]	chymotrypsin inhibitor-like [Apis mellifera]	84%	100.00%	1.00×10^{-25}
gi|328783471	Predicted: hypothetical protein LOC725114 isoform 1 [Apis mellifera]	chymotrypsin inhibitor [Apis mellifera]	100%	100.00%	2.00×10^{-59}
gi|328784821	Predicted: hypothetical protein LOC100577210 [Apis mellifera]	uncharacterized protein LOC413627 [Apis mellifera]	77%	100.00%	6.00×10^{-45}
		regucalcin [Apis cerana cerana]	77%	94.81%	4.00×10^{-45}
gi|328792767	Predicted: hypothetical protein LOC724993 [Apis mellifera]	uncharacterized protein LOC724993 [Apis mellifera]	83%	90.37%	$0.00 \times 10^{+00}$
		Predicted: kielin/chordin-like protein [Trachymyrmex zeteki]	82%	45.91%	2.00×10^{-85}
		kielin/chordin-like protein [Nasonia vitripennis]	81%	47.06%	3.00×10^{-89}
gi|380012917	Predicted: uncharacterized protein LOC100870850 [Apis florea]	uncharacterized protein LOC408608 [Apis mellifera]	99%	73.33%	2.00×10^{-84}
gi|380023404	Predicted: uncharacterized protein LOC100869599 [Apis florea]	uncharacterized protein LOC102654257 [Apis mellifera]	100%	94.71%	7.00×10^{-116}
		low-density lipoprotein receptor 1-like isoform X3 [Vespa mandarinia]	98%	55.88%	7.00×10^{-57}
gi|380026601	Predicted: uncharacterized protein LOC100863702 [Apis florea]	ataxin-2 [Apis cerana cerana]	77%	88.41%	5.00×10^{-30}
gi|380027252	Predicted: uncharacterized protein LOC100864410 [Apis florea]	odorant receptor Or1 [Apis mellifera]	98%	89.68%	0.0
gi|48094573	Predicted: hypothetical protein LOC408608 [Apis mellifera]	uncharacterized protein LOC408608 [Apis mellifera]	100%	100.00%	3.00×10^{-124}

For the identified gi\|110756431 protein (Predicted: hypothetical protein LOC725074 [Apis mellifera]), we found a significant alignment sequence referred to omega-conotoxin-like protein 1 [Apis mellifera]. This protein was also found in bee venom in our previous studies, based on HPLC, nanoLC-MALDI-TOF/TOF MS/MS, and LC–ESI–QToF methodology [56,57]. With a high degree of probability, we can conclude that the molecule we have identified corresponds entirely to the results obtained from the BLAST platform, as both query coverage and per cent identity equal 100%. It means that our identified sequence exactly corresponds to conotoxin-like protein 1. As far as we know, this protein has never been reported to be included in RJ before. Detection of conotoxin-like protein 1 in royal jelly is a particularly important discovery in terms of the safety of RJ and also its potential therapeutic use. Conotoxins are small, disulfide-rich, neurotoxic peptides isolated from the marine cone snails venom. There are many types of these molecules, differing in structure and function [58]. The mechanism of their action has not been utterly elucidated yet. However, it is known that they bind to ion channels, neurotransmitters, transporters, and neural receptors [59]. Because of their selective binding to neuronal targets, conotoxins may be used for the developments of new drugs. Moreover, these peptides may be potentially used in the treatment of neurodegenerative conditions, like Alzheimer or Parkinson diseases [60]. Results of our study suggest that conotoxin may also affect the therapeutic effect of RJ, especially since some studies revealed that RJ shows a neuroprotective effect [61–63]. What is interesting, is that it has been investigated whether the consumption of RJ can be beneficial in the treatment of behavioral deficits in Alzheimer's disease in rabbits [64]. Thus, the presence of conotoxin-like protein 1 in RJ may contribute to elucidate the neuroprotective effect caused by RJ. Interestingly, Kaplan et al. [65] found this protein to be mainly expressed in the bee brain. It's also worth noting that expression of a protein with an approximate mass of 8.2 kDa, referred to as omega-conotoxin-like protein 1, has been reported by Gätschenberger et al. [66] to be induced in the haemolymph of young drones as a response to a septic injury. It may suggest that this protein participates in an immune response.

The next protein, for which the sequence similarity was noticed, was ataxin-2. This protein has been found in the BLAST analysis of gi | 380026601 (Predicted: uncharacterised protein LOC100863702 [*Apis florea*]). However, database searching returned the BLAST scores under 100% (77% query coverage and 88.41% per cent identity). Therefore, the presence of this protein (or its homolog) in royal jelly should be confirmed in further research. Ataxin-2 is an RNA-binding protein, which regulates mRNA stability and translation. It participates in cell death, calcium homeostasis, and cellular metabolism [67,68]. In insects, ataxin-2 is involved in the regulation of circadian rhythms but also in the development of peripheral tissue [69–71]. This may explain the presence of ataxin-2 in RJ.

Other peptides that have been found to be similar to the sequences detected in RJ are odorant receptor Or1 and chymotrypsin inhibitor. These proteins play mainly regulatory functions. Odorant receptor Or1 targets in bee's periphery nervous system. It participates in the odorants transport and binding, and therefore, in the behavioral response to specific odorants [72]. Chymotrypsin inhibitor prevents proteolysis of proteins present in RJ. For the chymotrypsin inhibitor, 100% similarity to gi | 328783471 (Predicted: hypothetical protein LOC725114 isoform 1 [*Apis mellifera*]) was found. However, query coverage was lower (84%) for the analysis of gi | 328783362 (Predicted: hypothetical protein LOC725249 [*Apis mellifera*]). Regarding odorant receptor Or1, the similarity with Predicted: uncharacterised protein LOC100864410 [*Apis florea*] detected in our study was expressed with 98% query coverage and 89.68% per cent identity. Therefore, these proteomic features require further research.

Other peptides for which a partial similarity to the RJ-detected proteins was found were Chorion peroxidase, regucalcin, kielin/chordin-like protein, and low-density lipoprotein receptor 1-like isoform X3. All those proteins bear mainly regulatory functions. However, due to lower BLAST scores (see Table 2), a certain identification must be proved by additional analyses. For the remaining hypothetical and uncharacterized proteins found in royal jelly, no defined references were found in BLAST analysis. The structure, functions and properties of these proteins should be investigated in future research. However, gi | 380012917 (Predicted: uncharacterized protein LOC100870850 [*Apis florea*]), and gi | 48094573 (uncharacterized protein LOC408608 [*Apis mellifera*]) were found in *Hymenoptera* venom in our previous study [56].

As the number of fully sequenced honeybee proteins and peptides increases, omics and bioinformatics technologies provide a tremendous amount of information about sequential biomolecules that function in various biological aspects of the honeybee and that are found in its products. In the postgenomic era, the international project i5K (http://i5k.github.io/.2018, accessed on 16 February 2021) not only played an important role in understanding the honeybee genome but also resulted in extensive research on the bee's proteome and the identification of many active proteins and peptides in its products. For a number of new substances produced by honeybee, only one type of biological activity has been identified, and their wider physiological role is unknown. Because many substances often exhibit pleiotropic biological activity, there is a need to understand a wider spectrum of biological activity for newly discovered compounds. It is also necessary to develop new, specific, and sensitive physiological, pharmacological, or toxicological bioassays. In the last decade, we have been observing new trends in the development of modern biotechnology, where besides medical, industrial, food and plant protection biotechnology, the term "Yellow Biotechnology" was introduced as an alternative term for insect biotechnology, opening new horizons for multidisciplinary research in the field of experimental entomology [73]. Yellow biotechnology is defined as the use of biotechnology to transform insects, their molecules, cells or organs into products and services with great potential for applications in biomedicine, pharmacy, and industry, where the honeybee may be one of the candidates.

4. Materials and Methods

4.1. Royal Jelly

In this study, we analyzed three samples of fresh RJ, collected in June 2019. The non-commercial apiary, from which all fresh samples were collected, is located in Góry Złotnickie village (51°87′504″ N, 18°12′431″ E) in Poland (Greater Poland Voivodeship, west-central Poland). Right after the collection, samples were stored at −80 °C until analyses. All RJ samples were analyzed in three technical replicates.

4.2. Pre-Treatment of the Royal Jelly Samples

Samples of fresh royal jelly were first suspended in ultrapure water in a concentration of 330 mg/mL. The RJ-water suspensions were mixed with vortex for 1 min and then sonicated for 20 min. After sonication, suspensions were vortex-mixed for 1 min and subsequently spun with 13,000 RPM (9600 RCF) for 20 min. Collected supernatants (in this study called "extracts") were used in further steps.

4.3. Relative Protein Enrichment—ProteoMiner^{TM} Hexapeptide Ligand Library

For the presented research, we applied a commercial ProteoMiner^{TM} Sequential Elution Small Capacity Kit (Bio-Rad, Hercules, CA, USA). All enrichment steps were performed strictly according to the manufacturer's instructions. The only difference from the manufacturer's protocol was the type of the sample solution loaded onto the ProteoMiner^{TM} column. Instead of the 0.2 mL of serum (for which the protocol is optimized), we added 0.2 mL of RJ extract, prepared as described above (see Section 4.2). In brief, 0.2 mL of the RJ water extracts were loaded onto pre-conditioned ProteoMiner^{TM} columns containing slurry beads and incubated at room temperature (RT) on a platform shaker. Next the samples were washed with phosphate buffer saline (PBS), and sequentially eluted with solvents: 1 M sodium chloride, 20 mM HEPES, pH 7.4 (Elution Reagent 1), 200 mM glycine pH 2.4 (Elution Reagent 2), 60% ethylene glycol in water (Elution Reagent 3), and 33.3% 2-propanol, 16.7% acetonitrile (ACN), 0.1% trifluoroacetic acid (TFA) (Elution Reagent 4) In total, 4 separate fractions were obtained from each sample. These fractions were further trypsin digested and then purified with ZipTip.

4.4. Trypsin Proteolytic Digestion and ZipTip Concentration and Purification

All RJ fractions derived from the ProteoMiner^{TM} enrichment method and additionally crude RJ solutions (water suspended and sonicated, but not fractionated with ProteoMiner^{TM}; three technical repetitions) were subjected to digestion with trypsin

(Promega, Madison, WI, USA). We applied a modified protocol from pierce in-solution tryptic digestion kit. In brief, before adding trypsin, proteins in extracts (both ProteoMiner™ fractions and "crude" extracts non-enriched with ProteoMiner™) were denatured, reduced, and alkylated. Subsequently, after trypsin addition, RJ extracts were digested at 37 °C overnight. In the next step, ZipTip C18 (Millipore, Bedford, MA, USA) reverse phase chromatography micropipette tips were used to purify, desalt and concentrate digested RJ protein extracts before mass spectrometry analyses. Tips were conditioned using ACN and 0.1% TFA/water. Then, each solution of the digested proteins extracted from the RJ samples (the exact solutions formed due to trypsin digestion of the RJ extracts, without any additional reagents) was loaded onto ZipTip tips according to the manufacturer's instruction. Sample solutions were aspirated and dispensed in 10 cycles for maximum binding of proteins. After washing with 0.1% TFA, bound proteins and peptides were eluted with 50% ACN in 0.1% TFA.

4.5. NanoLC-MALDI-TOF/TOF MS/MS Analyses

For the identification of peptides and proteins included in RJ, the nanoLC-MALDI-TOF/TOF MS/MS technique was applied. Subsequent to ZipTip depletion, RJ protein extracts were subjected to nanoLC fractionation. The nanoLC set consisted of EASY-nLC II (Bruker Daltonics, Bremen, Germany) nanoflow HPLC system, and Proteineer-fc II (Bruker Daltonics, Bremen, Germany) collector of fractions. The nanoLC system parts were: NS-MP-10 BioSphere C18 (NanoSeparations, Nieuwkoop, the Netherlands) trap column (20 mm × 100 µm I.D., particle size 5 µm, pore size 120 Å), and an Acclaim PepMap 100 (Thermo Scientific, Sunnyvale, CA, USA) column (150 mm × 75 µm I.D., particle size 3 µm, pore size 100 Å). The gradient elution method was set on 2–50% of ACN in 96 min (mobile phase A—0.05% TFA in water, mobile phase B—0.05% TFA in 90% ACN). The flow rate for separation was 300 nL/min, and the volume of the sample eluent injected into the chromatography column was 4 µL. The nanoLC system was operated by HyStar 3.2 (Bruker Daltonics, Bremen, Germany) software. In total, 384 fractions were obtained in the nanoLC separation process. Fractions were mixed with a matrix solution consisted of 36 µL of HCCA saturated solution in 0.1% TFA and acetonitrile (90:10 v/v), 748 µL of acetonitrile and 0.1% TFA (95:5 v/v) mixture, 8 µL of 10% TFA, and 8 µL of 100 mM ammonium phosphate, and spotted automatically onto the AnchorChip 384 (Bruker Daltonics, Bremen, Germany) target plate by the collector of fractions. Fractions were subsequently subjected to MS/MS analysis. For this purpose, UltrafleXtreme (Bruker Daltonics, Bremen, Germany) mass spectrometer was used (mass range m/z of 700–3500, reflectron mode). External calibration was performed with a mixture of Peptide Calibration Standard (Bruker Daltonics, Bremen, Germany). The list of the precursor ions for the identification was established with WARP-LC (Bruker Daltonics, Bremen, Germany) software. For the spectra acquisition, processing and evaluation FlexControl 3.4, FlexAnalysis 3.4 and BioTools 3.2 (Bruker Daltonics, Bremen, Germany) software were used. For the identification of discriminative proteins and peptides, an NCBI database and Mascot 2.4.1 search engine with taxonomic restriction to *Apis* spp. were applied. The protein search parameters were as follows: fragment ion mass tolerance $m/z \pm 0.7$, precursor ion mass tolerance ± 50 ppm, peptide charge +1, monoisotopic mass.

5. Conclusions

The functional and health-promoting properties of RJ make it one of the most attractive healthy ingredients. However, because of an overly complicated composition, RJ is not yet fully characterized and may cause adverse effects. Therefore, in this study, we proposed an advanced strategy based on combinatorial hexapeptide ligand libraries (ProteoMiner™ kit) for the proteomic analysis of RJ water extracts. We identified a total of 86 proteins taxonomically classified to *Apis* spp. (bees). In addition to the well-known components of RJ, we identified 10 hypothetical proteins and 4 uncharacterized proteins. We also proved the utility of the ProteoMiner™ technique in the analysis of complex biological samples, as 74 out of all identified proteins were detected in RJ extracts pre-treated with ProteoMiner™ kit, and only 50 proteins were found in extracts that were not enriched with this technique. Broadening the knowledge of RJ composition may contribute to the development of standards and regulations, enhancing the quality of RJ, and consequently, the safety of its supplementation. Moreover, better characterization of proteins and peptides found in RJ may support the understanding of the functional properties of RJ regarding both humans and bees.

Author Contributions: Conceptualization, J.M. (Jan Matysiak) and G.R.; methodology, E.M. and J.M. (Jan Matysiak); software, E.M.; validation, E.M.; formal analysis, E.M. and J.M. (Joanna Matysiak); investigation, E.M.; resources, J.M. (Jan Matysiak); data curation, E.M.; Writing—Original draft preparation, E.M., J.M. (Joanna Matysiak), G.R., E.K., W.Z., J.Z. and J.M. (Jan Matysiak); Writing—Review and Editing, E.M., J.M. (Joanna Matysiak), G.R. and J.M. (Jan Matysiak); visualization, E.M.; supervision, J.M. (Jan Matysiak); project administration, J.M. (Jan Matysiak); funding acquisition, J.M. (Jan Matysiak). All authors have read and agreed to the published version of the manuscript.

Funding: This project received financial support from the Polish National Science Centre (grant number 2016/23/D/NZ7/03949).

Data Availability Statement: The data presented in this study are contained within the article.

Conflicts of Interest: The authors declare no conflict of interest.

References

1. Denisow, B.; Denisow-Pietrzyk, M. Biological and therapeutic properties of bee pollen: A review. *J. Sci. Food Agric.* **2016**, *96*, 4303–4309. [CrossRef]
2. Fratini, F.; Cilia, G.; Mancini, S.; Felicioli, A. Royal Jelly: An ancient remedy with remarkable antibacterial properties. *Microbiol. Res.* **2016**, *192*, 130–141. [CrossRef]
3. Taulavuori, K.; Julkunen-Tiitto, R.; Hyöky, V.; Taulavuori, E. Blue Mood for Superfood. *Nat. Prod. Commun.* **2013**, *8*. [CrossRef]
4. Ramadan, M.F.; Al-Ghamdi, A. Bioactive compounds and health-promoting properties of royal jelly: A review. *J. Funct. Foods* **2012**, *4*, 39–52. [CrossRef]
5. Wytrychowski, M.; Chenavas, S.; Daniele, G.; Casabianca, H.; Batteau, M.; Guibert, S.; Brion, B. Physicochemical characterisation of French royal jelly: Comparison with commercial royal jellies and royal jellies produced through artificial bee-feeding. *J. Food Compos. Anal.* **2013**, *29*, 126–133. [CrossRef]
6. Sabatini, A.G. Quality and standardisation of Royal Jelly. *J. ApiProd. ApiMed. Sci.* **2009**, *1*, 16–21. [CrossRef]
7. Garcia-Amoedo, L.H.; De Almeida-Muradian, L.B. Physicochemical composition of pure and adulterated royal jelly. *Quim. Nova* **2007**, *30*, 257–259. [CrossRef]
8. Alvarez-Suarez, J.M. *Bee Products-Chemical and Biological Properties*; Springer International Publishing: New York, NY, USA, 2017; pp. 1–306.
9. Priomorac, L.; Bilić Rajs, B.; Puškadija, Z.; Kovačić, M.; Vukadin, I.; Flanjak, I. Physicochemical characteristics of Croatian royal jelly. *Croat. J. Food Sci. Technol.* **2019**, *11*, 266–271.
10. Balkanska, R.; Crenguta, P. Comparison of physicochemical parameters in royal jelly from Romania and Bulgaria. *Bull. Univ. Agric. Sci. Vet. Med. Cluj-Napoca Anim. Sci. Biotechnol.* **2013**, *70*, 117–121.
11. Scarselli, R.; Donadio, E.; Giuffrida, M.G.; Fortunato, D.; Conti, A.; Balestreri, E.; Felicioli, R.; Pinzauti, M.; Sabatini, A.G.; Felicioli, A. Towards royal jelly proteome. *Proteomics* **2005**, *5*, 769–776. [CrossRef]
12. Lercker, G.; Caboni, M.F.; Vecchi, M.A.; Sabatini, A.G.; Nanetti, A. Caratterizzazione dei principali costituenti della gelatina reale. *Apicoltura* **1992**, *8*, 27–37.
13. Sano, O.; Kunikata, T.; Kohno, K.; Iwaki, K.; Ikeda, M.; Kurimoto, M. Characterization of Royal Jelly Proteins in both Africanized and European Honeybees (*Apis mellifera*) by Two-Dimensional Gel Electrophoresis. *J. Agric. Food Chem.* **2004**, *52*, 15–20. [CrossRef]
14. Buttstedt, A.; Moritz, R.F.A.; Erler, S. Origin and function of the major royal jelly proteins of the honeybee (*Apis mellifera*) as members of the yellow gene family. *Biol. Rev.* **2014**, *89*, 255–269. [CrossRef] [PubMed]
15. Kamakura, M. Royalactin induces queen differentiation in honeybees. *Nature* **2011**, *473*, 478–483. [CrossRef]
16. Buttstedt, A.; Ihling, C.H.; Pietzsch, M.; Moritz, R.F.A. Royalactin is not a royal making of a queen. *Nature* **2016**, *537*, E10–E12. [CrossRef]
17. Fujita, T.; Kozuka-Hata, H.; Ao-Kondo, H.; Kunieda, T.; Oyama, M.; Kubo, T. Proteomic Analysis of the Royal Jelly and Characterization of the Functions of its Derivation Glands in the Honeybee. *J. Proteome Res.* **2013**, *12*, 404–411. [CrossRef] [PubMed]
18. Schönleben, S.; Sickmann, A.; Mueller, M.J.; Reinders, J. Proteome analysis of Apis mellifera royal jelly. *Anal. Bioanal. Chem.* **2007**, *389*, 1087–1093. [CrossRef] [PubMed]

19. Han, B.; Li, C.; Zhang, L.; Fang, Y.; Feng, M.; Li, J. Novel royal jelly proteins identified by gel-based and gel-free proteomics. *J. Agric. Food Chem.* **2011**, *59*, 10346–10355. [CrossRef] [PubMed]
20. Furusawa, T.; Rakwal, R.; Nam, H.W.; Shibato, J.; Agrawal, G.K.; Kim, Y.S.; Ogawa, Y.; Yoshida, Y.; Kouzuma, Y.; Masuo, Y.; et al. Comprehensive royal jelly (RJ) proteomics using one- and two-dimensional proteomics platforms reveals novel RJ proteins and potential phospho/glycoproteins. *J. Proteome Res.* **2008**, *7*, 3194–3229. [CrossRef] [PubMed]
21. Li, J.; Wang, T.; Zhang, Z.; Pan, Y. Proteomic analysis of royal jelly from three strains of western honeybees (*Apis mellifera*). *J. Agric. Food Chem.* **2007**, *55*, 8411–8422. [CrossRef]
22. Hu, F.L.; Bíliková, K.; Casabianca, H.; Daniele, G.; Salmen Espindola, F.; Feng, M.; Guan, C.; Han, B.; Krištof Kraková, T.; Li, J.K.; et al. Standard methods for Apis mellifera royal jelly research. *J. Apic. Res.* **2017**, *58*, 1–68. [CrossRef]
23. Collazo, N.; Carpena, M.; Nuñez-Estevez, B.; Otero, P.; Simal-Gandara, J.; Prieto, M.A. Health Promoting Properties of Bee Royal Jelly: Food of the Queens. *Nutrients* **2021**, *13*, 543. [CrossRef] [PubMed]
24. Zhang, L.; Han, B.; Li, R.; Lu, X.; Nie, A.; Guo, L.; Fang, Y.; Feng, M.; Li, J. Comprehensive identification of novel proteins and N-glycosylation sites in royal jelly. *BMC Genom.* **2014**, *15*, 135. [CrossRef] [PubMed]
25. Zhang, L.; Fang, Y.; Li, R.; Feng, M.; Han, B.; Zhou, T.; Li, J. Towards posttranslational modification proteome of royal jelly. *J. Proteom.* **2012**, *75*, 5327–5341. [CrossRef] [PubMed]
26. Han, B.; Fang, Y.; Feng, M.; Lu, X.; Huo, X.; Meng, L.; Wu, B.; Li, J. In-depth phosphoproteomic analysis of royal jelly derived from western and eastern honeybee species. *J. Proteome Res.* **2014**, *13*, 5928–5943. [CrossRef] [PubMed]
27. Greco, V.; Piras, C.; Pieroni, L.; Ronci, M.; Putignani, L.; Roncada, P.; Urbani, A. Applications of MALDI-TOF mass spectrometry in clinical proteomics. *Expert Rev. Proteom.* **2018**, *15*, 683–696. [CrossRef]
28. Hajduk, J.; Matysiak, J.; Kokot, Z.J. Challenges in biomarker discovery with MALDI-TOF MS. *Clin. Chim. Acta* **2016**, *458*, 84–98. [CrossRef]
29. Li, S.; He, Y.; Lin, Z.; Xu, S.; Zhou, R.; Liang, F.; Wang, J.; Yang, H.; Liu, S.; Ren, Y. Digging More Missing Proteins Using an Enrichment Approach with ProteoMiner. *J. Proteome Res.* **2017**, *16*, 4330–4339. [CrossRef]
30. Righetti, P.G.; Boschetti, E.; Lomas, L.; Citterio, A. Protein Equalizer™ Technology: The quest for a "democratic proteome". *Proteomics* **2006**, *6*, 3980–3992. [CrossRef]
31. Boschetti, E.; Righetti, P.G. The ProteoMiner in the proteomic arena: A non-depleting tool for discovering low-abundance species. *J. Proteom.* **2008**, *71*, 255–264. [CrossRef]
32. Drapeau, M.D.; Albert, S.; Kucharski, R.; Prusko, C.; Maleszka, R. Evolution of the Yellow/Major Royal Jelly Protein family and the emergence of social behavior in honey bees. *Genome. Res.* **2006**, *16*, 1385–1394. [CrossRef]
33. Santos, K.S.; Delazari dos Santos, L.; Mendes, M.; Monson de Souza, B.; Malaspina, O.; Palma, M.S. Profiling the proteome complement of the secretion from hypopharyngeal gland of Africanized nurse-honeybees (*Apis mellifera* L.). *Insect Biochem. Mol. Biol.* **2005**, *35*, 85–91. [CrossRef]
34. Schmitzová, J.; Klaudiny, J.; Albert, Š.; Schröder, W.; Schreckengost, W.; Hanes, J.; Júdová, J.; Šimúth, J. A family of major royal jelly proteins of the honeybee *Apis mellifera* L. *Cell. Mol. Life Sci.* **1998**, *54*, 1020–1030. [CrossRef]
35. Hanes, J.; Šimuth, J. Identification and partial characterisation of the major royal jelly protein of the honey bee (*Apis mellifera* L.). *J. Apic. Res.* **1992**, *31*, 22–26. [CrossRef]
36. Ramanathan, A.N.K.G.; Nair, A.J.; Sugunan, V.S. A review on Royal Jelly proteins and peptides. *J. Funct. Foods* **2018**, *44*, 255–264. [CrossRef]
37. Buttstedt, A.; Mureşan, C.I.; Lilie, H.; Hause, G.; Ihling, C.H.; Schulze, S.H.; Pietzsch, M.; Moritz, R.F.A. How Honeybees Defy Gravity with Royal Jelly to Raise Queens. *Curr. Biol.* **2018**, *28*, 1095–1100.e3. [CrossRef]
38. Pirk, C.W.W. Honeybee Evolution: Royal Jelly Proteins Help Queen Larvae to Stay on Top. *Curr. Biol.* **2018**, *28*, R350–R351. [CrossRef] [PubMed]
39. Mureşan, C.I.; Buttstedt, A. pH-dependent stability of honey bee (*Apis mellifera*) major royal jelly proteins. *Sci. Rep.* **2019**, *9*, 3–8. [CrossRef]
40. Ragland, S.A.; Criss, A.K. From bacterial killing to immune modulation: Recent insights into the functions of lysozyme. *PLoS Pathog.* **2017**, *13*, e1006512. [CrossRef]
41. Oliver, W.T.; Wells, J.E. Lysozyme as an alternative to growth promoting antibiotics in swine production. *J. Anim. Sci. Biotechnol.* **2015**, *6*, 35. [CrossRef]
42. McCleskey, C.S.; Melampy, R.M. Bactericidal Properties of Royal Jelly of the Honeybee. *J. Econ. Entomol.* **1939**, *32*, 581–587. [CrossRef]
43. Bíliková, K.; Huang, S.-C.; Lin, I.-P.; Šimuth, J.; Peng, C.-C. Structure and antimicrobial activity relationship of royalisin, an antimicrobial peptide from royal jelly of Apis mellifera. *Peptides* **2015**, *68*, 190–196. [CrossRef] [PubMed]
44. Shen, L.; Liu, D.; Li, M.; Jin, F.; Din, M.; Parnell, L.D.; Lai, C.-Q. Mechanism of Action of Recombinant Acc-Royalisin from Royal Jelly of Asian Honeybee against Gram-Positive Bacteria. *PLoS ONE* **2012**, *7*, e47194. [CrossRef] [PubMed]
45. Yang, Y.C.; Chou, W.M.; Widowati, D.A.; Lin, I.P.; Peng, C.C. 10-Hydroxy-2-Decenoic Acid of Royal Jelly Exhibits Bactericide and Anti-Inflammatory Activity in Human Colon Cancer Cells. *BMC Complement. Altern. Med.* **2018**, *18*, 202. [CrossRef] [PubMed]
46. Kurek-Górecka, A.; Górecki, M.; Rzepecka-Stojko, A.; Balwierz, R.; Stojko, J. Bee Products in Dermatology and Skin Care. *Molecules* **2020**, *25*, 556. [CrossRef]

47. Pavel, C.I.; Mărghitaş, L.A.; Bobiş, O.; Dezmirean, D.S.; Şapcaliu, A.; Radoi, I.; Mădaş, M.N. Biological Activities of Royal Jelly—Review. *Sci. Pap. Anim. Sci. Biotechnol.* **2011**, *44*, 108–118.
48. Spillner, E.; Blank, S.; Jakob, T. Hymenoptera Allergens: From Venom to "Venome". *Front. Immunol.* **2014**, *5*, 77. [CrossRef]
49. Tomsitz, D.; Brockow, K. Component Resolved Diagnosis in Hymenoptera Anaphylaxis. *Curr. Allergy Asthma Rep.* **2017**, *17*, 38. [CrossRef] [PubMed]
50. Hoffman, D.R. Hymenoptera venom allergens. *Clin. Rev. Allergy Immunol.* **2006**, *30*, 109–128. [CrossRef]
51. Takahama, H.; Shimazu, T. Food-induced anaphylaxis caused by ingestion of royal jelly. *J. Dermatol.* **2006**, *33*, 424–426. [CrossRef]
52. Katayama, M.; Aoki, M.; Kawana, S. Case of anaphylaxis caused by ingestion of royal jelly. *J. Dermatol.* **2008**, *35*, 222–224. [CrossRef]
53. Mizutani, Y.; Shibuya, Y.; Takahashi, T.; Tsunoda, T.; Moriyama, T.; Seishima, M. Major royal jelly protein 3 as a possible allergen in royal jelly-induced anaphylaxis. *J. Dermatol.* **2011**, *38*, 1079–1081. [CrossRef] [PubMed]
54. Leung, R.; Lam, C.W.K.; Ho, A.; Chan, J.K.W.; Choy, D.; Lai, C.K.W. Allergic sensitisation to common environmental allergens in adult asthmatics in Hong Kong. *Hong Kong Med. J.* **1997**, *3*, 211–217.
55. Hata, T.; Furusawa-Horie, T.; Arai, Y.; Takahashi, T.; Seishima, M.; Ichihara, K. Studies of royal jelly and associated cross-reactive allergens in atopic dermatitis patients. *PLoS ONE* **2020**, *15*, e0233707. [CrossRef] [PubMed]
56. Matysiak, J.; Hajduk, J.; Światły, A.; Naskret, N.; Kokot, Z.J. Proteomic analysis of *Apis mellifera* venom determined by liquid chromatography (LC) coupled with nano-LC-MALDI-TOF/TOF MS. *Acta. Pol. Pharm. Drug Res.* **2017**, *74*, 53–65.
57. Matysiak, J.; Hajduk, J.; Mayer, F.; Hebeler, R.; Kokot, Z.J. Hyphenated LC–MALDI–ToF/ToF and LC–ESI–QToF approach in proteomic characterisation of honeybee venom. *J. Pharm. Biomed. Anal.* **2016**, *121*, 69–76. [CrossRef] [PubMed]
58. Robinson, S.D.; Norton, R.S. Conotoxin gene superfamilies. *Mar. Drugs* **2014**, *12*, 6058–6101. [CrossRef]
59. Daly, N.; Craik, D. Structural Studies of Conotoxins. *IUBMB Life* **2009**, *61*, 144–150. [CrossRef]
60. Dao, F.-Y.; Yang, H.; Su, Z.-D.; Yang, W.; Wu, Y.; Hui, D.; Chen, W.; Tang, H.; Lin, H. Recent Advances in Conotoxin Classification by Using Machine Learning Methods. *Molecules* **2017**, *22*, 1057. [CrossRef] [PubMed]
61. Ahmad, S.; Campos, M.G.; Fratini, F.; Altaye, S.Z.; Li, J. New insights into the biological and pharmaceutical properties of royal jelly. *Int. J. Mol. Sci.* **2020**, *21*, 382. [CrossRef]
62. Minami, A.; Matsushita, H.; Ieno, D.; Matsuda, Y.; Horii, Y.; Ishii, A.; Takahashi, T.; Kanazawa, H.; Wakatsuki, A.; Suzuki, T. Improvement of neurological disorders in postmenopausal model rats by administration of royal jelly. *Climacteric* **2016**, *19*, 568–573. [CrossRef]
63. Mohamed, A.A.-R.; Galal, A.A.A.; Elewa, Y.H.A. Comparative protective effects of royal jelly and cod liver oil against neurotoxic impact of tartrazine on male rat pups brain. *Acta Histochem.* **2015**, *117*, 649–658. [CrossRef]
64. Pan, Y.; Xu, J.; Jin, P.; Yang, Q.; Zhu, K.; You, M.; Hu, F.; Chen, M. Royal Jelly Ameliorates Behavioral Deficits, Cholinergic System Deficiency, and Autonomic Nervous Dysfunction in Ovariectomized Cholesterol-Fed Rabbits. *Molecules* **2019**, *24*, 1149. [CrossRef]
65. Kaplan, N.; Morpurgo, N.; Linial, M. Novel Families of Toxin-like Peptides in Insects and Mammals: A Computational Approach. *J. Mol. Biol.* **2007**, *369*, 553–566. [CrossRef] [PubMed]
66. Gätschenberger, H.; Gimple, O.; Tautz, J.; Beier, H. Honey bee drones maintain humoral immune competence throughout all life stages in the absence of vitellogenin production. *J. Exp. Biol.* **2012**, *215*, 1313–1322. [CrossRef] [PubMed]
67. Ostrowski, L.A.; Hall, A.C.; Mekhail, K. Ataxin-2: From RNA Control to Human Health and Disease. *Genes* **2017**, *8*, 157. [CrossRef] [PubMed]
68. Carmo-Silva, S.; Nobrega, C.; Pereira de Almeida, L.; Cavadas, C. Unraveling the Role of Ataxin-2 in Metabolism. *Trends Endocrinol. Metab.* **2017**, *28*, 309–318. [CrossRef]
69. Zhang, Y.; Ling, J.; Yuan, C.; Dubruille, R.; Emery, P. A role for Drosophila ATX2 in activation of PER translation and circadian behavior. *Science* **2013**, *340*, 879–882. [CrossRef]
70. Lim, C.; Allada, R. ATAXIN-2 Activates PERIOD Translation to Sustain Circadian Rhythms in Drosophila. *Science* **2013**, *340*, 875–879. [CrossRef]
71. Vianna, M.C.B.; Poleto, D.C.; Gomes, P.F.; Valente, V.; Paçó-Larson, M.L. Drosophila ataxin-2 gene encodes two differentially expressed isoforms and its function in larval fat body is crucial for development of peripheral tissues. *FEBS OpenBio* **2016**, *6*, 1040–1453. [CrossRef]
72. Venthur, H.; Zhou, J.-J. Odorant Receptors and Odorant-Binding Proteins as Insect Pest Control Targets: A Comparative Analysis. *Front. Physiol.* **2018**, *9*, 1163. [CrossRef] [PubMed]
73. Vilcinskas, A. *Biotechnology I Insect Biotechnologie in Drug Discovery*, 1st ed.; Vilcinskas, A., Ed.; Springer: Berlin/Heidelberg, Germany, 2013; p. 201.

Article

Effectiveness of Different Sample Treatments for the Elemental Characterization of Bees and Beehive Products

Maria Luisa Astolfi [1],*, Marcelo Enrique Conti [2], Elisabetta Marconi [3], Lorenzo Massimi [1] and Silvia Canepari [1]

[1] Department of Chemistry, Sapienza University, Piazzale Aldo Moro 5, I-00185 Rome, Italy; l.massimi@uniroma1.it (L.M.); silvia.canepari@uniroma1.it (S.C.)
[2] Department of Management, Sapienza University, Via del Castro Laurenziano 9, I-00161 Rome, Italy; marcelo.conti@uniroma1.it
[3] Department of Public Health and Infectious Diseases, Sapienza University, Piazzale Aldo Moro 5, I-00185 Rome, Italy; elisabetta.marconi@uniroma1.it
* Correspondence: marialuisa.astolfi@uniroma1.it; Tel./Fax: +39-06-4991-3384

Academic Editors: Gavino Sanna, Yolanda Picò, Marco Ciulu, Carlo Tuberoso and Nadia Spano
Received: 17 August 2020; Accepted: 15 September 2020; Published: 17 September 2020

Abstract: Bee health and beehive products' quality are compromised by complex interactions between multiple stressors, among which toxic elements play an important role. The aim of this study is to optimize and validate sensible and reliable analytical methods for biomonitoring studies and the quality control of beehive products. Four digestion procedures, including two systems (microwave oven and water bath) and different mixture reagents, were evaluated for the determination of the total content of 40 elements in bees and five beehive products (beeswax, honey, pollen, propolis and royal jelly) by using inductively coupled plasma mass and optical emission spectrometry. Method validation was performed by measuring a standard reference material and the recoveries for each selected matrix. The water bath-assisted digestion of bees and beehive products is proposed as a fast alternative to microwave-assisted digestion for all elements in biomonitoring studies. The present study highlights the possible drawbacks that may be encountered during the elemental analysis of these biological matrices and aims to be a valuable aid for the analytical chemist. Total elemental concentrations, determined in commercially available beehive products, are presented.

Keywords: sample preparation; trace element; toxic element; spectroanalytical technique; biomonitoring

1. Introduction

Various natural and anthropogenic emission sources of toxic elements may cause air pollution [1–6]. Among different air pollution monitoring techniques, biomonitoring has recently become one of the most widely used technique, due to its ease of operation, low cost, efficiency and specificity [7–10]. In fact, several living organisms, known as biomonitors, can accumulate toxic elements, allowing the monitoring of pollutants concentrations in the environment for integrated measurements over time [11–13]. The use of apis mellifera and beehive products for biomonitoring studies has been widely investigated [14–20] and reviewed [21–23]. Honeybees and the associated matrices are often considered as efficient sentinels for environmental biomonitoring [7,17]. Trace elements can be transferred to honeybees and beehive products from all the environmental compartments (soil, vegetation, air and water) in the areas covered by forager honeybees [24], within which honeybees and beehive products may supply integrated representative samples [25]. Measurements of element concentrations in honey samples are relevant for the healthiness assessment of the honey in terms of the presence of essential metals and for ensuring the human health safety by assessing the admissible levels of toxic elements [21]. Moreover, the assessment of element concentrations in honey is also useful for its classification based on its genuineness and its botanical and geographical origins [21].

Among the different instrumental methods used for the determination of elements in honeybee and beehive products, atomic and mass spectrometry are considered as the most sensitive, accurate, and robust techniques, thus being routinely and customarily applied [21,26,27]. Flame atomic absorption spectrometry (F-AAS) has been often employed for the low-cost and rapid determination of high concentrations of metals [21,22,28]. To control the quality of honey and other beehive products in terms of contamination by heavy metals, many sensitive techniques are required, including graphite furnace atomic absorption spectrophotometry (GF-AAS) [17,19], electrothermal atomic absorption spectrometry (ET-AAS) [29,30], microwave plasma technique atomic emission spectrometry (MP-AES) [31], inductively coupled plasma optical emission spectrometry (ICP-OES) [14,16,24,32–34], and inductively coupled plasma mass spectrometry (ICP-MS) [14,18,20,35–39]. In addition, other atomic techniques, such as atomic analyzer mercury (AMA), hydride generation-atomic absorption spectrometry (HG-AAS) and cold vapor atomic fluorescence spectrometry (CV-AFS), have been employed in environmental studies for the determination of Hg, As and Se [40,41].

The analysis of biological samples by atomic and mass spectrometry techniques is a difficult and challenging task [21]. Honey and other beehive products are very complex organic matrices with problems related to the sample heterogeneity, selection of sample treatment, and decomposition as well chemical interferences during measurements [21,22,38]. Moreover, bees and beehive products are matrices with high C contents [42–44] and their incomplete sample decomposition may cause a residual carbon content (RCC) in the final digests. During ICP analysis, the element signals with a similar ionization potential to that of C is enhanced due to C charge transfer reactions [45,46]. The residual acidity in the final digests is also important for the polyatomic interference (for ICP-MS), nebulization efficiency and for reducing the instrument interface damage [47–49]. Samples of honey and other beehive products are commonly decomposed using high temperature dry ashing [17,50] or wet digestion procedures, in order to destroy the carbohydrate-rich sample matrix and minimize the matrix-based interferences [21,22]. To obtain an efficient honeybee and beehive products digestion, various reagents or mixtures are used, including concentrated HNO_3 or other acids (such as HCl, $HClO_4$ and H_2SO_4), frequently mixed with 30% H_2O_2, using open vessels or sealed quartz or polytetrafluoroethylene (PTFE) vessels in microwave-assisted systems [21,22,35,38,39]. Unfortunately, to date the validation of methods and procedures used for the analysis of bees and beehive products is difficult because no certified reference material (CRM) of these matrices is available. Instead, the trueness of the methods and procedures applied is commonly checked by the analysis of other CRMs containing high levels of carbohydrates, such as Antarctic krill (MURST-ISSA2), apple leaves (NIST 1515), brown bread (BCR 191), corn (NBS 8413), mixed Polish herbs (INCT-MPH-2), tea leaves (INCT-TL-1) and wheat (IPE 684) or whole meal flour (BCR 189) [21,22]. For the same purpose, the recovery tests are carried out on the chosen samples spiked with known amounts of selected elements [21,22,35].

Generally, among all beehive products, only honey [14,20,21,23,32,34,51–53] or honey and pollen [17,24,35] are considered in the optimization approach for elemental determination with ICP techniques. To our knowledge, in the literature, few studies are focused on the elemental content of other beehive products such as propolis [19,31,38] or royal jelly and beeswax [15].

Therefore, the aim of this study is to compare different methods of sample preparation, including two systems (microwave oven and water bath) and different mixture reagents, which will allow the assessment of the exposure of bees to different element concentrations and the quality control of beehive products in terms of contamination made by toxic elements in both routine and large-scale investigations. The optimized methods were employed to determine 40 elements (Al, As, B, Ba, Be, Bi, Ca, Cd, Co, Cr, Cs, Cu, Fe, K, Li, Mg, Mn, Mo, Na, Ni, P, Pb, Rb, Sb, Se, Si, Sn, Sr, Te, Ti, Tl, U, V, and Zn) with ICP techniques. The analytical performance and quality control of the optimized procedures were evaluated on CRMs and field samples of bees, beeswax, honey, honeydew, pollen, propolis and royal jelly. For this purpose, different commercially available beehive products were analyzed.

2. Results and Discussion

2.1. Preliminary Evaluation of Digestion Efficiency

Preliminary tests were carried out in order to evaluate the effects of the reagent mixture, temperature and pressure on sample digestion, which was evaluated in digests by the RCC and residual acidity determination. HNO_3 and H_2O_2 (methods A and B) were preferred to other reagents (HCl, $HClO_4$, or H_2SO_4) because they allow the oxidation of almost all organic compounds and cause minor spectral interferences or problems in ICP-MS [49]. The aqua regia digestion procedures (methods B1 and B2) were chosen for assessing the total recoverable elements in all samples [54–56]. The total elemental content is important information to evaluate and control the quality of honey and other beehive products in terms of contamination. A mixture of aqua regia has been widely used for the digestion of various solid wastes such as ashes, sludge, sediments and soils [54,56]. In addition, HF is effective in extracting silicate-bound elements [54,56]. To improve element recovery and to increase the reaction kinetics, oxidizing agents such as H_2O_2 were added in the digestion procedures [56,57]. The comparison of the results obtained by using the different sample digestion procedures (methods A, B, B1 and B2) allows for the evaluation of the elements that have not been completely recovered in the various considered matrices.

A selected tolerance level of the RCC in solution, lower than 200 or 2000 mg L^{-1}, was considered appropriate for the subsequent analyses by ICP-MS or ICP-OES [58,59]. Preliminary experiments were performed using a fixed amount of 200 mg of the sample to evaluate the minimum HNO_3 amount that was sufficient to obtain suitable values of the RCC in the digests with methods A and B. In methods A and B, 67% HNO_3 was varied in order to achieve an efficient organic matter digestion, using an acid solution with a concentration as low as possible. Thus, digestion using two different 67% HNO_3 amounts (1 or 2 mL) was tested. Using 1 mL HNO_3, final digests presented a yellow color with solid residues remaining as suspended particles. All final digests obtained using 2 mL HNO_3 presented a colorless aspect, except for beeswax, which presented solid residues for both methods A and B. Thus, the RCC in digests (Figure 1) was lower than 231 ± 2 mg L^{-1} in the pollen digests by method A, and 155 ± 12 mg L^{-1} in the propolis digests by method B. The digestion efficiency was high when using both methods B1 and B2 with an RCC lower than 121 ± 8 and 99 ± 21 mg L^{-1} (in the propolis digest), respectively.

The residual acidity was also determined in the final digests obtained from the digestion procedure performed with the HNO_3/H_2O_2 mixture. The residual acidity was in the range 0.429 ± 0.016–0.909 ± 0.007 mol L^{-1} using microwave-assisted digestion, which showed a good digestion efficiency, while the activity was in the range 0.599 ± 0.016–0.829 ± 0.018 mol L^{-1} using water bath-assisted digestion. The results suggested that all the digests obtained using the four treatments were suitable for ICP-OES and ICP-MS analyses, except for pollen digests obtained by method A for ICP-MS analysis. Comparing the element concentrations in the digests obtained with all four digestion treatments, it is possible to verify if there is polyatomic ion interference due to the C concentrations in the digested pollen by method A and analyzed with ICP-MS.

2.2. Selection of ICP Instrument

ICP-MS analysis is generally more susceptible to interferences than ICP-OES. Spectral interference in ICP-MS may be polyatomic and isobaric due to the presence of nonanalyte elements of a similar mass, such as species produced by plasma gas (Ar), atmosphere, nebulizer gas and matrix or combinations thereof [58,60]. Elemental analyses of biological matrices are commonly subjected to interferences caused by major constituents such as C, Ca, Cl, Mg, N, Na, and S [61,62]. The occurrence of spectral effects during the complex sample analysis seriously interferes with the determination of many isotopes, mainly up to 100 amu [60–64]. Therefore, the collision-reaction interface (CRI) mode was used for the determination of As, Ca, Co, Cr, Fe, Mn, Ni, S, and Se in bees and beehive product samples, using H_2 and He as cell gases. The first high ionization potential of S (10.357 eV) leads to a relatively low ionization efficiency in an Ar-based plasma [65]. The 34S isotope was selected for the determination of total S because of its lower spectral interferences [65]. The main drawback of the use of the CRI is the reduction in sensitivity when the collision or reaction gases are employed [65]. The best gas flow rates selected for S were 90 mL min^{-1} H$_2$ and 0 mL min^{-1} He to the skimmer and sampler cone, respectively. For this element, a comparison with ICP-OES was made. As shown in Figure S1, the correlation between the S data obtained with ICP-MS and ICP-OES is not good. Hence, the ICP-OES determination of S was preferable because it not affected by spectral interferences. The best compromise was obtained for As, Ca, Cr, Fe, Mn, and Se with a mixture of 30 mL min^{-1} He and 70 mL min^{-1} H$_2$ to the sampler and skimmer cones, respectively, in agreement with previously reported methods [60]. There were few differences between the standard and CRI mode measurements of 59Co and 60Ni; therefore, these elements were analyzed in the standard mode.

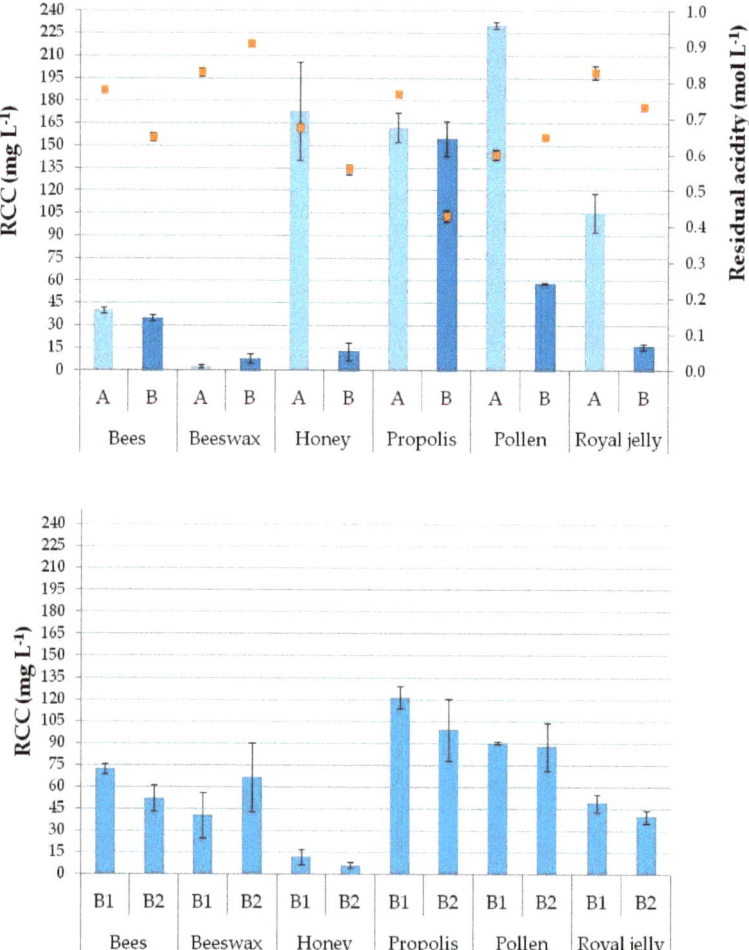

Figure 1. Sample treatment (A = open-vessel digestion heated with water bath, B = closed-vessel microwave-assisted digestion with HNO$_3$/H$_2$O$_2$ mixture, B1 = closed-vessel microwave-assisted digestion with aqua regia/H$_2$O$_2$ mixture, and B2 = closed-vessel microwave-assisted digestion with aqua regia/HF mixture) effect on bees and beehive products digestion efficiency. Bars represent residual carbon content (RCC; left Y axis; $n = 3$) and line (-□-) represents residual acidity (right Y axis; $n = 3$).

The wide differences in the potential interferences and concentration ranges of the elements revealed that the use of both ICP techniques increases the accuracy for some elements. Specifically, the As, Be, Bi, Ce, Ga, Li, Nb, Pb, Sb, Se, Sn, Te, Tl, U, W, and Zr content of most samples were below the limit of detection (LOD) for the ICP-OES analysis (Tables S1–S6); whereas very high concentrations of some macroelements, especially K, caused signal saturation in the ICP-MS analysis and required at least one additional dilution level for the analysis. Elements in the digests obtained by methods A and B were analyzed by both ICP-OES and ICP-MS instruments, while elements in the digests obtained by methods B1 and B2 were analyzed by ICP-OES only, to avoid the effect of interference in ICP-MS due to the presence of HCl or HF.

2.3. Analytical Performances

2.3.1. Linearity

The linearity ranges for ICP-MS and ICP-OES analyses are shown in Table 1 and Table S7, respectively. The obtained correlation coefficients of all calibration curves were >0.99 and a good linearity was confirmed by a Mandel test [66].

2.3.2. Limit of Detection and Quantification

The LOD and limit of quantification (LOQ) for each element are shown in Table 1 and Table S7. LOD and LOQ values were of a similar magnitude to those previously reported [67,68]. With Regulation No. 2015/1005 [69], the European Union fixed a maximum level of 0.10 mg kg^{-1} for Pb in honey. No regulated standards are available to evaluate the other element levels in honey samples; therefore, Codex Alimentarius [70] stated that "honey shall be free from heavy metals in amounts which may represent a hazard to human health". Consequently, there is the need to determine very low concentrations of elements that may be present in honey in trace and ultratrace levels. The tested analytical methods using ICP-MS were sufficiently sensitive to quantify all the selected elements in bees and beehive products, including Pb that has an LOD value (0.001 mg kg^{-1}) 100 times lower than the maximum accepted level for honey [69].

2.3.3. Precision, Trueness and Recovery Study

An apple leaf CRM was used to evaluate the trueness and precision under the repeatability of the tested methods (Tables S8 and S9). This CRM was used because there are no suitable reference materials for bees and beehive products and because it is frequently used [21]. However, the matrix of the apple leaf CRM is a powder and has a chemical and elemental composition different from the bees and beehive products. According to Pohl et al. (2009) [21], the use of reference materials that contain large amounts of C or carbohydrates can be very useful for the validation of the honey analysis method. Certified values of As, Be, Bi, Cs, Ga, Li, Nb, Se, Si, Sn, Te, Ti, Tl, and Zr concentrations were not available for the CRM used so further testing is required for validation. As shown in Table S9, the trueness bias percentage and repeatability obtained by method B1 were estimated in the range from −9.7 (La) to 12.6% (Co) and from 2.6 (S) to 25% (Sb), respectively, and were improved from those obtained by the other tested methods (A, B and B2). The results of As, Cr and Se in the digests obtained by method B1 and analyzed by ICP-MS are higher because they are strongly affected by the interference due to Cl. For this reason, we have chosen to report the data obtained with ICP-OES. To compare the observed results with the certified concentration of the CRM, the Z-scores were calculated [71]. Table S8 shows that the Z-scores of all elements excluding Cd, Rb and S obtained by method B1 were smaller than 2, thus the results are considered acceptable. For Cd and Rb, the results obtained by method B1 were underestimated compared to the certified values (Z-score < −2), whereas they rather tend to be overestimated for S.

Table 1. Limit of determination [a] (mg kg^{-1}) and linearity range for each element in bees and beehive products by inductively coupled plasma spectrometry.

Isotope/Element [b]	Internal Standard [c]	LOD$_A$	LOD$_B$	LOD$_{B1}$	N [e]	LLOQ–ULOQ [d] mg kg^{-1} Honey and Beeswax	N [e]	Bees, Pollen, Propolis, and Royal Jelly
27Al	45Sc	0.06	0.02	0.02	4	0.5–5	5	1.05–21
75As [f]	79Y	0.001	0.001	-	6	0.05–2	6	0.05–2
11B	45Sc	0.2	0.1	0.1	5	2.75–55	5	2.75–55
137Ba	115In	0.4	0.2	0.2	5	0.2–5	4	7–100
9Be	45Sc	0.00002	0.00002	0.00002	7	0.05–5	7	0.05–5
209Bi	232Th	0.00004	0.00006	0.00005	6	0.05–2	6	0.05–2
44Ca	79Y	11	5	14	4	100–1000	4	1000–10000
112Cd	115In	0.00004	0.00001	0.00001	7	0.05–110	7	0.05–110
140Ce	115In	0.00006	0.00004	0.0001	5	0.25–5	5	0.25–5
59Co	45Sc	0.0005	0.0004	0.0008	6	0.25–10	6	0.25–10
52Cr[f]	79Y	0.001	0.0008	-	7	0.05–5	4	1–10
133Cs	115In	0.00004	0.00002	0.00001	7	0.05–5	7	0.05–5
65Cu	79Y	0.003	0.002	0.002	6	0.05–2	9	0.1–100
56Fe	79Y	0.02	0.01	0.01	4	1–10	9	2.5–5000
71Ga	79Y	0.00005	0.00001	0.00001	4	0.1–1	4	0.1–1
39K	45Sc	5	3	0.7	6	50–2500	6	50–2500
139La	115In	0.00004	0.00006	0.00005	7	0.05–5	7	0.05–5
7Li	45Sc	0.0008	0.0008	0.0005	7	0.05–5	7	0.05–5
24Mg	45Sc	0.9	0.3	1.0	5	100–2500	5	50–1000
55Mn	79Y	0.002	0.002	0.003	4	0.2–2	6	5.2–200
98Mo	103Rh	0.0003	0.0001	0.0001	7	0.05–110	7	0.05–110
23Na	45Sc	0.6	0.4	0.3	5	100–2500	7	25–2000
93Nb	103Rh	0.00001	0.000007	0.00002	5	0.05–1	5	0.05–1
60Ni	45Sc	0.002	0.003	0.003	6	0.05–2	7	0.2–20
31P	45Sc	0.7	1	0.7	5	5–100	4	1100–5000
208Pb	232Th	0.001	0.001	0.001	7	0.05–5	4	7–100
85Rb	79Y	0.0003	0.0003	0.00007	5	0.05–1	4	20–200
121Sb	115In	0.0004	0.0002	0.0001	7	0.05–5	7	0.05–5
76Se	79Y	0.02	0.007	0.01	4	0.2–2	4	0.2–2
28Si	45Sc	9	8	10	4	10–100	4	50–500
118Sn	115In	0.0002	0.0001	0.0001	7	0.05–5	7	0.05–5
88Sr	79Y	0.008	0.02	0.02	5	5.5–110	5	5.5–110
125Te	115In	0.0003	0.0003	0.0002	5	0.05–1	5	0.05–1
49Ti	45Sc	0.002	0.003	0.0007	9	0.05–50	9	0.05–50
205Tl	232Th	0.00006	0.00003	0.00001	5	0.05–1	5	0.05–1
238U	232Th	0.00001	0.00001	0.00001	7	0.05–5	7	0.05–5
51V	79Y	0.0003	0.0006	0.00003	6	0.1–5	4	1–10
182W	232Th	0.0003	0.0002	0.00006	7	0.05–5	7	0.05–5
66Zn	79Y	0.09	0.04	0.09	5	2–50	4	70–1000
90Zr	79Y	0.0001	0.00009	0.0001	6	0.05–2	6	0.05–2

[a] LOD$_A$, LOD$_B$, and LOD$_{B1}$ are the limits of determination for methods A, B, and B1, respectively. [b] Isotopes are reported for all the analyzed elements by inductively coupled plasma mass spectrometry (ICP-MS). [c] Rh replaces Y for propolis samples. [d] LLOQ, lower limit of quantification; ULOQ upper limit of quantification. [e] N, number of calibration points for honey and beeswax or bees, pollen, propolis and royal jelly. [f] As and Cr in digests obtained by method B1 are determined with inductively coupled plasma optical emission spectrometry (ICP-OES).

With the lack of a suitable CRM, spiked samples were also used to determine the elemental recoveries by methods A and B, in accordance with previous studies [35]. It is worth mentioning that the study of recoveries does not allow for the assessment of the efficiency of the digestion procedure in decomposing the sample matrix but only enables the evaluation of matrix effects and losses or increases in a concentration of elements compared to the added amounts. All six sample matrices were spiked with two concentrations (the third and fifth instrumental calibration standards) before sample digestion. Recoveries for all elements fell within 20% of the expected value, with many of the elements recovering within 10%, excluding Al, Ca, Ce and Zn in beeswax, Ce in royal jelly, Ba, Cs, Ga, K, Mn, Na, Nb, P, Ti and Zn in bees, and Al, As, Si, Ti and Tl in propolis by both methods A and B, which fell within 30% (Table 2). The within-run precision for all the elements in honey and royal jelly and for most of the elements in other matrices was less than 10%. The intermediate precision was less than 15% for most of the elements in all matrices excluding As and Se in the digests by method A, and As, Ba, Be, Bi, Cd, Cr, Pb and Te in the digests by method B (Table 2).

Table 2. Summary of precision (repeatability as percent coefficient of variation intrarun (%CVr) and reproducibility as %CVR inter-run), and percent recovery (%R) ranges for each element in working bee and beehive products applying two different digestion methods prior to analysis by ICP-MS and ICP-OES (for S).

Isotope/Element [a]	Method A			Method B		
	%CVr Intraday (n = 3)	%CVR Interday (n = 9)	%R (n = 3)	%CVr Intraday (n = 3)	%CVR Interday (n = 9)	%R (n = 3)
^{27}Al	1.8–24	1.5–25	96–126	5.4–27	15–25	97–124
^{75}As	1.8–21	17–27	88–124	5.0–29	17–30	90–119
^{11}B	0.2–24	6.5–19	84–102	2.2–32	7.6–25	86–100
^{137}Ba	0.6–23	11–25	103–122	1.2–21	17–30	101–112
^{9}Be	0.9–13	8.2–24	86–97	14–22	21–30	90–101
^{209}Bi	1.8–17	11–21	85–107	3.1–25	19–31	86–110
^{44}Ca	1.4–12	7.2–25	79–127	5.7–26	14–26	82–127
^{112}Cd	0.7–19	9.1–25	83–100	1.5–22	19–30	83–104
^{140}Ce	1.1–19	8.7–26	96–123	1.0–25	10–30	98–119
^{59}Co	0.8–13	1.1–25	81–105	0.4–27	9.2–27	84–102
^{52}Cr	0.7–19	2.0–22	91–101	8.9–26	20–31	94–107
^{133}Cs	0.7–6.4	9.1–25	99–123	3.3–14	3.3–30	99–119
^{65}Cu	1.3–23	4.6–23	82–121	1.1–18	5.4–28	87–111
^{56}Fe	0.5–7.7	3.6–13	88–120	0.9–25	5.9–25	92–116
^{71}Ga	1.2–23	13–24	83–130	1.1–25	7.4–31	86–130
^{39}K	0.9–22	1.3–21	87–124	0.8–10	6.7–13	87–121
^{139}La	1.2–10	5.0–24	90–114	0.5–24	8.1–29	92–114
^{7}Li	0.2–11	3.8–24	90–118	1.5–23	14–21	90–120
^{24}Mg	0.01–9.0	5.2–25	82–108	0.5–19	7.0–23	82–110
^{55}Mn	0.4–17	12–25	84–125	0.7–24	3.7–24	86–115
^{98}Mo	0.3–6	10–25	95–117	1.9–25	1.9–25	94–114
^{23}Na	0.6–6.7	4.9–21	80–125	2.2–17	6.5–26	84–122
^{93}Nb	1.5–20	8.0–24	85–122	3.6–22	11–27	86–116
^{60}Ni	0.9–17	13–22	83–97	0.4–23	8.6–29	82–99
^{31}P	0.5–11	4.2–20	80–122	0.5–23	5.7–25	81–123
^{208}Pb	1.6–23	12–24	80–121	2.7–24	16–27	86–116
^{85}Rb	3.6–11	4.5–25	96–120	0.8–23	7.3–30	92–117
S	7.5–9.1	6.8–11	96–110	1.7–10	4.1–11	97–111
^{121}Sb	0.2–17	11–26	81–100	13–23	15–27	84–98
^{76}Se	0.4–18	16–26	83–108	3.4–24	13–30	82–102
^{28}Si	1.4–15	4.2–26	101–126	1.8–21	4.1–28	98–122
^{118}Sn	2.0–14	5.7–26	81–101	2.7–23	15–31	86–101
^{88}Sr	1.3–15	0.6–25	89–121	0.5–24	15–34	84–119
^{125}Te	0.7–15	7.0–26	81–89	3.0–17	27–30	81–90
^{49}Ti	1.2–22	6.8–18	95–128	0.2–19	3.5–30	96–118
^{205}Tl	1.6–19	13–20	94–124	1.0–19	11–24	96–120
^{238}U	1.7–21	9.8–29	87–110	1.9–21	15–26	87–110
^{51}V	0.3–19	9.4–25	83–111	1.2–22	10–30	86–106
^{182}W	1.8–7	19–26	84–113	0.6–15	11–30	84–116
^{66}Zn	0.3–24	5.2–25	88–126	1.3–21	10–29	86–124
^{90}Zr	1.2–21	11–25	90–112	2.3–17	13–31	92–114

[a] Isotopes were reported for all the analyzed elements by ICP-MS. Sulfur was determined by ICP-OES.

2.3.4. Mixture Reagent Digestion

The method's accuracy is very important for biomonitoring studies to control the quality of beehive products and to compare the results obtained from different samples. The variation of the elemental concentrations for each matrix must depend only on the variations of the environmental contamination and not on the selected method. The accuracy of the results allows for the assessment of the contamination in each matrix. For this purpose, a single homogeneous sample for each considered matrix was analyzed with the different digestion methods in order to estimate recovery and precision. Comparing the native data in the matrices (Figures 2–4) normalized with respect to method B, the greatest differences observed between the results of Al, B, Ba, K, Mn, P, S, Ti, V, and Zn in bees; Al, Cr, Fe, Mg, and Ti in beeswax, Al, B, Ba, K, and Mg in honey, Al, B, Ba, Cr, Na, P, Si, Ti, and Zn in pollen, Al, B, Ba, Be, Ca, Cd, Ce, Co, Cr, Cu, Li, Mn, P, Pb, S, Si, Ti, V, Zn, and Zr in propolis, and Al, Na, and S in royal jelly were due to the use of different digestion mixtures and not to the use of different digestion systems. Se, and Te in bees, B, Ba, Bi, Ga, Nb, Ni, Se, Si, Te, V, and W in beeswax, Be, Bi, Cd, Nb, Ni, Pb, Se, Te, V, and W in honey, Bi, Se, Te, and W in pollen, As, Ba, Bi, Cd, Ga, Nb, Ni, Pb, Te, V, and W in royal jelly were not considered since they were always <LOD. Considering all the methods with the use of microwave oven, the reagent mixtures used in methods B1 and B2 allow a greater extraction of the elements in the different matrices compared to the method B. The extraction capacity of the digestion reagent mixtures depends on the chemical compounds present in the various analyzed matrices. The higher temperatures and pressure that occur with the use of the microwave oven compared to the water bath affected the content of Ba, Be, Bi, Ga, Mn, Nb, P, Pb, S, Sn, and Sr in bees, Al, Be, Cd, Co, Cr, Cu, Fe, Li, Rb, Sb, Ti, Tl, and Zr in beeswax, As, Ba, Cr, Li, Mo, and Zr in honey, Al, As, Be, Ce, Cr, La, Li, Rb, Ti, U, V, and Zn in pollen, Al, Ba, Ga, Li, Se, Si, Sn, Ti, V, W, and Zr in propolis, and Al, Cr, Cs, Sb, Si, and U in royal jelly. Tables S1–S6 show that methods A and B gave similar results for all the elements excluding Ga in bees, La, Mo, and Ti in honey, Be, Ce, La, Li, U, and V in pollen, Se in propolis, and Cr and Si in royal jelly. The total content of the following elements in the selected matrices can only be obtained by using aqua regia mixtures: Al, B, Ba, Cr, P, and S in bees; Ba, P, and Ti in beeswax; Al, and Ba in honey; Al, B, Ba, Cr, Cu, Fe, Na, S, Si, and Ti in pollen; Al, B, Ba, Be, Ca, Cd, Co, Cr, Li, Mg, Si, Sn, Ti, V, Zn, and Zr in propolis; Na, and P in royal jelly. The improvements and optimal recoveries of Al, Ba, Fe, and Sb from a variety of matrices upon the addition of HCl have been demonstrated [56,72]. However, treatment with HNO_3 favors Cl elimination as nitrosyl chloride and minimizes the isobaric interferences in the case of some elements (As, Cr, Fe, Mn, Ni, Se and V) analyzed by ICP-MS [60,73].

In agreement with other studies [35], the procedure A is a good alternative to the procedure B because it limits sample manipulation, does not require the cleaning of sample vessels between analyses and allows 120 samples to be processed in 30 min. However, the content of some elements in specific matrices can be underestimated using the reactive digestion mixture HNO_3/H_2O_2. The data reported in the present study aim to be a valuable help for choosing the most appropriated methods and analytical techniques for the determination of each element in bees and beehive products.

2.4. Analysis of Commercial Beehive Products

Commercial beehive samples were analyzed in order to demonstrate the applicability of the optimized methods. Concentrations were above the LODs for most of the elements, including Pb, showing that both methods A (Table 3) and B (Tables S10 and S11) can be used to determine the elemental composition of beehive products. Both methods A and B provided similar concentration data for the elements above the LODs. However, the average levels of some elements (La, Mo, and Ti in honey; Be, Ce, La, Li, U, and V in pollen; Se in propolis; Cr and Si in royal jelly) obtained with method B were higher. Methods A and B allow the elemental characterization of bees and beehive products, except for some elements that were underestimated with respect to the data acquired with methods B1 and B2, as reported in Section 2.3.4.

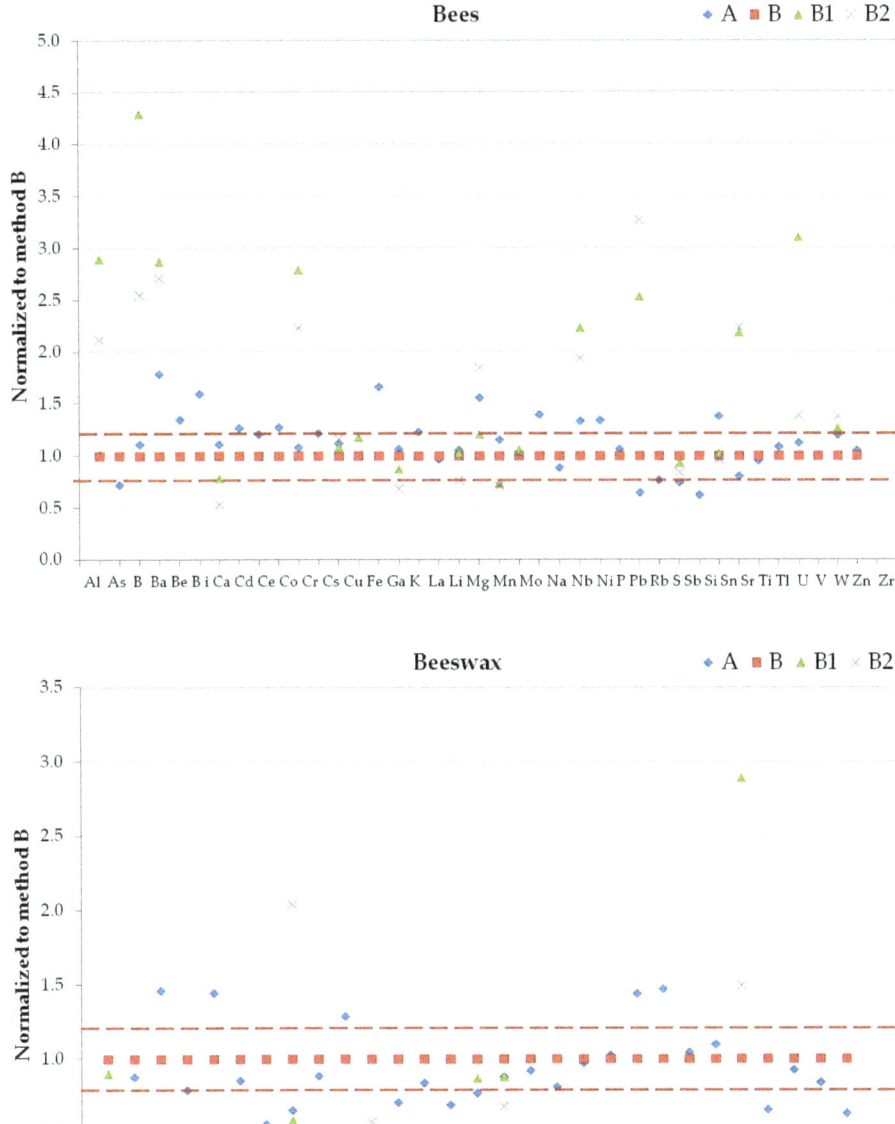

Figure 2. Normalized concentrations to results obtained by method B1 of detected elements in bees (top panel) and beeswax (bottom panel) samples by methods A, B, B1 and B2. Dashed lines are ±20%.

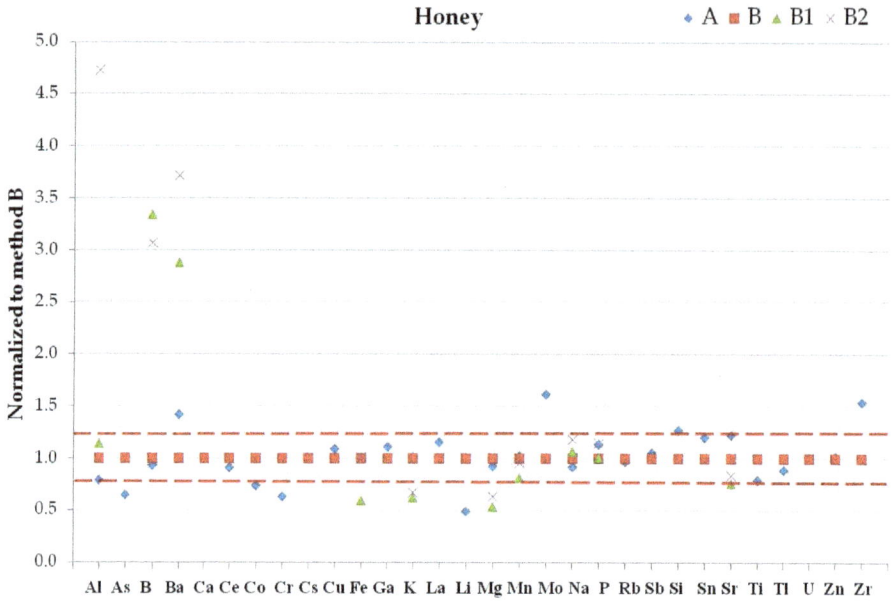

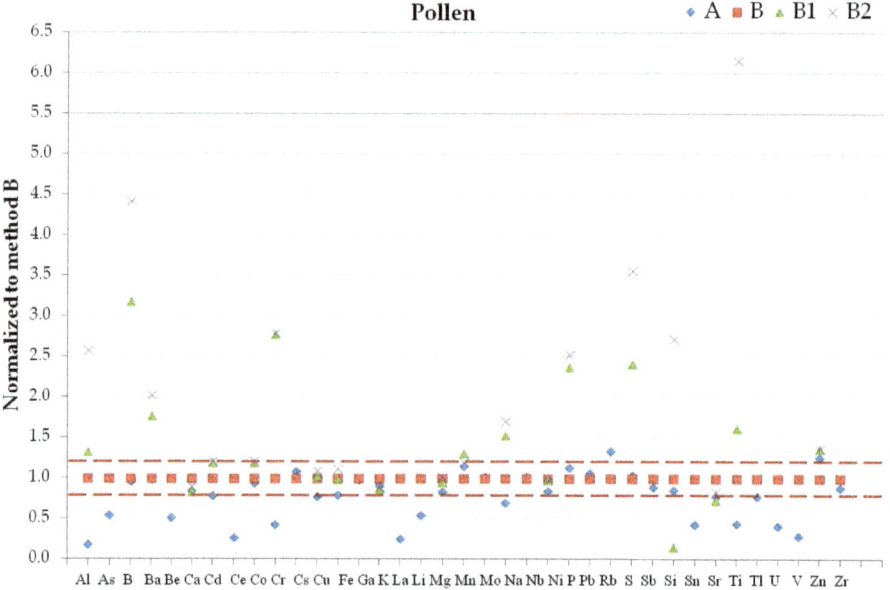

Figure 3. Normalized concentrations to results obtained by method B1 of detected elements in honey (top panel) and pollen (bottom panel) samples by methods A, B, B1 and B2. Dashed lines are ±20%.

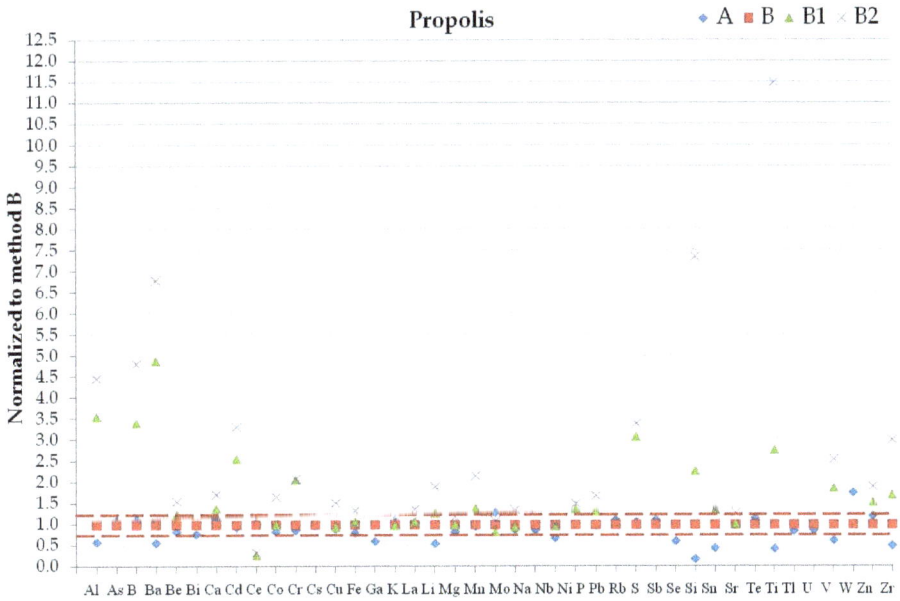
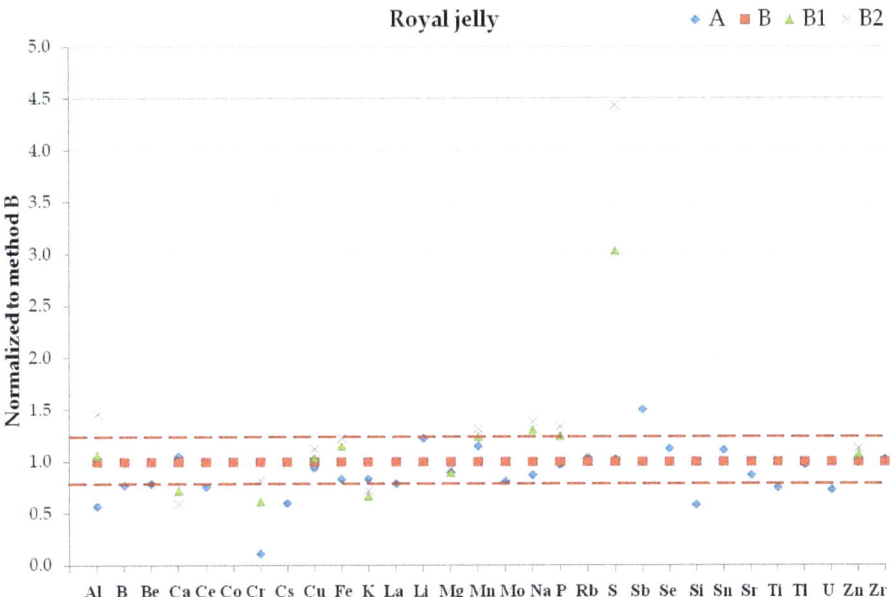

Figure 4. Normalized concentrations to results obtained by method B1 of detected elements in propolis (top panel) and royal jelly (bottom panel) samples by methods A, B, B1 and B2. Dashed lines are ±20%.

Table 3. Concentrations (mg Kg^{-1}) of each element in some commercial apiary products obtained by method A.

Element	Honey 1		Honey 2		Honey 3		Honey 4		Honey 5		Honeydew 1		Honeydew 2		Beeswax		Pollen		Royal Jelly	
	M	SD	M	SD	M	SD	M	SD	M	SD	M	SD	M	SD	M	SD	M	SD	M	SD
Al	0.09	0.02	0.187	0.078	6.33	0.43	2.48	0.22	0.153	0.027	0.439	0.035	1.65	0.03	<LOD	-	3.68	0.29	0.054	0.025
As	<LOD	-	<LOD	-	<LOD	-	<LOD	-	<LOD	-	<LOD	-	<LOD	-	<LOD	-	0.012	0.011	<LOD	-
B	6.56	0.32	0.697	0.045	3.60	0.25	3.99	0.03	7.75	0.16	4.54	0.10	7.33	0.46	<LOD	-	5.73	0.67	1.37	0.16
Ba	0.083	0.007	<LOD	-	1.93	0.08	0.627	0.043	0.0522	0.0053	0.113	0.011	0.891	0.061	<LOD	-	2.15	0.42	<LOD	-
Be	<LOD	-	<LOD	-	0.00091	0.00016	0.00117	0.00019	<LOD	-	<LOD	-	0.00183	0.00032	<LOD	-	0.00052	0.00037	<LOD	-
Bi	<LOD	-	<LOD	-	0.000516	0.000042	<LOD	-	<LOD	-	0.00057	0.00013	0.000270	0.000069	<LOD	-	<LOD	-	<LOD	-
Ca	91	6	38.6	8.7	181	24	154	21	51.1	6.8	50	12	118	8	<LOD	-	547	103	165	43
Cd	0.000218	0.000058	<LOD	-	0.00104	0.00034	<LOD	-	<LOD	-	0.00115	0.00022	0.000087	0.00010	<LOD	-	0.0642	0.0063	<LOD	-
Ce	0.00117	0.00015	0.000372	0.000087	0.0145	0.0012	0.0187	0.0012	0.00096	0.00025	0.00234	0.00017	0.0419	0.0018	0.00098	0.00023	0.0055	0.0026	0.000850	0.0002
Co	0.00630	0.00007	0.00130	0.00031	0.0107	0.0005	0.0107	0.0005	0.00174	0.00030	0.0176	0.0008	0.0241	0.0020	<LOD	-	0.158	0.017	0.00123	0.00023
Cr	0.00561	0.00074	0.0063	0.0017	0.0123	0.0016	0.0127	0.0033	<LOD	-	0.0146	0.0022	0.0587	0.0055	0.0204	0.0060	0.0547	0.0061	0.0340	0.0026
Cs	0.00443	0.00034	0.00062	0.00010	0.787	0.045	0.565	0.029	0.00245	0.00065	0.00365	0.00013	0.547	0.037	0.000292	0.000091	0.0142	0.0034	0.00063	0.00013
Cu	0.261	0.013	0.0623	0.0073	0.747	0.039	1.01	0.02	0.153	0.010	1.97	0.02	3.30	0.14	<LOD	-	5.82	0.89	4.26	0.41
Fe	2.93	0.41	1.81	0.19	2.68	0.31	4.28	0.96	0.58	0.13	2.54	0.25	4.50	0.35	0.31	0.15	24.0	8.3	9.23	0.75
Ga	0.00247	0.00035	0.00240	0.00029	0.042	0.011	0.0140	0.0012	0.00173	0.00035	0.00224	0.00048	0.0201	0.0025	<LOD	-	0.0436	0.0072	<LOD	-
K	1340	24	89	2	4661	341	3708	36	732	8	6520	46	2399	157	<LOD	-	3703	592	2520	260

3. Materials and Methods

3.1. Instrumentation

A Bruker 820-MS quadrupole ICP-MS spectrometer (Bremen, Germany) equipped with a collision-reaction interface (CRI) and an Analytik Jena AG MicroMistTM glass nebulizer (0.4 mL min^{-1}; Jena, Germany) was used for all the measurements. A radiofrequency power of 1.4 kW, plasma gas flow rate of 18.0 L min^{-1}, auxiliary gas flow rate of 1.8 L min^{-1} and nebulizer gas flow rate of 1.0 L min^{-1} were used. The monitored isotopes (m/z) are shown in Table 1. The CRI with He and H_2 (99.9995% purity; SOL Spa, Monza, Italy) as cell gases was used to quantify and remove polyatomic- and argon-based interference for As, Cr, Fe, Mn, Se, and V.

A Varian Vista MPX CCD Simultaneous ICP-OES spectrometer (Victoria, Mulgrave, Australia) in an axial configuration equipped with inert components (demountable torch with alumina injector, 1.8 mm, and PTFE injector holder; Sturman-Masters inert spray chamber, double pass, white Ertalyte; Agilent, Santa Clara, CA, United States) was used to determine the residual C content (RCC) of the final digests and all the selected elements. A radiofrequency power of 1.0 kW, plasma gas flow rate of 15.0 L min^{-1}, auxiliary gas flow rate of 1.5 L min^{-1}, and nebulizer gas flow rate of 0.75 L min^{-1} were used as operational conditions. Elements were detected at the wavelength that maximized the signal intensity and minimized spectral overlaps (Table 1)

Ar gas (99.9995% purity; SOL Spa, Monza, Italy) was used for plasma generation.

Analytical reagent-grade water with a resistivity of 18.2 MΩ cm was obtained with an Arioso Power I RO-UP Scholar UV water purification system from Human Corporation (Songpa-Ku, Seoul, Korea).

An Argo Lab WB12 water bath (Modena, Italy) with an electronic temperature control was used for open-vessel digestion (method A), as described in Section 3.3.1. A maximum temperature and pressure up to 95 °C (± 0.2 °C) and ~1 bar, respectively, were used for this system.

A Milestone Ethos1 Touch Control microwave system (Sorisole, Bergamo, Italy) equipped with six PTFE or 20 quartz vessels was used for closed-vessel microwave digestion (methods B, B1 and B2), as described in Section 3.3.2. The vessels were irradiated with a maximum power of 1000 W and all the experiments were carried out at the maximum temperature (180 °C) and pressures of ≤40 bar.

A Heto Power Dry LL1500 freeze dryer from Thermo Electron Corporation (Waltham, MA, USA) was employed for drying bee samples.

3.2. Reagents

Yttrium at 0.2 mg L^{-1} was used as the internal standard for ICP-OES and it was prepared from a standard stock solution (1000 ± 2 mg L^{-1}; Panreac Química, Barcelona, Spain) [58]. Yttrium, Sc, Rh, In, and Th (1000 ± 5 mg L^{-1}; Merck KGaA, Darmstadt, Germany) at 0.01 mg L^{-1} in a 1% (v/v) HNO_3 multistandard solution were employed as internal standards for ICP-MS [58,74].

HNO_3 (67–70%; super pure) from Carlo Erba Reagents S.r.l. (Milan, Italy), HCl (assay >36%; residue <3 mg L^{-1}), HF (assay >40%; residue <2 mg L^{-1}) and H_2O_2 (assay >30%) from Promochem, LGC Standards GmbH (Wesel, Germany) were used to prepare the standard and sample solution. All the reagents used were of analytical grade.

An apple leaf NIST 1515 was employed to evaluate the accuracy of the methods. The certified standard material was purchased from the National Institute of Standards and Technology (Gaithersburg, MD, USA).

ICP multielemental standard solutions of As, Al, Ba, Be, Bi, Cd, Cr, Cs, Cu, Ga, La, Li, Mn, Mo, Nb, Ni, Pb, Rb, Sb, Se, Sn, Te, Ti, Tl, U, V, W, and Zr at 1.000 ± 0.005 mg L^{-1}, Ce and Co at 5.00 ± 0.03 mg L^{-1}, Fe and Zn at 10.00 ± 0.05 mg L^{-1}, P and Si at 50.00 ± 0.25 mg L^{-1}, B and Sr at 55.00 ± 0.25 mg L^{-1}, K, Mg, and Na at 500.0 ± 2.5 mg L^{-1}, and Ca at 1000 ± 5 mg L^{-1} in 3% (v/v) HNO_3, from Ultra Scientific/Agilent Technologies (North Kingstown, RI, USA), were used for the calibration procedure and spiked samples.

A multielemental standard solution of Ba, Be, Ce, Co, In, Pb, Mg, Tl, and Th (10.00 ± 0.05 mg L^{-1} in 2% HNO$_3$) from Spectro Pure, Ricca Chemical Company (Arlington, TX, USA) was employed in order to select the best operating parameters for the ICP-MS analysis.

For the RCC assessment in the digested samples, a reference solution of 10,000 mg L^{-1} in C was prepared from anhydrous citric acid (assay >99.5%, ACS reagent; Sigma-Aldrich Chemie GmbH, Steinheim, Germany) in boiled deionized water, according to Muller et al., 2015 [49]. A standardized NaOH solution (0.5 mol L^{-1}; assay >98% sodium hydroxide anhydrous pellets, RPE for analysis, ACS and ISO; Carlo Erba Reagents, Milan, Italy) was prepared for the determination of residual acidity in final digested samples by acid–base titration.

3.3. Sample Preparation Methods

Ten samples of commercially available beehive products were collected from local supermarkets in Rome (Central Italy). The set of samples comprised several brands and consisted of five multifloral honeys, two honeydews, one pollen, one royal jelly and one beeswax cream.

The bee samples used for the recovery experiments were collected directly from the hive. At least 10 bees per sample with a wet weight of ~2 g were brought to a stable dry weight after freeze-drying for 48 h and were finely grounded in a glass mortar. Specific portions of honeybees (30, 50, 100 and 200 mg) were analyzed to assess whether the weight of honeybees samples might be reduced. A mass of 200 mg was selected for the subsequent analyses.

A volume of 1 or 2 mL of concentrated HNO$_3$ for each digestion method was selected considering the maximum volume of the digestion vessels and the minimum dilution of the samples, in order to have a final acidity of <5%, as recommended by the ICP-MS manual. The bee samples with mass of 200 mg and a reagent mixture ratio 1:2 of H$_2$O$_2$ and HNO$_3$ were kept at maximum constant temperature of 95 °C for the water bath and 180 °C for the microwave oven, in accordance with previous studies [15,35,51]. Two commonly used digestion procedures with strong reagent mixtures [54–56], microwave aqua regia + H$_2$O$_2$ and microwave aqua regia + HF, were used for the total digestion of elements in bees and beehive products samples. The sample analyses were carried out in duplicate. Certified reference material (NIST SRM 1515; three replicates) and blank digests (ten replicates) were subjected to the same sample preparation. All tested analytical procedures are described below.

3.3.1. Open-Vessel Water Bath-Assisted Digestion

A mass of ~200 mg for all samples was measured directly in an autosampler tube, 1 mL of 67% HNO$_3$ and 0.5 mL of 30% H$_2$O$_2$ were then added in the same tube. Subsequently, tubes were heated to 95 ± 5 °C in a water bath for 30 min (method A). After digestion, the mixture was left to cool and the contents of the tubes were diluted to 10 or 20 mL with deionized water for ICP-OES or ICP-MS analyses, respectively.

3.3.2. Closed-Vessel Microwave-Assisted Digestion

Weighed amounts (~200 mg) of all samples were transferred into the microwave vessels. Then, 1 mL 67% HNO$_3$, 0.5 mL 30% H$_2$O$_2$, and 1.5 mL of deionized water (method B) were added to the quartz vessels, and 1 mL 30% H$_2$O$_2$ and 4 mL aqua regia (method B1) or 1 mL 40% HF and 4 mL aqua regia (method B2) were added to the PTFE vessels. Subsequently, vessels were heated to 180 °C with microwave energy (at a power of 1000 W) for 40 min. Cooled digests were transferred to the autosampler tubes. All digests by methods B and B1 were diluted to 10 or 20 mL with deionized water for the ICP-OES or ICP-MS analyses, respectively. The digests obtained with method B2 were diluted to a volume of 10 mL with deionized water for ICP-OES analysis.

3.4. Quality Assurance and Control

For the method validation, the selectivity, linearity, accuracy, LOD and LOQ were tested. According to the Eurachem Guide [75], accuracy, the closeness of agreement between a test result and the accepted reference value, is a parameter described by two contributions: the trueness bias or spike recovery and the precision. The accuracy was studied using a CRM and spiked samples for all the digestion methods or only for methods A and B, respectively.

The comparison between the obtained results and the certified values of the CRM was carried out using Z-scores [71]. These were calculated according to the following formula:

$$(\overline{X}_{found} - \overline{X}_{certified}/(SD/\sqrt{n}) \tag{1}$$

where $\overline{X}_{found}$ is the result found by the analyst, $\overline{X}_{certified}$ is the certified value, SD is the standard deviation, and n is the number of independent replicates. Z-scores smaller than 2 are usually considered acceptable, Z-scores between 2 and 3 are questionable, and Z-scores larger than 3 are not satisfactory.

For recovery experiments, a spike solution was added to the samples at two levels before the sample digestion. Each digestion batch contained a reagent blank (matrix samples) to allow background correction.

The LODs were calculated with three times the relative standard deviation percentage (RSD%) of ten method blanks multiplied by the background equivalent concentration (BEC)/100 [58], and the dilution factor used for sample preparation (Table S7). The LOQs were the lowest standard curve points that could be used for quantification (LLOQs). Together, the LLOQ and upper LOQ (ULOQ) define the linearity range.

At regular intervals (every 20 samples) during all analyses, an intermediate calibration standard was analyzed as a sample to monitor the instrument drift. A maximum percentage drift of ± 10% was considered acceptable for all the elements. Furthermore, calibration blanks (3% HNO_3) were frequently analyzed alongside samples to check any loss or cross contamination. Blanks were prepared by performing the full analytical procedure without samples.

The matrix effects on the sample uptake and nebulization were monitored by an internal standardization with Y for ICP-OES, and Y, Sc, Rh, In and Th for ICP-MS, and measurements were automatically corrected by the respective ICP software. Yttrium was not used as an internal standard for the ICP-MS analyses of the propolis samples because it is contained in the propolis in a concentration of ~0.25 mg kg^{-1} (~0.0025 mg L^{-1} in digests). This concentration was negligible for ICP-OES analyses because Y was added to the samples as an internal standard in a concentration of 0.2 mg L^{-1}; therefore, in this case, Y was used as the internal standard for the propolis samples.

3.5. Statistical Analysis

All statistical calculations were made by the SPSS software package (IBM SPSS Statistics 25 software; IBM Corp., Armonk, NY, USA). Values < LOD were designated as LOD/2 [76]. The studies of significant differences were carried out by Mann–Whitney U or Kruskal–Wallis tests with pairwise post-hoc tests [77]. The difference in the results was considered statistically significant for p-values < 0.05.

4. Conclusions

Unfortunately, the validation of methods and procedures used for the analysis of bees and beehive products is difficult because no CRMs of these matrices are available. Instead, the trueness of the methods and procedures applied is commonly checked by the analysis of different CRMs or by the recovery tests with samples spiked with known amounts of elements. It is worth noting that the study of elemental recovery in bees and beehive products may not allow for the evaluation of the efficiency of digestion procedures in decomposing the sample matrix. This in fact may not have the ability to readily absorb aqueous solutions containing known amounts of elements that should be released with only a complete digestion of the sample. In the present study, it is highlighted how each element in a specific matrix responds differently to the different sample treatment procedures used, thus allowing one to choose a specific method accordingly for the needs and purpose of the analysis. In screening analyses and biomonitoring studies, method A was a faster and a good alternative compared to microwave-assisted acid digestion for the determination of all the analyzed elements. Both the methods with mixture HNO_3/H_2O_2 (methods A and B) showed an acceptable accuracy for all the analyzed elements, and low levels of detection for trace elements including Pb. Considering all the tested methods, to have total levels of some elements (such as Al, B, Ba, Cr, P, and S in bees; Ba, P, and Ti in beeswax; Al, and Ba in honey; Al, B, Ba, Cr, Cu, Fe, Na, S, Si, and Ti in pollen; Al, B, Ba, Be, Ca, Cd, Co, Cr, Li, Mg, Si, Sn, Ti, V, Zn, and Zr in propolis; Na, and P in royal jelly) is necessary to use an aqua regia mixture. This is very important for evaluating the quality of products and for preserving human health and the environment.

Supplementary Materials: The following are available online, Figure S1, Tables S1–S11.

Author Contributions: Conceptualization, M.L.A.; validation, M.L.A.; formal analysis, M.L.A.; investigation, M.L.A. and E.M.; resources, S.C. and M.E.C.; data curation, M.L.A.; writing—original draft preparation, M.L.A.; writing—review and editing, M.L.A., S.C., M.E.C. and L.M.; visualization, M.L.A. and E.M.; supervision, M.L.A. All authors have read and agreed to the published version of the manuscript.

Funding: This research received no external funding.

Acknowledgments: The authors wish to thank Massimo Marcolini and Marco Papi (Associazione Apicoltori Roma e Provincia) for their kind support in the bee sample collection.

Conflicts of Interest: The authors declare no conflict of interest.

References

1. Massimi, L.; Ristorini, M.; Astolfi, M.L.; Perrino, C.; Canepari, S. High resolution spatial mapping of element concentrations in PM10: A powerful tool for localization of emission sources. *Atmos. Res.* **2020**, *244*, 105060. [CrossRef]
2. Manigrasso, M.; Protano, C.; Astolfi, M.L.; Massimi, L.; Avino, P.; Vitali, M.; Canepari, S. Evidences of copper nanoparticle exposure in indoor environments: Long-term assessment, high-resolution field emission scanning electron microscopy evaluation, in silico respiratory dosimetry study and possible health implications. *Sci. Total Environ.* **2019**, *653*, 1192–1203. [CrossRef]
3. Canepari, S.; Castellano, P.; Astolfi, M.L.; Materazzi, S.; Ferrante, R.; Fiorini, D.; Curini, R. Release of particles, organic compounds, and metals from crumb rubber used in synthetic turf under chemical and physical stress. *Environ. Sci. Pollut. Res.* **2018**, *25*, 1448–1459. [CrossRef]
4. Canepari, S.; Astolfi, M.L.; Marcovecchio, F.; Maretto, M.; Perrino, C. Seasonal variations in the concentration and solubility of elements in atmospheric particulate matter: A case study in Northern Italy, E3S Web of Conferences. *EDP Sci.* **2013**, *1*, 20002. [CrossRef]
5. Marconi, E.; Canepari, S.; Astolfi, M.L.; Perrino, C. Determination of Sb (III), Sb (V) and identification of Sb-containing nanoparticles in airborne particulate matter. *Procedia Environ. Sci.* **2011**, *4*, 209–217. [CrossRef]
6. Astolfi, M.L.; Canepari, S.; Catrambone, M.; Perrino, C.; Pietrodangelo, A. Improved characterisation of inorganic components in airborne particulate matter. *Environ. Chem. Lett.* **2006**, *3*, 186–191. [CrossRef]
7. Al-Alam, J.; Chbani, A.; Faljoun, Z.; Millet, M. The use of vegetation, bees, and snails as important tools for the biomonitoring of atmospheric pollution—A review. *Environ. Sci. Pollut. Res.* **2019**, *26*, 9391–9408. [CrossRef]
8. Protano, C.; Canepari, S.; Astolfi, M.L.; de Meo, S.D.O.; Vitali, M. Urinary reference ranges and exposure profile for lithium among an Italian paediatric population. *Sci. Total Environ.* **2018**, *619*, 58–64. [CrossRef]
9. Astolfi, M.L.; Protano, C.; Marconi, E.; Massimi, L.; Piamonti, D.; Brunori, M.; Vitali, M.; Canepari, S. Biomonitoring of Mercury in Hair among a Group of Eritreans (Africa). *Int. J.Environ. Res. Public Health* **2020**, *17*, 1911. [CrossRef]
10. Astolfi, M.L.; Protano, C.; Schiavi, E.; Marconi, E.; Capobianco, D.; Massimi, L.; Ristorini, M.; Baldassarre, M.E.; Laforgia, N.; Vitali, M.; et al. A prophylactic multi-strain probiotic treatment to reduce the absorption of toxic elements: In-vitro study and biomonitoring of breast milk and infant stools. *Environ. Int.* **2019**, *130*, 104818. [CrossRef]
11. Ristorini, M.; Astolfi, M.L.; Frezzini, M.A.; Canepari, S.; Massimi, L. Evaluation of the efficiency of Arundo donax L. leaves as biomonitors for atmospheric element concentrations in an urban and industrial area of central Italy. *Atmosphere* **2020**, *11*, 226. [CrossRef]

12. Vitali, M.; Antonucci, A.; Owczarek, M.; Guidotti, M.; Astolfi, M.L.; Manigrasso, M.; Avino, P.; Bhattacharya, B.; Protano, C. Air quality assessment in different environmental scenarios by the determination of typical heavy metals and Persistent Organic Pollutants in native lichen Xanthoria parietina. *Environ. Pollut.* **2019**, *254*, 113013. [CrossRef]
13. Massimi, L.; Conti, M.E.; Mele, G.; Ristorini, M.; Astolfi, M.L.; Canepari, S. Lichen transplants as indicators of atmospheric element concentrations: A high spatial resolution comparison with PM10 samples in a polluted area (Central Italy). *Ecol. Indic.* **2019**, *101*, 759–769. [CrossRef]
14. Conti, M.E.; Canepari, S.; Finoia, M.G.; Mele, G.; Astolfi, M.L. Characterization of Italian multifloral honeys on the basis of their mineral content and some typical quality parameters. *J. Food Compos. Anal.* **2018**, *74*, 102–113. [CrossRef]
15. Losfeld, G.; Saunier, J.B.; Grison, C. Minor and trace-elements in apiary products from a historical mining district (Les Malines, France). *Food Chem.* **2014**, *146*, 455–459. [CrossRef]
16. Dżugan, M.; Wesołowska, M.; Zaguła, G.; Kaczmarski, M.; Czernicka, M.; Puchalski, C. Honeybees (*Apis mellifera*) as a biological barrier for contamination of honey by environmental toxic metals. *Environ. Monit. Assess.* **2018**, *190*, 101. [CrossRef]
17. Lambert, O.; Piroux, M.; Puyo, S.; Thorin, C.; Larhantec, M.; Delbac, F.; Pouliquen, H. Bees, honey and pollen as sentinels for lead environmental contamination. *Environ. Pollut.* **2012**, *170*, 254–259. [CrossRef]
18. Zhou, X.; Taylor, M.P.; Davies, P.J.; Prasad, S. Identifying sources of environmental contamination in European honey bees (*Apis mellifera*) using trace elements and lead isotopic compositions. *Environ. Sci. Technol.* **2018**, *52*, 991–1001. [CrossRef]
19. Matin, G.; Kargar, N.; Buyukisik, H.B. Bio-monitoring of cadmium, lead, arsenic and mercury in industrial districts of Izmir, Turkey by using honey bees, propolis and pine tree leaves. *Ecol. Eng.* **2016**, *90*, 331–335. [CrossRef]
20. Smith, K.E.; Weis, D.; Amini, M.; Shiel, A.E.; Lai, V.W.M.; Gordon, K. Honey as a biomonitor for a changing world. *Nat. Sustain.* **2019**, *2*, 223–232. [CrossRef]
21. Pohl, P.; Sergiel, I.; Stecka, H. Determination and fractionation of metals in honey. *Crit. Rev. Anal. Chem.* **2009**, *39*, 276–288. [CrossRef]
22. Herrero-Latorre, C.; Barciela-García, J.; García-Martín, S.; Peña-Crecente, R.M. The use of honeybees and honey as environmental bioindicators for metals and radionuclides: A review. *Environ. Rev.* **2017**, *25*, 463–480. [CrossRef]
23. Pascual-Maté, A.; Osés, S.M.; Fernández-Muiño, M.A.; Sancho, M.T. Methods of analysis of honey. *J. Apic. Res.* **2018**, *57*, 38–74. [CrossRef]
24. Álvarez-Ayuso, E.; Abad-Valle, P. Trace element levels in an area impacted by old mining operations and their relationship with beehive products. *Sci. Total Environ.* **2017**, *599*, 671–678. [CrossRef] [PubMed]
25. van der Steen, J.J.; de Kraker, J.; Grotenhuis, T. Spatial and temporal variation of metal concentrations in adult honeybees (*Apis mellifera* L.), Environ. *Monit. Assess.* **2012**, *184*, 4119–4126. [CrossRef]
26. Ajtony, Z.; Bencs, L.; Haraszi, R.; Szigeti, J.; Szoboszlai, N. Study on the simultaneous determination of some essential and toxic trace elements in honey by multi-element graphite furnace atomic absorption spectrometry. *Talanta* **2007**, *71*, 683–690. [CrossRef]
27. Packer, A.P.; Giné, M.F. Analysis of undigested honey samples by isotope dilution inductively coupled plasma mass spectrometry with direct injection nebulization (ID-ICP-MS). *Spectrochim. Acta B* **2001**, *56*, 69–75. [CrossRef]
28. Madejczyk, M.; Baralkiewicz, D. Characterization of Polish rape and honeydew honey according to their mineral contents using ICP-MS and F-AAS/AES. *Anal. Chim. Acta* **2008**, *617*, 11–17. [CrossRef]
29. Conti, M.E.; Botrè, F. Honeybees and their products as potential bioindicators of heavy metals contamination. *Environ. Monit. Assess.* **2001**, *69*, 267–282. [CrossRef]
30. Gutiérrez, M.; Molero, R.; Gaju, M.; van der Steen, J.; Porrini, C.; Ruiz, J.A. Assessment of heavy metal pollution in Córdoba (Spain) by biomonitoring foraging honeybee. *Environ. Monit. Assess.* **2015**, *18*, 651. [CrossRef]
31. Sajtos, Z.; Herman, P.; Harangi, S.; Baranyai, E. Elemental analysis of Hungarian honey samples and bee products by MP-AES method. *Microchem. J.* **2019**, *149*, 103968. [CrossRef]
32. Mračević, S.Đ.; Krstić, M.; Lolić, A.; Ražić, S. Comparative study of the chemical composition and biological potential of honey from different regions of Serbia. *Microchem. J.* **2020**, *152*, 104420. [CrossRef]

33. Bazeyad, A.Y.; Al-Sarar, A.S.; Rushdi, A.I.; Hassanin, A.S.; Abobakr, Y. Levels of heavy metals in a multifloral Saudi honey. *Environ. Sci. Pollut. Res.* **2019**, *26*, 3946–3953. [CrossRef]
34. Oliveira, S.S.; Alves, C.N.; Morte, E.S.B.; Júnior, A.D.F.S.; Araujo, R.G.O.; Santos, D.C.M.B. Determination of essential and potentially toxic elements and their estimation of bioaccessibility in honeys. *Microchem. J.* **2019**, *151*, 104221. [CrossRef]
35. Grainger, M.N.; Hewitt, N.; French, A.D. Optimised approach for small mass sample preparation and elemental analysis of bees and bee products by inductively coupled plasma mass spectrometry. *Talanta* **2020**, *214*, 120858. [CrossRef]
36. Sadowska, M.; Gogolewska, H.; Pawelec, N.; Sentkowska, A.; Krasnodębska-Ostręga, B. Comparison of the contents of selected elements and pesticides in honey bees with regard to their habitat. *Environ. Sci. Pollut. Res.* **2019**, *26*, 371–380. [CrossRef]
37. Giglio, A.; Ammendola, A.; Battistella, S.; Naccarato, A.; Pallavicini, A.; Simeon, E.; Tagarelli, A.; Giulianini, P.G. Apis mellifera ligustica, Spinola 1806 as bioindicator for detecting environmental contamination: A preliminary study of heavy metal pollution in Trieste, Italy. *Environ. Sci. Pollut. Res.* **2017**, *24*, 659–665. [CrossRef]
38. González-Martín, M.I.; Revilla, I.; Betances-Salcedo, E.V.; Vivar-Quintana, A.M. Pesticide residues and heavy metals in commercially processed propolis. *Microchem. J.* **2018**, *143*, 423–429. [CrossRef]
39. Voica, C.; Iordache, A.M.; Ionete, R.E. Multielemental characterization of honey using inductively coupled plasma mass spectrometry fused with chemometrics. *J. Mass Spectrom.* **2020**, *55*, e4512. [CrossRef]
40. Raeymaekers, B. A prospective biomonitoring campaign with honey bees in a district of Upper-Bavaria (Germany). *Environ. Monit. Assess.* **2006**, *116*, 233–243. [CrossRef]
41. Vieira, H.P.; Nascentes, C.C.; Windmöller, C.C. Development and comparison of two analytical methods to quantify the mercury content in honey. *J. Food Compos. Anal.* **2014**, *34*, 1–6. [CrossRef]
42. Morgano, M.A.; Martins, M.C.T.; Rabonato, L.C.; Milani, R.F.; Yotsuyanagi, K.; Rodriguez-Amaya, D.B. A comprehensive investigation of the mineral composition of Brazilian bee pollen: Geographic and seasonal variations and contribution to human diet. *J. Braz. Chem. Soc.* **2012**, *23*, 727–736. [CrossRef]
43. da Silva, P.M.; Gauche, C.; Gonzaga, L.V.; Costa, A.C.O.; Fett, R. Honey: Chemical composition, stability and authenticity. *Food Chem.* **2016**, *196*, 309–323. [CrossRef] [PubMed]
44. Bastos, D.H.M.; Bastos, M.; Bartli, O.M.; Rocha, C.I.; Cunha, I.B.d.S.; de Carvalho, P.; Torres, E.A.S.; Michelan, M. Fatty acid composition and palynological analysis of bee (Apis) pollen loads in the states of São Paulo and Minas Gerais. *Brazil. J. Apic. Res.* **2004**, *43*, 35–39. [CrossRef]
45. Pettine, M.; Casentini, B.; Mastroianni, D.; Capri, S. Dissolved inorganic carbon effect in the determination of arsenic and chromium in mineral waters by inductively coupled plasma-mass spectrometry. *Anal. Chim. Acta* **2007**, *599*, 191–198. [CrossRef]
46. Allain, P.; Jaunault, L.; Mauras, Y.; Mermet, J.M.; Delaporte, T. Signal enhancement of elements due to the presence of carbon-containing compounds in inductively coupled plasma mass spectrometry. *Anal. Chem.* **1991**, *63*, 1497–1498. [CrossRef]
47. US EPA. *SW-846 Test Method 6020B: Inductively Coupled Plasma—Mass Spectrometry*; United States Environmental Protection Agency: Cincinnati, OH, USA, 1994.
48. Nóbrega, J.A.; Pirola, C.; Fialho, L.L.; Rota, G.; de Campos Jordão, C.E.K.M.A.; Pollo, F. Microwave-assisted digestion of organic samples: How simple can it become? *Talanta* **2012**, *98*, 272–276. [CrossRef]
49. Muller, A.L.; Oliveira, J.S.; Mello, P.A.; Muller, E.I.; Flores, E.M. Study and determination of elemental impurities by ICP-MS in active pharmaceutical ingredients using single reaction chamber digestion in compliance with USP requirements. *Talanta* **2015**, *136*, 161–169. [CrossRef]
50. Latimer, G.W.; Association of Official Analytical Chemists International (AOAC). *Official Methods of Analysis of AOAC International*; AOAC International: Gaithersburg, MD, USA, 2012.
51. Bilandžić, N.; Đokić, M.; Sedak, M.; Kolanović, S.B.; Varenina, I.; Končurat, A.; Rudan, N. Determination of trace elements in Croatian floral honey originating from different regions. *Food Chem.* **2011**, *128*, 1160–1164. [CrossRef]
52. Leme, A.B.; Bianchi, S.R.; Carneiro, R.L.; Nogueira, A.R. Optimization of sample preparation in the determination of minerals and trace elements in honey by ICP-MS. *Food Anal. Methods* **2014**, *7*, 1009–1015. [CrossRef]

53. Döker, S.; Aydemir, O.; Uslu, M. Evaluation of digestion procedures for trace element analysis of Cankiri, Turkey honey by inductively coupled plasma mass spectrometry. *Anal. Lett.* **2014**, *47*, 2080–2094. [CrossRef]
54. Chen, M.; Ma, L.Q. Comparison of three aqua regia digestion methods for twenty Florida soils. *Soil Sci. Soc. Am. J.* **2001**, *65*, 491–499. [CrossRef]
55. Abbruzzini, T.F.; Silva, C.A.; Andrade, D.A.D.; Carneiro, W.J.D.O. Influence of digestion methods on the recovery of iron, zinc, nickel, chromium, cadmium and lead contents in 11 organic residues. *Rev. Bras. Ciênc. Sol.* **2014**, *38*, 166–176. [CrossRef]
56. Das, S.; Ting, Y.P. Evaluation of wet digestion methods for quantification of metal content in electronic scrap material. *Resources* **2017**, *6*, 64. [CrossRef]
57. Edgell, K. *USEPA Method Study 37 SW-846 Method 3050 Acid Digestion of Sediments, Sludges, and Soils*; US Environmental Protection Agency, Environmental Monitoring Systems Laboratory: Washington, DC, USA, 1989.
58. Astolfi, M.L.; Protano, C.; Marconi, E.; Massimi, L.; Brunori, M.; Piamonti, D.; Migliara, G.; Vitali, M.; Canepari, S. A new treatment of human hair for elemental determination by inductively coupled mass spectrometry. *Anal. Methods* **2020**, *12*, 1906–1918. [CrossRef]
59. Bizzi, C.A.; Barin, J.S.; Garcia, E.E.; Nóbrega, J.A.; Dressler, V.L.; Flores, E.M. Improvement of microwave-assisted digestion of milk powder with diluted nitric acid using oxygen as auxiliary reagent. *Spectrochim. Acta B* **2011**, *66*, 394–398. [CrossRef]
60. Astolfi, M.L.; Marconi, E.; Protano, C.; Vitali, M.; Schiavi, E.; Mastromarino, P.; Canepari, S. Optimization and validation of a fast digestion method for the determination of major and trace elements in breast milk by ICP-MS. *Anal. Chim. Acta* **2018**, *1040*, 49–62. [CrossRef]
61. Phan-Thien, K.Y.; Wright, G.C.; Lee, N.A. Inductively coupled plasma-mass spectrometry (ICP-MS) and-optical emission spectroscopy (ICP–OES) for determination of essential minerals in closed acid digestates of peanuts (Arachis hypogaea L.). *Food Chem.* **2012**, *134*, 453–460. [CrossRef]
62. Pick, D.; Leiterer, M.; Einax, J.W. Reduction of polyatomic interferences in biological material using dynamic reaction cell ICP-MS. *Microchem. J.* **2010**, *95*, 315–319. [CrossRef]
63. D'Ilio, S.; Petrucci, F.; D'Amato, M.; di Gregorio, M.; Senofonte, O.; Violante, N. Method validation for determination of arsenic, cadmium, chromium and lead in milk by means of dynamic reaction cell inductively coupled plasma mass spectrometry. *Anal. Chim. Acta* **2008**, *624*, 59–67. [CrossRef]
64. May, T.W.; Wiedmeyer, R.H. A table of polyatomic interferences in ICP-MS. *At. Spectrosc.* **1998**, *19*, 150–155.
65. Martínez-Sierra, J.G.; San Blas, O.G.; Gayón, J.M.; Alonso, J.G. Sulfur analysis by inductively coupled plasma-mass spectrometry: A review. *Spectrochim. Acta B* **2015**, *108*, 35–52. [CrossRef]
66. Mandel, J. *The Statistical Analysis of Experimental Data*; John Wiley & Sons: New York, NY, USA, 1964; pp. 1–410.
67. Cubadda, F.; Raggi, A.; Testoni, A.; Zanasi, F. Multielemental analysis of food and agricultural matrixes by inductively coupled plasma-mass spectrometry. *J. AOAC Int.* **2002**, *85*, 113–121. [CrossRef]
68. Nardi, E.P.; Evangelista, F.S.; Tormen, L.; Pierre, T.D.S.; Curtius, A.J.; de Souza, S.S.; Junior, F.B. The use of inductively coupled plasma mass spectrometry (ICP-MS) for the determination of toxic and essential elements in different types of food samples. *Food Chem.* **2009**, *112*, 727–732. [CrossRef]
69. European Commission. Commission Regulation (EU) 2015/1005 of 25 June 2015 amending Regulation (EC) No 1881/2006 as regards maximum levels of lead in certain foodstuffs. *Off. J. Eur. Union* **2015**, *161*, 9–13.
70. Codex Alimentarius. *Joint FAO/WHO Food Standards Programme*; Twenty-fourth Session; Codex Alimentarius Commission: Geneva, Switzerland, 2001.
71. Jorhem, L.; Engman, J.; Schröder, T. Evaluation of results derived from the analysis of certified reference materials–a user-friendly approach based on simplicity. *Fresenius J. Anal. Chem.* **2001**, *370*, 178–182. [CrossRef]
72. U.S. EPA. *Method 3051A (SW-846): Microwave Assisted Acid Digestion of Sediments, Sludges, and Oils*; Revision 1; United States Environmental Protection Agency: Washington, DC, USA, 2007.
73. Cava-Montesinos, P.; Cervera, M.L.; Pastor, A.; de la Guardia, M. Room temperature acid sonication ICP-MS multielemental analysis of milk. *Anal. Chim. Acta* **2005**, *531*, 111–123. [CrossRef]
74. Astolfi, M.L.; Marconi, E.; Protano, C.; Canepari, S. Comparative elemental analysis of dairy milk and plant-based milk alternatives. *Food Control* **2020**, *116*, 107327. [CrossRef]

75. Magnusson, B.; Örnemark, U. (Eds.) *Eurachem Guide: The Fitness for Purpose of Analytical Methods—A Laboratory Guide to Method Validation and Related Topics*, 2nd ed.; Eurachem: Teddington, UK, 2014; ISBN 978-91-87461-59-0. Available online: www.eurachem.org (accessed on 7 September 2020).
76. Hornung, R.W.; Reed, L.D. Estimation of average concentration in the presence of nondetectable values. *Appl. Occup. Environ. Hyg.* **1990**, *5*, 46–51. [CrossRef]
77. Siegel, S.; Castellan, N.J., Jr. *Non Parametric Statistics for the Behavioral Sciences*, 2nd ed.; McGraw-Hill: Milan, Italy, 1992; p. 477.

Sample Availability: Not available.

© 2020 by the authors. Licensee MDPI, Basel, Switzerland. This article is an open access article distributed under the terms and conditions of the Creative Commons Attribution (CC BY) license (http://creativecommons.org/licenses/by/4.0/).

Article

Multielemental Analysis of Bee Pollen, Propolis, and Royal Jelly Collected in West-Central Poland

Eliza Matuszewska [1], Agnieszka Klupczynska [1], Krzysztof Maciołek [2], Zenon J. Kokot [3] and Jan Matysiak [1,*]

[1] Department of Inorganic and Analytical Chemistry, Poznan University of Medical Sciences, Grunwaldzka 6 Street, 60-780 Poznań, Poland; eliza.matuszewska@ump.edu.pl (E.M.); aklupczynska@ump.edu.pl (A.K.)
[2] Aquanet Laboratory Ltd., 126 Dolna Wilda Street, 61-492 Poznań, Poland; krzysztof.maciolek@aquanet-laboratorium.pl
[3] Faculty of Health Sciences, Calisia University, 13 Street, 62-800 Kalisz, Poland; z.kokot@akademiakaliska.edu.pl
* Correspondence: jmatysiak@ump.edu.pl

Abstract: Beehive products possess nutritional value and health-promoting properties and are recommended as so-called "superfoods". However, because of their natural origin, they may contain relevant elemental contaminants. Therefore, to assess the quality of bee products, we examined concentrations of a broad range of 24 selected elements in propolis, bee pollen, and royal jelly. The quantitative analyses were performed with inductively coupled plasma-mass spectrometry (ICP-MS) and inductively coupled plasma optical emission spectrometry (ICP-OES) techniques. The results of our research indicate that bee products contain essential macronutrients (i.e., K, P, and S) and micronutrients (i.e., Zn and Fe) in concentrations depending on the products' type. However, the presence of toxic heavy metals makes it necessary to test the quality of bee products before using them as dietary supplements. Bearing in mind that bee products are highly heterogenous and, depending on the environmental factors, differ in their elemental content, it is necessary to develop standards regulating the acceptable levels of inorganic pollutants. Furthermore, since bees and their products are considered to be an effective biomonitoring tool, our results may reflect the environment's condition in west-central Poland, affecting the health and well-being of both humans and bees.

Keywords: bee products; multielemental analysis; biomonitoring; ICP-MS; ICP-OES; inorganic contaminants; heavy metals

1. Introduction

Macroelements and trace elements (including heavy metals) play an essential role in human nutrition. Since they are involved in many biochemical processes, they are crucial for life processes [1,2]. Moreover, minerals are components of biological structures, i.e., bones, nerves, and muscles. They are also included in enzymes, hormones, and pigments, such as oxygen-carrying hemoglobin [3]. Although a precise distinction between macronutrients and micronutrients is still under discussion, at least 23 of the minerals are defined as essential for humans [4,5]. They are the macronutrients: sodium (Na), potassium (K), magnesium (Mg), calcium (Ca), chlorine (Cl), phosphorus (P), and sulfur (S), and micronutrients: manganese (Mn), iron (Fe), copper (Cu), zinc (Zn), selenium (Se), cobalt (Co), molybdenum (Mo), and iodine (I). Some other elements are also discussed to be considered essential, such as vanadium (V), nickel (Ni), and silicon (Si) [4].

Deficiencies in macroelements and microelements cause serious health problems [6,7]. On the other hand, if consumed at excessively high levels for a long time, minerals can be toxic and trigger adverse effects [8]. They may disrupt the metabolic pathways by accumulating in the body, replacing the appropriate ions, or binding to enzymes responsible for controlling the metabolic reactions [9]. Moreover, by increasing the production of reactive oxygen species (ROS), mineral elements may also cause cancer, degenerative diseases, and damage to the central nervous system [10,11]. Therefore, consumers should be aware of the mineral content of foods. However, not all food products have been thoroughly investigated yet, including bee products.

Bee products are complex natural mixtures produced by a honey bee (*Apis mellifera*). The products are rich in nutrients and biologically active compounds, such as carbohydrates, proteins, amino acids, lipids, vitamins, phenolics, and minerals [12,13]. Due to their health-supporting properties, bee products are willingly used as dietary supplements within the branch of alternative medicine called apitherapy [14]. The most recognizable bee products are honey, bee venom, propolis, bee pollen, and royal jelly. In this research, we examined the selected minerals' levels in three of them: bee pollen, propolis, and royal jelly.

Bee pollen is a mixture of flower pollen from different species gathered by forager bees, formed into pellets with nectar and secretions of bees' salivary glands. In Poland, this beehive product's main plant sources are *Brassica napus, Taraxacum officinale, Robinia pseudoacacia, Tilia cordata*, and *Trifolium repens*. Ethanol extract of Polish bee pollen was reported to be rich in flavonoids and polyphenols as well as responsible for anti-atherogenic activity [15]. Moreover, due to its nutritional and energetic properties, bee pollen supplementation is recommended for people leading active lifestyles and during recuperation [16]. Propolis, a resinous product with an aromatic smell, is produced by mixing bees' saliva, beeswax, and pollen with secretions collected from plants [17]. In Poland's temperate climate, its primary plant source is *Populus spp., Betula verucosa, Pinus sylvestris, Aesculus hippocastanum*, and *Acer pseudoplatanus* [18]. The botanical origin determines the main biologically active constituents of Polish propolis: flavonoids, phenolic acids, and esters. These compounds are responsible for bacteriostatic, antifungal, antioxidative, and antiproliferative properties [18–21]. Royal jelly is a milky secretion of worker bees' salivary glands [22]. Unlike bee pollen and propolis, it does not contain plant tissues. In addition to water, royal jelly contains mainly sugars, proteins (including enzymes), amino acids, and lipids [12]. This bee product is becoming increasingly popular, as its antibacterial, anti-inflammatory, antioxidative, and even antitumor properties have been reported [22–25].

Bee pollen, propolis, and royal jelly are recommended as so-called "superfoods", meaning foods of rich nutrient content having favorable effects on human health [26]. Therefore, they should be of the highest quality. However, the challenge is a variability of the composition of bee products, depending on the geographical region, climate, and season [16,27–29]. Differences may involve many groups of nutrients, including mineral contents [30,31]. Since excessive element intake may lead to adverse effects, the concentrations (along with the acceptable levels) of heavy metals and other mineral contaminants in bee products should be precisely defined.

The multielemental composition of bee products may also reflect environmental pollution, affecting both humans and bees. A continuous decline in the honeybee population has been reported in industrialized countries [32]. Several factors are supposed to contribute to honeybee declines, such as pests and diseases, global warming, and environmental pollutants associated with industrial and agricultural human activity, including heavy metals [32,33]. Due to the many vital roles of honeybees in the environment and agriculture, their health has become a public concern. Thus, it is imperative to assess the pollution to which bees are exposed in their natural habitat.

Therefore, to assess the safety of bee products for humans and bees, in this study, we examined concentrations of 24 selected elements (including macronutrients and micronutrients) in propolis, bee pollen, and royal jelly. We decided not to analyze bee venom, which was thoroughly examined in our previous study [34], and honey, which other researchers extensively studied [35–38]. The bee product samples used in this study were collected in Poland (Greater Poland region) in 2018 and 2019. The quantitative analyses were performed with inductively coupled plasma-mass spectrometry (ICP-MS) and inductively coupled plasma optical emission spectrometry (ICP-OES) techniques. Our research results contributed to assessing bees' exposure to various chemicals and evaluating their products' quality.

2. Results

Levels of 24 chemical elements were determined in all samples of bee products. We measured one of the broadest spectra of chemical elements (including micro- and macronutrients, heavy metals, and trace elements) in bee products compared to the available literature [39–49]. In this study, we examined various beehive matrices (bee pollen, propolis, and royal jelly) collected over two years from the same area. The majority of studies on the determination of elements (mainly heavy metals) in bee products included one selected product collected from different locations [31,39,46,48–50]. The analysis of different beehive products derived from the same area allowed us to compare the levels of inorganic contaminants between them and evaluate their role as bioindicators. On the other hand, the obtained results allowed comparing bee products in terms of micronutrient and macronutrient content. For the studied products, the determined concentrations of minerals, along with mean, standard deviations (SD), and relative standard deviation (%RSD) values, calculated for each year separately and all samples in total, are presented in Tables 1 and 2.

The highest content in bee pollen was measured for K (mean = 4233.33 mg/kg), P (mean = 4050.00 mg/kg), and S (mean = 2383.33 mg/kg). These elements are classified as macroelements, for which the dietary requirement exceeds 100 mg per day. On the other hand, the lowest concentration in the macronutrient group in bee pollen was found for Na (mean = 25.17 mg/kg). Concerning micronutrients, the highest content in bee pollen was recorded for Zn (mean = 31.3 mg/kg), Fe (mean = 114.5 mg/kg), and Mn (mean = 25.0 mg/kg), and the lowest was found for Co (mean = 0.038 mg/kg). However, a relatively high Al content was measured in a group of toxic metals in bee pollen (mean = 26.13 mg/kg). On the other hand, Sb (mean = 0.009 mg/kg) and As (mean = 0.02 mg/kg) levels were the lowest among the bee pollen's trace elements.

In propolis, the highest content among macronutrients was determined for K (mean = 706.67 mg/kg), Ca (mean = 373.33 mg/kg), and S (mean = 225.0 mg/kg). However, the concentrations of these elements in propolis were much lower than in bee pollen. The macroelement with the lowest content in propolis, similar to bee pollen, was Na (mean = 22.67 mg/kg). Among microelements, similar to bee pollen, the highest content was found for Zn (mean = 13.67 mg/kg). However, the concentration of Zn in bee pollen was over two times higher than in propolis. The micronutrient with the second-highest content in propolis was Fe (mean = 49.17 mg/kg) followed with Mn (mean = 7.2 mg/kg), but its concentration was more than three times lower than in bee pollen. The microelement with the lowest content in propolis was Se (mean = 0.047 mg/kg). Similar to bee pollen, the high Al content was measured in propolis (mean = 93.0 mg/kg), and this concentration was more than three times higher than in bee pollen. Among trace elements in propolis, Ag was determined at the lowest level (mean = 0.006 mg/kg). Summing up, mean levels of Mn, Ni, Co, Zn, Ag, Cd, Mo, Na, Mg, K, Ca, P, and S were higher in bee pollen than in propolis.

Table 1. Levels of 24 selected elements measured in bee pollen along with mean, standard deviations (SD), and relative standard deviation (%RSD) values, calculated for each year separately and all samples in total; 1, 2, 3—samples collected in 6 May, 23 June and 8 July 2018, respectively; 4, 5, 6—samples collected in 16 May, 2 June and 20 July 2019, respectively. Values of mean and SD are expressed in mg/kg.

Element	1	2	3	Year 2018 Mean + SD	%RSD	4	5	6	Year 2019 Mean + SD	%RSD	Years 2018 and 2019 Mean + SD	%RSD
	[mg/kg]						[mg/kg]					
Ag	0.18	0.28	0.04	0.17 ± 0.12	72.33	0.05	0.09	0.05	0.06 ± 0.02	36.46	0.12 ± 0.10	83.54
Al	53.00	25.00	11.00	29.67 ± 21.39	72.09	7.20	9.60	51.00	22.60 ± 24.62	108.96	26.13 ± 20.99	80.31
As	0.03	0.03	0.01	0.02 ± 0.01	48.42	0.01	0.01	0.03	0.02 ± 0.01	79.52	0.03 ± 0.01	57.55
Ba	0.84	0.57	0.51	0.64 ± 0.18	27.47	0.25	0.65	1.00	0.63 ± 0.38	59.25	0.64 ± 0.26	41.17
Ca	1800.00	1500.00	450.00	1250.00 ± 708.87	56.71	1500.00	1500.00	680.00	1226.67 ± 473.47	38.59	1238.33 ± 539.27	43.55
Cd	0.02	0.11	0.03	0.05 ± 0.05	91.14	0.02	0.04	0.12	0.06 ± 0.06	96.25	0.06 ± 0.05	84.12
Co	0.03	0.05	0.02	0.03 ± 0.02	49.01	0.04	0.05	0.04	0.042 ± 0.003	7.16	0.038 ± 0.012	30.99
Cr	0.12	0.08	0.04	0.08 ± 0.04	50.00	0.04	0.03	0.11	0.06 ± 0.04	76.82	0.07 ± 0.04	57.54
Cu	5.30	5.30	2.40	4.33 ± 1.67	38.64	5.40	5.70	4.40	5.17 ± 0.68	13.17	4.75 ± 1.23	25.91
Fe	75.00	57.00	22.00	51.33 ± 26.95	52.50	24.00	49.00	68.00	47.00 ± 22.07	46.95	49.17 ± 22.16	45.07
K	4500.00	4100.00	4300.00	4300.00 ± 200.00	4.65	3500.00	4600.00	4400.00	4166.67 ± 585.95	14.06	4233.33 ± 398.33	9.41
Mg	1100.00	720.00	620.00	813.33 ± 253.25	31.15	980.00	1000.00	520.00	833.33 ± 271.54	32.58	823.33 ± 235.09	28.55
Mn	16.00	21.00	13.00	16.67 ± 4.04	24.25	22.00	62.00	16.00	33.33 ± 25.07	75.02	25.00 ± 18.44	73.76
Mo	0.31	0.29	0.16	0.25 ± 0.08	32.15	0.18	0.30	0.16	0.21 ± 0.08	35.49	0.23 ± 0.07	31.57
Na	25.00	32.00	14.00	23.67 ± 9.07	38.34	24.00	24.00	32.00	26.67 ± 4.62	17.32	25.17 ± 6.65	26.41
Ni	0.63	0.83	0.37	0.61 ± 0.23	37.81	0.66	0.74	0.67	0.69 ± 0.04	6.32	0.65 ± 0.15	23.81
P	4600.00	4100.00	3200.00	3966.67 ± 709.46	17.89	4300.00	4600.00	3500.00	4133.33 ± 568.62	13.76	4050.00 ± 582.24	14.38
Pb	0.17	0.19	0.12	0.16 ± 0.04	22.53	0.14	0.09	0.22	0.15 ± 0.07	45.35	0.15 ± 0.05	31.59
S	2700.00	2600.00	1,800.00	2366.67 ± 493.29	20.84	2600.00	2700.00	1900.00	2400.00 ± 435.89	18.16	2383.33 ± 416.73	17.49
Sb	0.007	0.013	0.005	0.008 ± 0.004	49.033	0.004	0.008	0.016	0.009 ± 0.006	64.58	0.009 ± 0.005	52.45
Se	0.04	0.06	0.04	0.05±0.01	32.46	0.02	0.03	0.09	0.05 ± 0.04	82.58	0.05 ± 0.03	56.12
Si	75.00	45.00	20.00	46.67 ± 27.54	59.01	8.50	22.00	71.00	33.83 ± 32.89	97.20	40.25 ± 28.02	69.63
V	0.11	0.05	0.03	0.06 ± 0.045	70.84	0.05	0.02	0.11	0.06 ± 0.05	76.48	0.06 ± 0.04	65.88
Zn	27.00	28.00	35.00	30.00 ± 4.36	14.53	26.00	31.00	41.00	32.67 ± 7.64	23.38	31.33 ± 5.75	18.35

Table 2. Levels of 24 selected elements measured in propolis along with mean, standard deviations (SD), and relative standard deviation (%RSD) values, calculated for each year separately and all samples in total; 1, 2, 3—samples collected in 19 May, 23 June and 8 July 2018, respectively; 4, 5, 6—samples collected in 19 May, 1 June and 29 June 2019, respectively. Values of mean and SD are expressed in mg/kg, QL—quantification limit.

Element	1	2	3	Year 2018		4	5	6	Year 2019		Years 2018 and 2019	
	[mg/kg]			Mean + SD	%RSD	[mg/kg]			Mean + SD	%RSD	Mean + SD	%RSD
Ag	<QL	0.01	<QL	0.003 ± 0.006	173.21	<QL	0.07	<QL	0.009 ± 0.016	173.21	0.009 ± 0.016	177.76
Al	140.00	120.00	120.00	126.67 ± 11.55	9.12	93.00	97.00	69.00	86.33 ± 15.14	17.54	86.33 ± 15.14	23.63
As	0.11	0.06	0.08	0.09 ± 0.03	26.98	0.06	0.06	0.05	0.056 ± 0.009	16.20	0.056 ± 0.01	32.40
Ba	3.90	2.60	2.90	3.13 ± 0.68	21.72	1.50	1.80	1.77	1.67 ± 0.15	9.17	1.67 ± 0.15	38.19
Ca	500.00	560.00	300.00	453.33 ± 136.14	30.03	300.00	220.00	360.30	293.33 ± 70.24	23.94	293.33 ± 70.24	34.99
Cd	0.05	0.04	0.03	0.043 ± 0.01	23.26	0.03	0.08	0.03	0.032 ± 0.006	19.52	0.032 ± 0.0184	25.56
Co	0.13	0.10	0.20	0.14 ± 0.05	35.80	0.10	0.10	0.08	0.09 ± 0.01	13.45	0.09 ± 0.01	36.95
Cr	0.69	0.72	0.54	0.65 ± 0.10	14.84	0.28	0.41	0.25	0.31 ± 0.09	27.14	0.31 ± 0.09	41.84
Cu	2.00	1.40	1.40	1.6 ± 0.35	21.65	3.00	1.10	0.98	1.69 ± 1.13	66.92	1.69 ± 1.13	45.62
Fe	150.00	160.00	120.00	143.33 ± 20.81	14.52	97.00	84.00	76.00	85.67 ± 10.60	12.37	85.67 ± 10.60	30.45
K	730.00	1100.00	470.00	766.67 ± 316.60	41.30	610.00	600.00	730.00	646.67 ± 72.34	11.19	646.67 ± 72.34	30.52
Mg	110.00	140.00	76.00	108.67 ± 32.02	29.47	100.00	64.00	110.00	91.33 ± 24.19	26.49	91.33 ± 24.19	27.10
Mn	7.90	8.30	6.40	7.533 ± 1.002	13.30	5.00	9.70	5.90	6.87 ± 2.49	36.33	6.87 ± 2.49	24.15
Mo	0.07	0.13	0.04	0.08 ± 0.05	58.10	0.04	0.05	0.06	0.05 ± 0.01	26.01	0.05 ± 0.01	51.43
Na	30.00	20.00	29.00	26.33 ± 5.51	20.91	23.00	20.00	14.00	19.00 ± 4.58	24.12	19.00 ± 4.58	26.71
Ni	0.53	0.86	0.47	0.62 ± 0.21	33.87	0.25	0.40	0.36	0.34 ± 0.08	23.07	0.37 ± 0.08	43.92
P	230.00	160.00	110.00	166.67 ± 60.28	36.17	300.00	190.00	210.0	233.33 ± 58.59	25.11	233.33 ± 58.59	32.25
Pb	1.00	0.68	0.74	0.817 ± 0.17	21.09	0.50	0.60	0.44	0.51 ± 0.08	15.75	0.51 ± 0.08	30.30
S	270.00	320.00	160.00	250 ± 81.86	32.74	180.00	200.00	220.00	200.00 ± 20.00	10.00	200.00 ± 20.00	26.63
Sb	0.04	0.04	0.05	0.042 ± 0.002	5.46	0.04	0.03	0.08	0.031 ± 0.005	17.07	0.031 ± 0.005	19.64
Se	0.08	0.05	0.04	0.057 ± 0.0	37.96	0.03	0.05	0.04	0.034 ± 0.008	19.87	0.038 ± 0.008	37.12
Si	170.00	160.00	140.00	156.677 ± 15.28	9.75	97.00	110.00	96.00	101.00 ± 7.81	7.73	101 ± 7.815	25.12
V	0.39	0.27	0.26	0.317 ± 0.07	23.59	0.20	0.22	0.15	0.19 ± 0.04	18.98	0.19 ± 0.04	32.95
Zn	19.00	12.00	11.00	14 ± 4.36	31.13	15.00	13.00	12.00	13.33 ± 1.53	11.46	13.33 ± 1.53	21.54

Analysis of bee product samples collected from the same location but in different months and years allowed us to estimate the variability in elemental composition caused by the date of sample collection. We observed quite substantial differences in the levels of individual elements in the beehive products between months and years of sample collection (Tables 1 and 2). Among all determined elements in bee pollen, Cd, Al, and Ag showed the greatest variation in concentration between years (%RSD > 80%), whereas K, P, S, and Zn were present in a similar concentration throughout all analyzed years (%RSD < 20%). High %RSD values calculated for Ag and Cd can result from a very low level of that element determined in pollen samples. However, Al, Cd, and Ag constitute metals of anthropogenic origin. Thus, variation in their concentrations observed in bee pollen samples throughout the analyzed years can also reflect changes in the degree of pollution of the environment surrounding the apiary. The content of chemical elements determined in propolis samples did not vary so substantially as in the case of pollen, and the calculated %RSD values were noticeably lower (Table 2). The only exception was Ag, whose concentration occurred below the quantification limit (QL) in part of the samples.

To better assess variability in the elemental composition of different beehive products and different years of collection, univariate and multivariate statistical analyses were performed. Univariate statistical tests showed no significant differences in the elemental composition between bee pollen samples collected in two different years (Table 3). In the comparison of the elemental composition of propolis samples collected in two different years, significant differences were observed for six elements (Al, Ba, Cr, Fe, Sb, and Si). On the other hand, most of the observed differences in the levels of chemical elements determined in two various beehive products (pollen and propolis) were statistically significant. It should be emphasized that all studied samples were collected from the same location and during the same season; thus, the observed differences are not related to geographical region, climate, and season but are more associated with the process of the production of these bee products. Only for Cd, Na, Ni, and Se were differences not significant; thus, we can conclude that those elements occur at comparable levels in bee pollen and propolis. To better visualize how distinct the content of elements between pollen and propolis is, PCA analysis was conducted. A clear separation of the bee pollen samples and propolis samples was attained on the score plot (Figure 1). The results of PCA indicate that the differences in the elemental composition of pollen and propolis were strong enough to cause grouping of the samples according to the type of beehive products. It should be noted that the results of multivariate statistical analysis are in line with the results of univariate tests. Figure 1 demonstrates that differences in elements' concentrations are more pronounced between two beehive products than between years of collection of these products. Moreover, it is visible that propolis samples collected in 2018 are separated from propolis samples collected in 2019, whereas no grouping is visible in the case of pollen samples.

Table 3. Results of univariate tests performed to compare levels of chemical elements ($n = 24$) in two beehive products collected over two years from the same area: bee pollen and propolis. Bold text indicates a statistically significant difference with a p-value below 0.05.

Element	Pollen 2018 vs. 2019	Propolis 2018 vs. 2019	Pollen vs. Propolis
		p Value	
Ag	0.21856	NA *	**0.00508**
Al	0.72650	**0.02142**	**0.00015**
As	0.65361	0.10144	**0.00169**
Ba	0.97911	**0.02194**	**0.00430**
C	0.96446	0.14470	**0.01307**
Cd	0.93551	0.18140	0.81018
Co	0.34784	0.16963	**0.00537**
Cr	0.56601	**0.01054**	**0.00508**
Cu	0.46925	0.89808	**0.00824**

Table 3. Cont.

Element	Pollen 2018 vs. 2019	Propolis 2018 vs. 2019	Pollen vs. Propolis
		p Value	
Fe	0.83994	0.01289	0.00423
K	0.72807	0.55699	0.00508
Mg	0.93016	0.49600	0.00508
Mn	0.31806	0.68966	0.00508
Mo	0.56704	0.38378	0.00144
Na	0.63668	0.15094	0.51126
Ni	0.58672	0.09351	0.13817
P	0.76675	0.24152	0.00508
Pb	0.81004	0.05422	0.00123
S	0.93433	0.36214	0.00508
Sb	0.82629	0.02728	0.00003
Se	1.00000	0.23406	0.92881
Si	0.63166	0.00493	0.00052
V	0.95890	0.06677	0.00133
Zn	0.62719	0.81490	0.00021

* NA—no statistical tests were performed because Ag concentration was determined only in one sample per year.

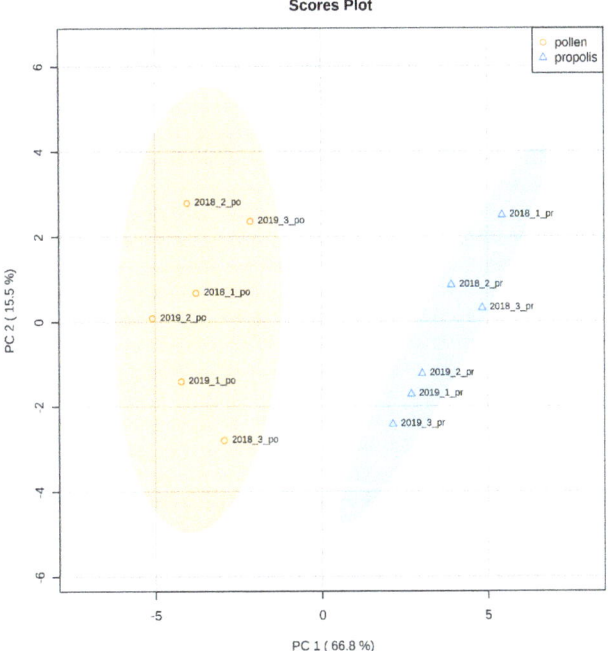

Figure 1. The principal component analysis (PCA) score plot using the first two principal components derived from elements' determination in two beehive products: bee pollen (orange circles; $n = 6$) and propolis (blue triangles; $n = 6$).

Additionally, we analyzed levels of elements in the royal jelly sample collected in 2019. The concentrations of chemical elements measured in royal jelly compared to mean values calculated for the bee pollen and propolis samples are shown in Table 4. Among the macroelements, the highest content in royal jelly was measured for P (1700 mg/kg) and S (1200 mg/kg), whereas the lowest concentration was found for Ca (35 mg/kg). Concerning microelements, the highest content in royal jelly, similarly to bee pollen and propolis, was measured for Zn (21 mg/kg), and the lowest was measured for Co (0.003 mg/kg), as in bee pollen. Among the other elements, the highest concentration was found for Al and the lowest was found for V (0.001 mg/kg). In general, most of the elements' levels were lower in royal jelly than in propolis or bee pollen. The exception is the content of Na, which was the highest in the royal jelly sample. Moreover, Cu, Zn, Ag, Mg, K, P, and S were at higher levels in royal jelly than in the propolis.

Table 4. Mean values of 24 selected chemical elements determined in bee pollen and propolis and measured levels of elements in the royal jelly sample. Values are expressed in mg/kg.

Chemical Element	Bee Pollen	Propolis	Royal Jelly
Ag	0.12	0.01	0.09
Al	26.13	106.50	1.00
As	0.02	0.07	0.01
Ba	0.64	2.40	0.12
Ca	1238.33	373.33	35.00
Cd	0.06	0.04	0.002
Co	0.04	0.12	0.003
Cr	0.07	0.48	0.02
Cu	4.75	1.65	4.20
Fe	49.17	114.50	3.90
K	4233.33	706.67	970.00
Mg	823.33	100.00	120.00
Mn	25.00	7.20	0.73
Mo	0.23	0.07	0.05
Na	25.17	22.67	41.00
Ni	0.65	0.48	0.25
P	4050.00	200.00	1700.00
Pb	0.15	0.66	0.07
S	2383.33	225.00	1200.00
Sb	0.01	0.04	0.003
Se	0.05	0.05	0.02
Si	40.25	128.83	0.88
V	0.06	0.25	0.001
Zn	31.33	13.67	21.00

Legend: the highest, medium, the lowest

3. Discussion

Beehive products are a rich source of nutrients and biologically active compounds. As reported in the available literature, they contain macro- and microelements essential for the human body's proper functioning [50–53]. In this study, we examined the concentrations of several macroelements, including Ca, K, Mg, Na, P, and S, and microelements, including Co, Cu, Fe, Mn, Mo, Se, and Zn in Polish bee pollen, propolis, and royal jelly.

Our study results indicate that bee pollen and royal jelly contain large amounts of K, P, and S. Additionally, bee pollen is rich in Ca. The same macronutrients are present at the highest level in propolis, but the concentrations are lower than in bee pollen and royal jelly. K, P, Ca, and S play a vital role in human health. K ensures energy metabolism, membrane transport, and cardiac contraction. This element is commonly found in food, so severe deficiencies are virtually non-existent [54]. However, the intake of K in modern diets is well below the current recommended nutritional requirements, which, according to the World Health Organization (WHO), is 3510 mg/day [55,56]. A sufficient intake of K is particularly important because K prevents cardiovascular disease and hypertension, kidney stones, and osteoporosis [55,57]. The second element, P, is a vital intracellular anion. It participates in various metabolic processes and serves as a component of cell membranes. In the body, it is usually found in the form of phosphate (PO_4^{3-}). P is essential in a variety of key biological molecules, including adenosine triphosphate (ATP) and 2,3-diphos-phoglycerate (2,3-DPG) [3,58]. Along with Ca, P is crucial for teeth and bone development [59]. Ca plays also a significant role in cell signaling and biochemical processes [60]. Disruption of Ca-P homeostasis leads to skeletal and cardiovascular disorders [61]. Therefore, the intake of these nutrients should be balanced. However, due to the use of phosphorus-based food additives, the phosphorus content of food has increased significantly [62,63] and in the United States, it far exceeds current recommendations for daily intake [64]. On the other hand, dietary intakes of the least abundant element, S, have decreased as a consequence of modern agricultural practice [65]. S is a component of important molecules, such as proteins, enzymes, and vitamins [66]. Hence, because of the high S, K, Ca and other essential macronutrients content proven in our study, bee products' consumption is beneficial for the human body.

Among determined micronutrients, we found Zn to be the most abundant in bee pollen, royal jelly, and propolis (Table 3). Zn is essential in all types of human tissues. It serves as a component or an activator of various enzymes involved in more than 300 enzymatic reactions [67]. Zn supplementation may prevent or facilitate the treatment of depressive illnesses, diarrhea, and, by enhancing immune responses, prevent pneumonia and viral diseases [68–71]. In addition to Zn, we measured relatively high Fe and Mn levels in all analyzed bee products. In the available literature, it has been already proven that bee pollen is an excellent source of Zn and Fe [42,72]. Fe possesses an essential role in several enzymatic functions, including oxygen transport and oxidative phosphorylation [73]. Fe deficiency, which is the leading cause of anemia worldwide, affects nearly one-third of the population [74]. Moreover, the deficit of Fe in infants results in learning and memory deficits [75]. Mn, similar to Fe, is associated with the activation and synthesis of various enzymes. It is involved in cellular metabolism regulation, bone mineralization, blood clotting, and cellular protection from reactive oxygen species (ROS) [76]. For example, Mn deficiency may result in skeletal abnormalities, dermatitis, deafness, or infertility [77]. To summarize, our study proved that all studied honeybee products (bee pollen, propolis, royal jelly) should be considered as a valuable source of both macro- and micronutrients. However, the levels of particular elements varied significantly between different types of products (Table S2).

Although bee products are rich in selected micro- and macronutrients, Pohl et al. [78] have recently reported that not all minerals are bioaccessible for humans after pollen in vitro digestion. According to the results of that research, Ca and Mg are among the most accessible elements contained in bee products. However, the bioaccessibility of Zn, Fe, and Cu may be lower than 40%. In the case of micro- and macroelements, low absorption after digestion is unfavorable. However, although it has not been investigated, the low elements' accessibility may also concern toxic heavy metals, and then, it is highly beneficial.

However, it should be kept in mind that elements occurring in natural products may have toxic effects on the body, even when consumed in small amounts. Taking into account the therapeutic and nourishing properties of bee products, it is important to know their contamination. Data from the available literature indicated that honeybee products are affected by environmental pollution [79]. However, there are scarce data that compare the inorganic contaminations (including heavy metals) between different bee products. All bee products contain potentially harmful trace elements and heavy metals, but our study results indicated significant differences in the degree of heavy metal contamination between them (Table S2). A clear distinction is observed, especially between the concentrations of heavy metals in the royal jelly compared to the bee pollen and propolis, which seem to be contaminated to a greater extent (Table 3). Among the elements with a potentially toxic effect on the human organism, Al was found at the highest levels in all studied beehive matrices. The highest Al concentration was observed in propolis, then bee pollen, and the lowest in royal jelly. Al is commonly found on Earth. However, due to anthropogenic activities, increasing Al content is observed in food [80]. Al accumulates in the brain, bones, kidney, and liver [81]. Prolonged exposure to even low levels of Al may lead to neurodegenerative disorders, such as Alzheimer's disease [82]. Al can also interfere with some essential elements, such as Ca, by replacing it and thus affecting the bones' mineralization [83].

Contamination of natural food is a serious problem, especially taking into account increasing environmental pollution. However, there is a lack of up-to-date regulations concerning bee products. Polish obsolete regulations specified only As, Cd, Cr, and Pb levels in honey (PN-88/A-77626: 1988), bee pollen (PN-R-78893: 1997), and propolis (PN-P-77627: 1997). Comparing our results to those regulations, we found that the studied samples of bee pollen fully met the requirements. However, levels of Cr and Pb in propolis were overdrawn. Excessive intake of Pb and Cr can lead to serious health consequences. Pb is toxic for children and adults, causing anemia, hypertension, cognitive deficits, immunodeficiency, infertility, bone and tooth development delays, vitamin D deficiency, and hepatic and renal effects [84,85]. The second element, Cr, although involved in human lipid and protein metabolism (in a form of Cr^{3+}) [86], is also connected with several pathologies, including carcinogenic (in a form of Cr^{6+}) [87,88]. Thus, finding these elements in propolis at excessive levels indicates that bee products must be tested before being released for consumption or medical use. However, proper regulations should be prepared, considering all potentially harmful trace elements and heavy metals. The results of our research may be the first step toward developing appropriate regulations.

The contaminants involved in bee products may arise both from beekeeping practice and from the environment. Trace elements and heavy metals can be transferred to honeybees and consequently to their products from all the environmental compartments (plants, soil, air, and water) in the areas adjacent to the hive [89,90]. The presence of toxic elements in bee pollen, propolis, and royal jelly is certainly associated with anthropogenic pollution around the apiaries. Hence, bees and their products can be used for effective environmental biomonitoring [91,92]. Road traffic and industry have a substantial impact on environmental pollution with trace elements and heavy metals. Elevated Cr content in the tested bee pollen samples may come from metal electroplating and metallurgical, paint, and leather tanning industries. The source of increased Pb levels can be metallurgical and glass industries. Additionally, Al can come from the cosmetic and pharmaceutical industries, Ni can come from the steel industry, and Cd can come from metal smelters [93,94]. Cd was also suggested to correlate with Pb contamination [46]. Moreover, heavy metal ions, such as Pb (II), Fe (III), Hg (II), Cu (II), and Cr (VI), are present in groundwater, which nowadays is one of the most severe environmental problems [95].

In addition to humans, bees are also affected by toxic heavy metal pollution. The data gathered from 2012 to 2014 in a pan-European epidemiological study on honeybee colony losses (EPILOBEE) revealed that bee colony mortality rates reached up to 36% [96]. A syndrome that is characterized by a sudden loss of worker bees in a colony has been named Colony Collapse Disorder [97]. A decline in bee colonies number affects all agricultural activities that rely on these pollinators' activity [98]. The origin of this loss is undoubtedly an anthropogenic activity, generating environmental contamination with elevated levels of toxic elements and heavy metals. In the study of Polish forager bees, Roman [99] reported that elemental contaminations accumulate in the bees' bodies to the degree depending on the industrialization of the region in which they live. Similar relationships were also observed by other researchers from different countries [100–103]. Bees' exposition to elevated heavy metals contaminants may affect the expression of genes encoding mainly enzymes involved in the detoxification metabolism, which indicates the physiological response of bees toward environmental pollutions [104]. In addition, bees' acetylcholinesterase levels have been suggested to correlate with inorganic contaminants [105]. Changes in enzymes expression may contribute to the disruption of homeostasis in bees, eventually causing their decline. Moreover, in bees exposed to heavy metals, alterations in feeding behavior have been observed [106]. Forager bees may reject the resources contaminated with some elements, such as Cu. On the other hand, they may prefer food moderately contaminated with Pb. That poses a significant risk to larvae, which are more sensitive to heavy metals than adult bees [79,106,107]. Feeding the larvae with contaminated resources may cause the accumulation of metals in their bodies, leading to an increased risk of brood survival decrease. A reduction in the number of bee colonies is a serious problem that requires immediate steps to prevent bees' extinction. As a first step, it is necessary to assess the factors influencing bee losses, including the pollution to which bees are exposed in their natural habitat. Thus, this study's results may contribute to broadening the knowledge of the quality of apiaries' surroundings in west-central Poland. Analysis of the data can help in deciding on possible ways to prevent bee decline.

Since the raw plant materials that bees use to make pollen may be contaminated to varying degrees, trace elements and heavy metals content in bee products may depend on the type of bee product and the kind of plant the bees use to produce it. In our study, we detected significantly lower concentrations of most elements in royal jelly compared to bee pollen and propolis. This is probably due to the fact that bee pollen and propolis are made directly from the pollen and resins of the plants flowering in the area covered by forager bees. Plants tend to accumulate heavy metals absorbed from the environment, mainly from water [108]. On the other hand, royal jelly is a pure secretion of bees' glands. Hence, the differences in elements content between royal jelly, bee pollen, and propolis may result from these products' origins. In central-west Poland, where the samples were collected, the primary pollen source is annual or biennial plants. Thus, samples of bee pollen used in our study contained mainly pollen of *Phacelia tanacetifolia*, *Achillea millefolium*, *Anthriscus sylvestris*, and *Brassica napus*. These plants, living just for a short period, perfectly reflect the dynamic changes in the environment's quality. Therefore, the results of our study may mirror the current environmental condition. Samples of the same bee product (collected from the same apiary) show some differences in contaminants levels, which are reflected in relatively high values of standard deviation (SD). Thus, it may be assumed that both bee pollen and propolis may be successfully used for biomonitoring.

Several reports presenting the elemental content in bee products collected in different regions of Earth can be found in the available literature. Temizer et al. [39] analyzed 20 elements in bee pollen collected in Turkey. In the study, the authors found the highest level of Fe in all pollen samples. In general, the Al, As, Cr, and Fe concentrations found in the reported research were higher than those in our study. On the other hand, Mn and Zn levels were higher in samples examined in our research (i.e., pollen samples of Polish origin). Turkish propolis was also analyzed by Altunatmaz et al. [40], who measured the concentrations of 12 chemical elements. The concentration values obtained in that study were in general higher than ours. Only levels of Ni, Pb, Si, and Zn were reported to be lower than in our research. The comparison of our results with data obtained by Formicki et al. [41], who analyzed Polish bee pollen and propolis, revealed that our studied samples contained, in general, lower levels of chemical elements. However, the authors of the above-mentioned article measured concentrations of only six elements: Cd, Ni, Pb, Fe, Mg, and Zn, whereas our study involved 24 elements. An explanation for the observed differences between beehive products may be the origin of the bee pollen and propolis used in the study. Although the samples in both studies were collected in Poland, our samples were collected in the Greater Poland Voivodeship in west-central Poland. The apiary where our samples were collected is located in Góry Złotnickie village. It is a typically agricultural region. There are fields within a radius of 20 km (mainly rapeseed, phacelia, and clusters of linden, acacia and fruit trees), as well as uncultivated lands and meadows. This area is several kilometers away from human settlements. The nearest bigger city (Kalisz, population c.a. 100,000) is about 20 km away from the apiary. In contrast, Formicki et al. used samples from Lesser Poland Voivodeship in southern Poland. The Lesser Poland region is an area of high industrial and agricultural activity. In this province, heavy metals are the most common and toxic contaminants [109]. Bee pollen was also analyzed by researchers from other countries, such as Serbia, New South Wales, and Greece (42–44). In these studies, the concentrations of the selected elements were generally higher than in Polish samples analyzed by us (see Supplementary Materials).

Polish propolis was also analyzed by Roman et al. [45] but only for Cd, Co, Pb, Zn, and As content. The concentration of the selected elements in that study was higher than in ours. This is also probably due to the location of the apiaries—Roman et al. analyzed propolis collected from a heavily industrialized region (Lower Silesia Voivodeship). Additionally, the apiaries were located in the area of Wrocław—a large provincial city. Comparisons of our study with the literature reports suggest that a higher pollutant content may be associated with higher industrial and agricultural activity. This observation supports the assumption that bee products are excellent matrices for biomonitoring.

Comparing our results with the elements' levels measured in Brazilian propolis [46,47] indicated lower levels of both nutritional and toxic elements in samples of propolis collected in Poland. In addition, Polish propolis seems to be less contaminated than samples collected in Spain [48]. However, samples of Macedonian propolis [49] contained lower concentrations of the selected chemical elements than our beehive products. These notable discrepancies between the literature reports and our study prove that bee products are characterized by a high variability (see Supplementary Materials). For this reason, quality testing of bee products should be performed routinely.

In the available literature, there are very limited data reporting concentrations of chemical elements in royal jelly. Stocker et al. [50] analyzed royal jelly samples collected in France. The authors showed higher elemental concentrations in royal jelly than measured in our study. However, the correlation between the levels of individual elements in the study of Stocker et al. [50] and our research was comparable. Royal jelly, which is the pure secretion of bee glands and contains any components taken directly from plants or the environment, can also serve as a biomonitoring tool. Bees are inextricably linked to the natural environment. Therefore, both bee secretions and the bee's whole body can reflect the degree of environmental pollution [110].

4. Materials and Methods

4.1. Sample Collection

Bee pollen, propolis, and royal jelly samples were harvested directly from hives of *Apis mellifera* bees in apiaries located in Góry Złotnickie village (N 51°87′504″, E 18°12′431″), Greater Poland Voivodeship in west-central Poland. Bee pollen and propolis samples were collected from May 2018 to July 2019 during the summer season. We analyzed three bee pollen samples collected in 2018 and three collected in 2019 as well as three propolis samples collected in 2018 and three in 2019. Moreover, pilot tests were conducted using a royal jelly sample collected in 2019. The samples were stored at -80 °C in the darkness until analysis.

4.2. Sample Preparation

A microwave-assisted digestion method was used for sample preparation. Mineralization was conducted in closed vessels made of PTFE. The weighed honeybee products (0.25–0.75 g) were mixed with 10 mL of suprapure grade 65% nitric acid (Merck, Darmstadt, Germany). The mineralization temperature was set at 180 °C, and the duration was 20 min. After cooling, the minerals were quantitatively transferred to centrifuge tubes with a capacity of 50 mL, and the volume of the solution was adjusted to the mark with deionized water. The obtained solutions were filtered through 0.45 μm polypropylene syringe filters and diluted with deionized water in an appropriate ratio. The pilot tests and subsequent validation procedures demonstrated that the proposed method of mineralization is effective enough to convert the elements into a form enabling their quantitative determination.

4.3. Elemental Analysis

A quadrupole ICP-MS 7800 (Agilent Technologies, Tokyo, Japan) was used for the determination of heavy metals and trace elements (Ag, Al, As, Ba, Cd, Co, Cr, Cu, Mn, Mo, Ni, Pb, Sb, Se, V, and Zn) in the collected honeybee products. Additionally, ICP-OES experiments were performed to determine the main elements' content (Ca, Fe, K, Mg, Na, P, S, and Si) in the studied beehive products. The assays were performed in accordance with the requirements of PN-EN ISO 17294-2: 2016-11 standard.

The methods used were validated in terms of linearity, the limit of detection, the limit of quantification, precision, and recovery. Calibration curves for each element were determined from seven calibration solutions analyzed in triplicate. Calibration solutions were prepared using commercial standard solutions (c = 1000 mg/kg, Merck Darmstadt, Germany), consistent with NIST standards, and the serial dilution method. The linearity of the calibration curves was found in the working range of the method for each of the analyzed elements. The instrumental limits of detection and quantification were estimated as three times and 10 times the standard deviation, respectively, from the measurements of a blank sample with the addition of a small amount of standard. Precision was determined by analyzing real samples representing three tested matrices (propolis, bee pollen, royal jelly). Each sample was analyzed in seven replicates (the entire analytical procedure starting from weighing the sample was conducted). Precision values expressed as coefficient of variation (%RSD) were determined for each type of matrix separately. In the case of unfortified samples, precision was evaluated only for analytes exceeding the limit of quantification in the tested samples. Due to some analytes' low content in real samples, an additional determination of precision was performed for fortified samples. Precision (%RSD) was calculated based on three repetitions of each fortified sample. The analysis of the fortified matrices proved that the applied analytical technique is characterized by high precision (Supplementary Materials, Table S3). However, higher %RSD values were obtained based on analysis of unfortified samples. This could be partially explained by some elements' content in the lower working range of the method (near the limit of quantification). Within the validation, recovery tests of the fortified samples were also performed. Honeybee products were fortified (before mineralization) at two concentration levels within the method's working range. The percent recovery was calculated according to the equation:

%R = (conc. in fortified sample − conc. in unfortified sample)/(amount of standard added) * 100%

The obtained results were in line with generally accepted requirements, according to which the recovery values for pollutants present in amounts (m/m) below 0.1% should be within the range of 75–125%. The detailed values of the method parameters determined during validation are contained in Supplementary Materials in Table S3. To sum up, the performed validation experiments showed good linearity, satisfactory precision, and acceptable accuracy of the applied method.

Elemental analysis of the beehive products was preceded by an analysis of calibration solutions, based on which a calibration curve is prepared. The matrix of the calibration solutions corresponded to the matrix of the analyzed samples. In order to eliminate interferences of matrix origin, an internal standard method was used. The internal standard solution was introduced in the same amount to all analyzed solutions (calibration solutions, real samples, quality control samples) using a separate peristaltic pump line. A solution of Sc, Y and Tb was used as an internal standard. Polyatomic interferences were eliminated by using a crash chamber with the use of helium as a reaction gas.

4.4. Statistical Analysis

The statistical analyses were conducted using Statistica 13.0 (TIBCO Software Inc., CA, USA) and the MetaboAnalyst 5.0 [111]. In the first step of univariate statistical analyses, the Shapiro–Wilk test of normality was applied. Then, to assess the equality of variances, Levene's test was used. Variables with normal distribution and equal variances were subjected to a t-test, whereas variables with normal distribution and not equal variances were subjected to Welch's test. For the analysis of variables that were not normally distributed, the Mann–Whitney U test was used. In all tests, a p-value ≤ 0.05 was considered statistically significant. Additionally, in order to visualize the differences in the elemental composition of the analyzed bee products, a multivariate statistical analysis - principal component analysis (PCA) was conducted. Prior to PCA, autoscaling of the elements' concentrations were performed.

5. Conclusions

In the current study, we analyzed one of the broadest ranges of chemical elements in the selected beehive products. Among the determined elements were macro- and micronutrients, trace elements, and heavy metals. Therefore, the obtained data contributed to assessing both nutritional value and levels of inorganic contamination of bee products collected in west-central Poland. Although bee pollen, propolis, and royal jelly are characterized by different micronutrient and macronutrient content, it is impossible to consider one product as the best dietary supplement. The choice of a bee product should be tailored to the condition whose treatment we want to support. It should be borne in mind that apart from micronutrients and macronutrients, bee products contain other valuable components, such as proteins, amino acids, lipids, vitamins, and minerals. However, the concentrations of heavy metals provide us information on the safety of using bee products. The levels of heavy metals measured in bee pollen and propolis were generally lower than in samples collected in Poland's highly industrialized regions. These results suggest that due to the high variability of the contaminants of bee products from the same country, measurements should be performed routinely. Moreover, bee pollen samples analyzed in this study seem to be less contaminated than those collected in Turkey, and propolis examined in this research contained lower inorganic pollutants than collected in Brazil and Spain. These results may indicate better environmental quality in the vicinity of apiaries in west-central Poland. It may also be concluded that honeybees and their products may serve as sensitive bioindicators, reflecting the degree and dynamic changes of environmental pollution.

Supplementary Materials: The following are available online. Tables S1 and S2. Comparison of the selected elements' levels in bee pollen to the available literature data; Table S3. Performance characteristics of the analytical method used; Figures S1–S17: Bar charts visualizing the differences between levels (mean±SD) of the selected heavy metals measured in bee pollen and propolis in our study and by other research teams.

Author Contributions: Conceptualization, E.M., A.K., Z.J.K. and J.M.; methodology, K.M.; software, E.M., A.K. and K.M.; validation, K.M.; formal analysis, E.M. and A.K.; investigation, K.M. resources, J.M.; data curation, E.M. and A.K.; writing—original draft preparation, E.M. and A.K.; writing—review and editing, E.M., A.K., K.M., Z.J.K. and J.M.; visualization, E.M. and A.K.; supervision, J.M.; project administration, J.M.; funding acquisition, J.M. All authors have read and agreed to the published version of the manuscript.

Funding: This project received financial support from the Polish National Science Centre (grant number 2016/23/D/NZ7/03949).

Institutional Review Board Statement: Not applicable.

Informed Consent Statement: Not applicable.

Data Availability Statement: The data presented in this study contained within the article and supplementary materials.

Conflicts of Interest: The authors declare no conflict of interest.

Sample Availability: Not available.

References

1. Gharibzahedi, S.M.T.; Jafari, S.M. The importance of minerals in human nutrition: Bioavailability, food fortification, processing effects and nanoencapsulation. *Trends Food Sci. Technol.* **2017**, *62*, 119–132. [CrossRef]
2. Al-fartusie, F.S.; Mohssan, S.N. Essential Trace Elements and Their Vital Roles in Human Body. *Indian J. Adv. Chem. Sci.* **2017**, *5*, 127–136.
3. Soetan, K.O.; Olaiya, C.O.; Oyewole, O.E. The importance of mineral elements for humans, domestic animals and plants: A review. *Afr. J. Food Sci.* **2010**, *4*, 200–222.
4. Zoroddu, M.A.; Aaseth, J.; Crisponi, G.; Medici, S.; Peana, M.; Nurchi, V.M. The essential metals for humans: A brief overview. *J. Inorg. Biochem* **2019**, *195*, 120–129. [CrossRef] [PubMed]
5. Quintaes, K.D.; Diez-Garcia, R.W. *The importance of minerals in the human diet. Handbook of Mineral Elements in Food*; John Wiley & Sons: New York, NY, USA, 2015; pp. 1–21, (Wiley Online Books).

6. Gupta, S.C. Sources and Deficiency Diseases of Mineral Nutrients in Human Health and Nutrition: A Review. *Pedosphere* **2014**, *24*, 13–38. [CrossRef]
7. Stein, A.J. Global impacts of human mineral malnutrition. *Plant. Soil* **2010**, *335*, 133–154. [CrossRef]
8. Verkaik-Kloosterman, J.; McCann, M.; Hoekstra, J.; Verhagen, H. Vitamins and minerals: Issues associated with too low and too high population intakes. *Food Nutr. Res.* **2012**, *56*, 5728. [CrossRef] [PubMed]
9. Fu, Z.; Xi, S. The effects of heavy metals on human metabolism. *Toxicol Mech. Methods* **2020**, *30*, 167–176. [CrossRef] [PubMed]
10. Matés, J.M.; Segura, J.A.; Alonso, F.J.; Márquez, J. Roles of dioxins and heavy metals in cancer and neurological diseases using ROS-mediated mechanisms. *Free Radic. Biol. Med.* **2010**, *49*, 1328–1341. [CrossRef]
11. Cicero, C.E.; Mostile, G.; Vasta, R.; Rapisarda, V.; Signorelli, S.S.; Ferrante, M.; Zappio, M.; Nicoletti, A. Metals and neurodegenerative diseases. A systematic review. *Env. Res.* **2017**, *159*, 82–94. [CrossRef]
12. Cornara, L.; Biagi, M.; Xiao, J.; Burlando, B. Therapeutic Properties of Bioactive Compounds from Different Honeybee Products. *Front. Pharmacol.* **2017**, *412*, 1–20. [CrossRef]
13. Alvarez-Suarez, J.M. *Bee Products-Chemical and Biological Properties*; Springer International Publishing: New York, NY, USA, 2017; pp. 1–306.
14. Fratellone, P.M.; Tsimis, F.; Fratellone, G. Apitherapy Products for Medicinal Use. *J. Altern. Complement Med.* **2016**, *22*, 1020–1022. [CrossRef] [PubMed]
15. Rzepecka-Stojko, A.; Stojko, J.; Jasik, K.; Buszman, E. Anti-Atherogenic Activity of Polyphenol-Rich Extract from Bee Pollen. *Nutrients* **2017**, *9*, 1369. [CrossRef] [PubMed]
16. Denisow, B.; Denisow-Pietrzyk, M. Biological and therapeutic properties of bee pollen: A review. *J. Sci. Food Agric.* **2016**, *96*, 4303–4309. [CrossRef] [PubMed]
17. Anjum, S.I.; Ullah, A.; Khan, K.A.; Attaullah, M.; Khan, H.; Ali, H.; Bashir, M.A.; Tahir, M.; Ansari, M.J.; Ghramh, H.A. Composition and functional properties of propolis (bee glue): A review. *Saudi. J. Biol. Sci* **2019**, *26*, 1695–1703. [CrossRef]
18. Popova, M.; Giannopoulou, E.; Skalicka-Woźniak, K.; Graikou, K.; Widelski, J.; Bankova, V.; Kalofonos, H.; Sivolapenko, G.; Gaweł-Bęben, K.; Antosiewicz, B.; et al. Characterization and Biological Evaluation of Propolis from Poland. *Molecules* **2017**, *22*, 1159. [CrossRef] [PubMed]
19. Pobiega, K.; Kraśniewska, K.; Derewiaka, D.; Gniewosz, M. Comparison of the antimicrobial activity of propolis extracts obtained by means of various extraction methods. *J. Food Sci. Technol.* **2019**, *56*, 5386–5395. [CrossRef] [PubMed]
20. Gucwa, K.; Kusznierewicz, B.; Milewski, S.; Van Dijck, P.; Szweda, P. Antifungal Activity and Synergism with Azoles of Polish Propolis. *Pathogens* **2018**, *7*, 56. [CrossRef] [PubMed]
21. Pobiega, K.; Kraśniewska, K.; Przybył, J.L.; Bączek, K.; Żubernik, J.; Witrowa-Rajchert, D.; Gniewosz, M. Growth Biocontrol of Foodborne Pathogens and Spoilage Microorganisms of Food by Polish Propolis Extracts. *Molecules* **2019**, *24*, 2965. [CrossRef]
22. Fratini, F.; Cilia, G.; Mancini, S.; Felicioli, A. Royal Jelly: An ancient remedy with remarkable antibacterial properties. *Microbiol Res.* **2016**, *192*, 130–141. [CrossRef]
23. Chen, Y.-F.; Wang, K.; Zhang, Y.-Z.; Zheng, Y.-F.; Hu, F.-L. In Vitro Anti-Inflammatory Effects of Three Fatty Acids from Royal Jelly. Yin, J.; editor. *Mediat. Inflamm* **2016**, *2016*, 3583684. [CrossRef]
24. Guo, H.; Kouzuma, Y.; Yonekura, M. Structures and properties of antioxidative peptides derived from royal jelly protein. *Food Chem* **2009**, *113*, 238–245. [CrossRef]
25. Nikokar, M.; Shirzad, M.; Kordyazdi, R.; Shahinfard, N. Does Royal jelly affect tumor cells? *J. Herbmed Pharm.* **2013**, *2*, 45–48.
26. Taulavuori, K.; Julkunen-Tiitto, R.; Hyöky, V.; Taulavuori, E. Blue Mood for Superfood. *Nat. Prod. Commun* **2013**, *8*, 1934578X1300800627. [CrossRef]
27. Lau, P.; Bryant, V.; Ellis, J.D.; Huang, Z.Y.; Sullivan, J.; Schmehl, D.R.; Cabrera, A.R.; Rangel, J. Seasonal variation of pollen collected by honey bees (Apis mellifera) in developed areas across four regions in the United States. *PLoS ONE* **2019**, *14*, 0217294. [CrossRef]
28. Sabatini, A.G. Quality and standardization of Royal Jelly. *J. Apiproduct Apimedical Sci* **2009**, *1*, 16–21. [CrossRef]
29. Chaillou, L.L.; Nazareno, M.A. Chemical variability in propolis from Santiago del Estero, Argentina, related to the arboreal environment as the sources of resins. *J. Sci. Food Agric.* **2009**, *89*, 978–983. [CrossRef]
30. Golubkina, N.A.; Sheshnitsan, S.S.; Kapitalchuk, M.V.; Erdenotsogt, E. Variations of chemical element composition of bee and beekeeping products in different taxons of the biosphere. *Ecol. Indic.* **2016**, *66*, 452–457. [CrossRef]
31. Morgano, M.A.; Teixeira Martins, M.C.; Rabonato, L.C.; Milani, R.F.; Yotsuyanagi, K.; Rodriguez-Amaya, D.B. A comprehensive investigation of the mineral composition of Brazilian bee pollen: Geographic and seasonal variations and contribution to human diet. *J. Braz. Chem. Soc.* **2012**, *23*, 727–736. [CrossRef]
32. Neumann, P.; Carreck, N.L. Honey bee colony losses. *J. Apic. Res.* **2010**, *49*, 1–6. [CrossRef]
33. Feldhaar, H.; Otti, O. Pollutants and Their Interaction with Diseases of Social Hymenoptera. *Insects* **2020**, *11*, 153. [CrossRef]
34. Kokot, Z.J.; Matysiak, J. Inductively coupled plasma mass spectrometry determination of metals in honeybee venom. *J. Pharm. Biomed. Anal.* **2008**, *48*, 955–959. [CrossRef]
35. Solayman, M.; Islam, M.A.; Paul, S.; Ali, Y.; Khalil, M.I.; Alam, N.; Gan, S.H. Physicochemical Properties, Minerals, Trace Elements, and Heavy Metals in Honey of Different Origins: A Comprehensive Review. *Compr Rev. Food Sci. Food Saf.* **2016**, *15*, 219–233. [CrossRef]

36. Adugna, E.; Hymete, A.; Birhanu, G.; Ashenef, A. Determination of some heavy metals in honey from different regions of Ethiopia. Yildiz, F.; editor. *Cogent. Food Agric.* **2020**, *6*, 1764182. [CrossRef]
37. Lazor, P.; Tomas, J.; Toth, T.; Toth, J.; Ceryova, S. Monitoring of air pollution and atmospheric deposition of heavy metals by analysis of honey. *J. Microbiol Biotechnol Food Sci.* **2012**, *1*, 522–533.
38. Bazeyad, A.Y.; Al-Sarar, A.S.; Rushdi, A.I.; Hassanin, A.S.; Abobakr, Y. Levels of heavy metals in a multifloral Saudi honey. *Env. Sci. Pollut. Res.* **2019**, *26*, 3946–3953. [CrossRef]
39. Temizer, İ.K.; Güder, A.; Temel, F.A.; Avcl, E. A comparison of the antioxidant activities and biomonitoring of heavy metals by pollen in the urban environments. *Env. Monit. Assess.* **2018**, *190*, 462. [CrossRef]
40. Altunatmaz, S.S.; Tarhan, D.; Aksu, F.; Barutçu, U.B.; Or, M.E. Mineral element and heavy metal (Cadmium, lead and arsenic) levels of bee pollen in Turkey. *Food Sci. Technol.* **2017**, *37*, 136–141. [CrossRef]
41. Formicki, G.; Greń, A.; Stawarz, R.; Zyśk, B.; Gał, A. Metal Content in Honey, Propolis, Wax, and Bee Pollen and Implications for Metal Pollution Monitoring. *Pol. J. Env. Stud.* **2013**, *22*, 99–106.
42. Kostić, A.Z.; Pešić, M.B.; Mosić, M.; Dojèinović, B.P.; Natić, M.M.; Trifković, J.D. Mineral content of bee pollen from Serbia. *Arh Hig. Rada. Toksikol* **2015**, *66*, 251–258. [CrossRef]
43. Somerville, D.C.; Nicol, H.I. Mineral content of honeybee-collected pollen from southern New South Wales. *Aust. J. Exp. Agric.* **2002**, *42*, 1131–1136. [CrossRef]
44. Liolios, V.; Tananaki, C.; Papaioannou, A.; Kanelis, D.; Rodopoulou, M.A.; Argena, N. Mineral content in monofloral bee pollen: Investigation of the effect of the botanical and geographical origin. *J. Food Meas. Charact.* **2019**, *13*, 1674–1682. [CrossRef]
45. Roman, A.; Madras-Majewska, B.; Popiela-Pleban, E. Comparative study of selected toxic elements in propolis and honey. *J. Apic. Sci.* **2011**, *55*, 97–106.
46. Finger, D.; Filho, I.K.; Torres, Y.R.; Quináia, S.P. Propolis as an indicator of environmental contamination by metals. *Bull. Env. Contam. Toxicol* **2014**, *92*, 259–264. [CrossRef] [PubMed]
47. Hodel, K.V.S.; Machado, B.A.S.; Santos, N.R.; Costa, R.G.; Menezes-Filho, J.A.; Umsza-Guez, M.A. Metal Content of Nutritional and Toxic Value in Different Types of Brazilian Propolis. *Sci. World. J.* **2020**, *2020*, 4395496. [CrossRef]
48. Bonvehí, J.S.; Bermejo, F.J.O. Element content of propolis collected from different areas of South Spain. *Env. Monit. Assess.* **2013**, *185*, 6035–6047. [CrossRef]
49. Popov, B.B.; Hristova, V.K.; Presilski, S.; Shariati, M.A.; Najman, S. Assessment of heavy metals in propolis and soil from the Pelagonia region, republic of Macedonia. *Maced, J. Chem. Chem. Eng.* **2017**, *36*, 1–11. [CrossRef]
50. Stocker, A.; Schramel, P.; Kettrup, A.; Bengsch, E. Trace and mineral elements in royal jelly and homeostatic effects. *J. Trace Elem. Med. Biol* **2005**, *19*, 183–189. [CrossRef] [PubMed]
51. Komosinska-Vassev, K.; Olczyk, P.; Kaźmierczak, J.; Mencner, L.; Olczyk, K. Bee pollen: Chemical composition and therapeutic application. *Evid. Based Complement. Altern. Med.* **2015**, *2015*, 297425. [CrossRef]
52. Kieliszek, M.; Piwowarek, K.; Kot, A.M.; Błażejak, S.; Chlebowska-Śmigiel, A.; Wolska, I. Pollen and bee bread as new health-oriented products: A review. *Trends Food Sci. Technol* **2018**, *71*, 170–180. [CrossRef]
53. Tosic, S.; Stojanovic, G.; Mitic, S.; Pavlovic, A.; Alagic, S. Mineral composition of selected serbian propolis samples. *J. Apic. Sci.* **2017**, *61*, 5–15. [CrossRef]
54. Demigné, C.; Sabboh, H.; Rémésy, C.; Meneton, P. Protective Effects of High Dietary Potassium: Nutritional and Metabolic Aspects. *J. Nutr.* **2004**, *134*, 2903–2906. [CrossRef] [PubMed]
55. Palmer, B.F.; Clegg, D.J. Achieving the Benefits of a High-Potassium, Paleolithic Diet, Without the Toxicity. *Mayo. Clin. Proc.* **2016**, *91*, 496–508. [CrossRef]
56. *Guideline: Potassium Intake for Adults and Children*; WHO Press: Geneva, Switzerland, 2012.
57. Adrogué, H.J.; Madias, N.E. The Impact of Sodium and Potassium on Hypertension Risk. *Semin. Nephrol.* **2014**, *34*, 257–272. [CrossRef] [PubMed]
58. Kraft, M.D. Phosphorus and Calcium. *Nutr Clin. Pr.* **2015**, *30*, 21–33. [CrossRef]
59. Loughrill, E.; Wray, D.; Christides, T.; Zand, N. Calcium to phosphorus ratio, essential elements and vitamin D content of infant foods in the UK: Possible implications for bone health. *Matern. Child. Nutr.* **2017**, *13*, 12368. [CrossRef]
60. Bazydlo, L.A.L.; Needham, M.; Harris, N.S. Calcium, Magnesium, and Phosphate. *Lab. Med.* **2014**, *45*, 44–50. [CrossRef]
61. Sun, M.; Wu, X.; Yu, Y.; Wang, L.; Xie, D.; Zhang, Z.; Chen, L.; Lu, A.; Zhang, G.; Li, F. Disorders of Calcium and Phosphorus Metabolism and the Proteomics/Metabolomics-Based Research. *Front. Cell. Dev. Biol.* **2020**, *8*, 576110. [CrossRef]
62. Gutiérrez, O.M. Sodium- and phosphorus-based food additives: Persistent but surmountable hurdles in the management of nutrition in chronic kidney disease. *Adv. Chronic. Kidney Dis.* **2013**, *20*, 150–156. [CrossRef]
63. Gutiérrez, O.M.; Luzuriaga-McPherson, A.; Lin, Y.; Gilbert, L.C.; Ha, S.-W.; Beck, G.R., Jr. Impact of Phosphorus-Based Food Additives on Bone and Mineral Metabolism. *J. Clin. Endocrinol Metab.* **2015**, *100*, 4264–4271. [CrossRef]
64. Calvo, M.S.; Uribarri, J. Public health impact of dietary phosphorus excess on bone and cardiovascular health in the general population. *Am. J. Clin. Nutr.* **2013**, *98*, 6–15. [CrossRef] [PubMed]
65. Parcell, S. Sulfur in human nutrition and applications in medicine. *Altern. Med. Rev.* **2002**, *7*, 22–44.
66. Hewlings, S.; Kalman, D. Sulfur and Human Health. *Ec. Nutr.* **2019**, *14*, 785–791.
67. Białek, M.; Zyska, A. The Biomedical Role of Zinc in the Functioning of the Human Organism. *Pol. J. Public Heal.* **2014**, *124*, 160–163. [CrossRef]

68. Penny, M.E. Zinc Supplementation in Public Health. *Ann. Nutr. Metab.* **2013**, *62*, 31–42. [CrossRef] [PubMed]
69. Patel, A.; Mamtani, M.; Dibley, M.J.; Badhoniya, N.; Kulkarni, H. Therapeutic Value of Zinc Supplementation in Acute and Persistent Diarrhea: A Systematic Review. *PLoS ONE* **2010**, *5*, 10386. [CrossRef]
70. Lai, J.; Moxey, A.; Nowak, G.; Vashum, K.; Bailey, K.; McEvoy, M. The efficacy of zinc supplementation in depression: Systematic review of randomized controlled trials. *J. Affect. Disord.* **2012**, *136*, 31–39. [CrossRef]
71. Wessels, I.; Rolles, B.; Rink, L. The Potential Impact of Zinc Supplementation on COVID-19 Pathogenesis. *Front. Immunol.* **2020**, *11*, 1712. [CrossRef]
72. Serra Bonvehí, J.; Escolà Jordà, R. Nutrient Composition and Microbiological Quality of Honeybee-Collected Pollen in Spain. *J. Agric. Food Chem.* **1997**, *45*, 725–732. [CrossRef]
73. Milic, S.; Mikolasevic, I.; Orlic, L.; Devcic, E.; Starcevic-Cizmarevic, N.; Stimac, D.; Kapovic, M.; Ristic, S. The Role of Iron and Iron Overload in Chronic Liver Disease. *Med. Sci. Monit.* **2016**, *22*, 2144–2151. [CrossRef]
74. Elstrott, B.; Khan, L.; Olson, S.; Raghunathan, V.; DeLoughery, T.; Shatzel, J.J. The role of iron repletion in adult iron deficiency anemia and other diseases. *Eur. J. Haematol.* **2020**, *104*, 153–161. [CrossRef] [PubMed]
75. Fretham, S.J.B.; Carlson, E.S.; Georgieff, M.K. The Role of Iron in Learning and Memory. *Adv. Nutr.* **2011**, *2*, 112–121. [CrossRef]
76. Attar, T. A mini-review on importance and role of trace elements in the human organism. *Chem. Rev. Lett.* **2020**, *3*, 117–130.
77. Dutta, T.K.; Mukta, V. Trace elements. *Med. Updat.* **2012**, *22*, 353–357.
78. Pohl, P.; Dzimitrowicz, A.; Lesniewicz, A.; Welna, M.; Szymczycha-Madeja, A.; Cyganowski, P.; Jamroz, P. Room temperature solvent extraction for simple and fast determination of total concentration of Ca, Cu, Fe, Mg, Mn, and Zn in bee pollen by FAAS along with assessment of the bioaccessible fraction of these elements using in vitro gastrointestinal digestion. *J. Trace Elem. Med. Biol.* **2020**, *60*, 126479. [CrossRef]
79. Hladun, K.R.; Di, N.; Liu, T.X.; Trumble, J.T. Metal contaminant accumulation in the hive: Consequences for whole-colony health and brood production in the honey bee (*Apis mellifera* L.). *Environ. Toxicol. Chem.* **2016**, *35*, 322–329. [CrossRef]
80. Paz, S. Aluminium Exposure through the Diet. *Food Sci. Nutr.* **2017**, *3*, 1–10. [CrossRef]
81. Verstraeten, S.V.; Aimo, L.; Oteiza, P.I. Aluminium and lead: Molecular mechanisms of brain toxicity. *Arch. Toxicol.* **2008**, *82*, 789–802. [CrossRef] [PubMed]
82. Bondy, S.C. Prolonged exposure to low levels of aluminum leads to changes associated with brain aging and neurodegeneration. *Toxicology* **2014**, *315*, 1–7. [CrossRef] [PubMed]
83. Malluche, H.H. Aluminium and bone disease in chronic renal failure. *Nephrol. Dial. Transplant.* **2002**, *17*, 21–24. [CrossRef] [PubMed]
84. Obeng-Gyasi, E. Sources of lead exposure in various countries. *Rev. Environ. Health* **2019**, *34*, 25–34. [CrossRef]
85. Mitra, P.; Sharma, S.; Purohit, P.; Sharma, P. Clinical and molecular aspects of lead toxicity: An update. *Crit. Rev. Clin. Lab. Sci.* **2017**, *54*, 506–528. [CrossRef]
86. Bielicka, A.; Bojanowska, I.; Wiśniewski, A. Two Faces of Chromium - Pollutant and Bioelement. *Polish. J. Environ. Stud.* **2005**, *14*, 5–10.
87. Pavesi, T.; Moreira, J.C. Mechanisms and individuality in chromium toxicity in humans. *J. Appl. Toxicol* **2020**, *40*, 1183–1197. [CrossRef] [PubMed]
88. Achmad, R.T.; Auerkari, E.I. Effects of chromium on human body. *Annu. Res. Rev. Biol.* **2017**, *13*, 1–8. [CrossRef]
89. Álvarez-Ayuso, E.; Abad-Valle, P. Trace element levels in an area impacted by old mining operations and their relationship with beehive products. *Sci Total Environ.* **2017**, *599–600*, 671–678. [CrossRef] [PubMed]
90. Astolfi, M.L.; Conti, M.E.; Marconi, E.; Massimi, L.; Canepari, S. Effectiveness of Different Sample Treatments for the Elemental Characterization of Bees and Beehive Products. *Molecules* **2020**, *25*, 4263. [CrossRef]
91. AL-Alam, J.; Chbani, A.; Faljoun, Z.; Millet, M. The use of vegetation, bees, and snails as important tools for the biomonitoring of atmospheric pollution—a review. *Environ. Sci. Pollut. Res.* **2019**, *26*, 9391–9408. [CrossRef]
92. Lambert, O.; Piroux, M.; Puyo, S.; Thorin, C.; Larhantec, M.; Delbac, F.; Pouliquen, H. Bees, honey and pollen as sentinels for lead environmental contamination. *Environ. Pollut.* **2012**, *170*, 254–259. [CrossRef]
93. Zwolak, A.; Sarzyńska, M.; Szpyrka, E.; Stawarczyk, K. Sources of Soil Pollution by Heavy Metals and Their Accumulation in Vegetables: A Review. *Water Air Soil Pollut* **2019**, *230*, 164. [CrossRef]
94. Stahl, T.; Taschan, H.; Brunn, H. Aluminium content of selected foods and food products. *Environ. Sci. Eur.* **2011**, *23*, 37. [CrossRef]
95. Liu, W.; Yang, L.; Xu, S.; Chen, Y.; Liu, B.; Li, Z.; Jiang, C. Efficient removal of hexavalent chromium from water by an adsorption-reduction mechanism with sandwiched nanocomposites. *RSC. Adv.* **2018**, *8*, 15087–15093. [CrossRef]
96. Chauzat, M.-P.; Laurent, M.; Ribiere-Chabert, M.; Hendrikx, P. A Pan-European Epidemiological Study on Honeybee Colony Losses 2012–2014. 2016:44. Available online: https://ec.europa.eu/food/sites/food/files/animals/docs/la_bees_epilobee-report_2012-2014.pdf (accessed on 21 April 2021).
97. vanEngelsdorp, D.; Evans, J.D.; Saegerman, C.; Mullin, C.; Haubruge, E.; Nguyen, B.K.; Frazier, M.; Frazier, J.; Cox-Foster, D.; Chen, Y. Colony Collapse Disorder: A Descriptive Study. *PLoS ONE* **2009**, *4*, 6481. [CrossRef]
98. Stokstad, E. The Case of the Empty Hives. *Science* **2007**, *316*, 970–972. [CrossRef] [PubMed]
99. Roman, A. Levels of Copper, Selenium, Lead, and Cadmium in Forager Bees. *Polish J. Environ. Stud.* **2010**, *19*, 663–669.
100. Zhelyazkova, I.; Atanasova, S.; Barakova, V.; Mihaylova, G. Content of heavy metals and metalloids in bees and bee products from areas with different degree of anthropogenic impact. *Agric. Sci. Technol.* **2010**, *3*, 136–142.

101. Perugini, M.; Manera, M.; Grotta, L.; Abete, M.C.; Tarasco, R.; Amorena, M. Heavy Metal (Hg, Cr, Cd, and Pb) Contamination in Urban Areas and Wildlife Reserves: Honeybees as Bioindicators. *Biol. Trace Elem. Res.* **2011**, *140*, 170–176. [CrossRef]
102. Gutiérrez, M.; Molero, R.; Gaju, M.; van der Steen, J.; Porrini, C.; Ruiz, J.A. Assessment of heavy metal pollution in Córdoba (Spain) by biomonitoring foraging honeybee. *Environ. Monit. Assess.* **2015**, *187*, 651. [CrossRef]
103. Gliga, O. The content of heavy metals in the bees body depending on location area of hives. *Sci Pap. Anim. Sci. Ser.* **2016**, *65*, 169–175.
104. Gizaw, G.; Kim, Y.; Moon, K.; Choi, J.B.; Kim, Y.H.; Park, J.K. Effect of environmental heavy metals on the expression of detoxification-related genes in honey bee *Apis. Mellifera Apidologie* **2020**, *51*, 664–674. [CrossRef]
105. Abdelhameed, K.M.A.; Khalifa, M.H.; Aly, G.F. Heavy metal accumulation and the possible correlation with acetylcholinesterase levels in honey bees from polluted areas of Alexandria, Egypt. *African. Entomol.* **2020**, *28*, 385–393.
106. Burden, C.M.; Morgan, M.O.; Hladun, K.R.; Amdam, G.V.; Trumble, J.J.; Smith, B.H. Acute sublethal exposure to toxic heavy metals alters honey bee (Apis mellifera) feeding behavior. *Sci. Rep.* **2019**, *9*, 4253. [CrossRef]
107. Di, N.; Hladun, K.R.; Zhang, K.; Liu, T.-X.; Trumble, J.T. Laboratory bioassays on the impact of cadmium, copper and lead on the development and survival of honeybee (*Apis mellifera* L.) larvae and foragers. *Chemosphere* **2016**, *152*, 530–538. [CrossRef] [PubMed]
108. Turkyilmaz, A.; Sevik, H.; Cetin, M.; Saleh, E.A.A. Changes in Heavy Metal Accumulation Depending on Traffic Density in Some Landscape Plants. *Polish. J. Environ. Stud.* **2018**, *27*, 2277–2284. [CrossRef]
109. WIOŚ Kraków. Report on the State of the Environment in Lesser Poland in 2008. Kraków. 2009. Available online: https://www.euro.who.int/__data/assets/pdf_file/0005/95333/E92584.pdf (accessed on 21 April 2021).
110. Skorbiłowicz, E.; Skorbiłowicz, M.; Ciesluk, I. Bees as bioindicators of environmental pollution with metals in an urban area. *J. Ecol. Eng.* **2018**, *19*, 229–234. [CrossRef]
111. Chong, J.; Wishart, D.S.; Xia, J. Using MetaboAnalyst 4.0 for Comprehensive and Integrative Metabolomics Data Analysis. *Curr. Protoc. Bioinforma* **2019**, *68*, 86. [CrossRef] [PubMed]

Article

Characterizing the Volatile and Sensory Profiles, and Sugar Content of Beeswax, Beebread, Bee Pollen, and Honey

Małgorzata Starowicz [1], Paweł Hanus [2], Grzegorz Lamparski [3] and Tomasz Sawicki [4,*]

1. Department of Chemistry and Biodynamics of Food, Institute of Animal Reproduction and Food Research of the Polish Academy of Sciences, 10 Tuwima Street, 10-748 Olsztyn, Poland; m.starowicz@pan.olsztyn.pl
2. Department of Technology and Plant Product Quality Assessment, University of Rzeszów, 4 Zelwerowicza Street, 35-601 Rzeszów, Poland; phanus@ur.edu.pl
3. Sensory Laboratory, Institute of Animal Reproduction and Food Research of the Polish Academy of Sciences, 10 Tuwima Street, 10-748 Olsztyn, Poland; g.lamparski@pan.olsztyn.pl
4. Department of Human Nutrition, Faculty of Food Sciences, University of Warmia and Mazury in Olsztyn, 45f Słoneczna Street, 10-718 Olsztyn, Poland
* Correspondence: tomasz.sawicki@uwm.edu.pl

Abstract: Bee products are a well-known remedy against numerous diseases. However, from the consumers' perspective, it is essential to define factors that can affect their sensory acceptance. This investigation aimed to evaluate the volatile and sensory profiles, and sugar composition of beeswax, beebread, pollen, and honey. According to the HS-SPME/GC-MS results, 20 volatiles were identified in beeswax and honey, then 32 in beebread, and 33 in pollen. Alkanes were found to dominate in beeswax, beebread, and pollen, while aldehydes and monoterpenes in honey. In the case of sugars, a higher content of fructose was determined in beebread, bee pollen, and honey, whereas the highest content of glucose was assayed in beeswax. In the QDA, the highest aroma intensity characterized as honey-like and sweet was found in honey, while the acid aroma was typical of beebread. Other odor descriptors, including waxy, pungent, and plant-based aromas were noted only in beeswax, honey, and pollen, respectively.

Keywords: aroma composition; bee products; PLS; sugar content; QDA profile

1. Introduction

Honeybee products, including beeswax, beebread, bee pollen, and honey, have been intensively studied for their efficacy against some diseases [1–4]. However, aroma and flavor determination is nowadays also an important aspect considering the consumers' acceptance of varied food products, and thus the potential industrial application of raw materials in food design. Kaškoniene and Venskutonis [5] collected examples of research proving that the volatile composition of honey is strictly related to its botanical origin.

Honey is the best-known bee product. It is broadly used in every household and its consumption has been declared by over 95% of the Polish consumers [6]. Consumers typically prefer light-colored honey and often choose multiflorous, linden, acacia, rapeseed, and honeydew honeys [6]. Honey has a very pleasant aroma and sweet taste [7], therefore it is used to sweeten meals (as declared by about 70% of consumers). An increasing number of consumers appreciate the therapeutic effects of other bee products with the wide application in apitherapy, but also their sensorial properties. Bee pollen contains about 250 substances, two-thirds of which are carbohydrates, and ~94% of these carbohydrates are monosaccharides (mainly fructose, glucose, and sucrose) [8,9]. Moreover, beebread is a unique product, while beeswax is synthetized from honey sugars and represents a complex mixture of esters, hydrocarbons, free fatty acids, alcohols, and other substances [10]. Beeswax is used in the European Union as a glazing agent, food additive labeled as E901, and also widely as a carrier for flavors and coatings [11,12]. The bee products, especially honey, are often used as meal sweeteners, due to a high content of carbohydrates. The sugar composition of honey depends mainly on its botanical and geographical origins, and is affected by climate, processing, and storage [13]. Moreover, the concentration of fructose and glucose, as well as the ratio between them are useful indicators in the classification of honey types [14].

Honey and other bee products are used in food technology due to their valuable chemical composition and sensory characteristics. According to Tomczyk et al. [15], spray-dried honey can be an additive to isotonic drinks, improving their taste and colour. In addition to its wide use in the cosmetics industry, beeswax is also used in food technology to protect ripened cheeses as a polishing agent or edible coating mixed with polylactic-acid (PLA) [16]. Moreover, the addition of bee pollen improves the nutritional, functional, and sensory properties of food [17]. The use of bee pollen in white wine production increases the number of volatile compounds in wine. The addition of up to 1 g/L of multifloral pollen significantly affects the perception of fruit and floral notes in wines [18].

The collected literature allows us to conclude that beeswax and beebread are the least characterized of bee products. Therefore, the aim of this research was to study the volatile and sensorial profiles, and sugar content of beeswax, beebread, bee pollen, and honey of the same origin and from the same batch.

2. Results and Discussion

2.1. The Profile of Volatiles Compounds in Bee Products

There is only scarce information about the profiles of volatile compounds of bee products. Most studies dedicated to the volatile profiles refer to honey [19,20]. While, the research on the characteristics of volatiles in beeswax, bee pollen, and beebread mainly concern their use as additives, which allows enriching the flavor and aroma profile, and bioactive properties of various final products [21,22]. Thus, this is the first research presenting the profile of volatile compounds of four different bee products (honey, beebread, bee pollen, and beeswax).

A total of 55 volatile compounds identified in this study are presented in Table 1. They include 15 alkanes (volatile nos. X1, X2, X3, X4, X5, X6, X7, X8, X9, X11, X16, X17, X18, 25, and 27; refer to Table 1), eight compounds from monoterpenes (nos. 10, 20, 22, 30, 39, 41, X47, and X50), seven alcohols (nos. X23, X29, X33, X38, X43, X51, and X52), also seven acids (nos. X31, X45, X46, X48, X49, X53, and X55), five aldehydes (nos. X13, X21, X28, X34, and X37), two esters (X26 and X54), two ketones (24 and 36), and two compounds from benzenes (nos. X15 and X19), as well as disulfides (compound no. X12), sulfoxides (X14), furans (X32), oxygenated hydrocarbon (X36), pyrroles (X40), and lactones (X44), represented by one molecule.

Nonanal (X28), furfural (X32), and benzaldehyde (X37) were detected in all samples of been products (Table 1). In addition, 2,4-Dimethyl-heptane (X1), 4-methyl-octane (X2), 2-methyl-nonane (X5), 4-methyl-decane (X7), dodecane (X17), octanal (X21), and tetradecane (X27) were found in beeswax, beebread, and bee pollen, while, acetic acid (X31) and hexanoic acid (X49) were present only in beebread, bee pollen, and honey.

As shown in Table 1, the total sum of volatiles was detected within the range of 138.29–267.13 ppm (presented as peak areas). The highest relative content of these compounds was detected in bee pollen (267.13 ± 8.32). However, no statistical differences were found between bee pollen and beebread ($p < 0.05$). In contrast, the lowest sum of volatiles was determined in beeswax (138.29 ± 5.73). The sum of volatiles could be ranked as follows: Bee pollen ≥ beebread > honey > beeswax. However, the order of the number of volatiles in bee products was: Bee pollen (33 compounds) > beebread (32) > honey = beeswax (20). In addition, 2,4-dimethyl-heptane (X1) was the dominant compound in beeswax (18.9% of total volatile amount) and bee pollen (13.7%). Furthermore, 4-methyl-octane (X2; 13.8%; $p < 0.05$) also dominated in bee pollen, while benzaldehyde (X37) had a significantly higher percent contribution in the volatile profile in honey (26% of total volatiles). In case of the beebread, four compounds were prevailing, including three from the group of alkenes (X6,

X9, and X25) and one from the group of furans (X32). These major compounds accounted for approximately 38% of the total peak area in the beebread. As mentioned above, several volatiles were found only in one bee product (Table 1). Nine volatiles (X20, X22, X30, X39, X41, X42, X48, X50, and X53) were only present in honey, six (X14, X23, X26, X34, X47, and X54) in bee pollen, while, five (X10, X29, X35, X38, and X43) volatiles were only present in beeswax.

The richest profile of volatiles was detected in the bee pollen (33 volatiles), therefore they are listed in Table 1. In contrast, a study conducted by Keskin and Özkök [23], showed only 25 compounds in the bee pollen from Turkey. Tetradecane (X27) was the common compound identified in the Turkish bee pollen and bee products analyzed in our study. A small number of volatiles (propanedioic acid, benzoic acid, phenylacetic acid, 1-dodecene, hexadecenoic acid, 9,12-octadecadienoic acid, 9,12,15-octadecatrienoic acid, and nonadecanoic acid) were also detected in the bee pollen from Brazil [24]. On the other hand, the largest number of volatiles (137 compounds) was detected in the bee pollen from the mid-north region of Brazil [25]. Identification of such a large number of volatile compounds was feasible due to the use of many extraction methods (micro-hydrodistillation, dynamic headspace, ultrasound-assisted extraction, and solid phase microextraction). Only the headspace solid phase microextraction method alone (HS-SPME, a method that was also used in our study) allowed determining a high number of volatiles (84) in the bee pollen samples [25]. On the other hand, the compound number detected may be related to the region from which the samples were collected. The previous studies have also shown a relationship between the profile of volatile compounds in bee products and their region of origin [19]. Kaškonienė et al. [26] also found a high number of volatiles (103) in bee pollen collected in Lithuania. However, only 26 of them were in the amounts exceeding 1% of total volatiles in the sample. The following volatiles detected by these authors in bee pollen: Hexanal (X13), styrene (X19), benzaldehyde (X37), 6-methyl-5-hepten-2-one (X24), octanal (X21), (E,E)-3,5-octadien-2-one (X36), nonanal (X28), dodecane (X17), and 1-tridecene (X25) were also identified in our research. One of the main compounds in the bee pollen from Lithuania was styrene, with the content ranging from 19.6 to 27% of total volatiles [26]. In our study, its relative content was only 1% of the total volatiles of bee pollen, which may indicate that bee pollen from Poland may be less contaminated. According to the available literature, styrene in bee pollen can come from plastic bags [27], but it may also exist naturally or could be produced during enzymatic synthesis since bee pollen is produced by bees with the use of their saliva [28].

The bee pollen possessed the richest profile of chemical classes among other bee products (12 from 13 chemical groups). Its volatile profile included 13 alkanes, five aldehydes, two acids, two benzene derivatives, two ketones, two esters, and one compound from other groups (sulfoxides, alcohols, pyrroles, furans, lactones, and monoterpenes). The main chemical group in bee pollen was represented by alkanes (65.5% of total volatiles; Figure 1). Furthermore, the second most dominant chemical group included acids (11.8%), followed by sulfoxides (9.3%), and aldehydes (5.4%), whereas the percent contribution of other groups was below 2%. In comparison, the bee pollen from Brazil contained only two groups of compounds, i.e., esters (seven compounds) and alkane [24]. The other Brazilian bee pollen was characterized by compounds from six chemical groups, while thirteen groups were identified in our study. Brazilian bee pollen was characterized by higher numbers of esters, ketones, and hydrocarbons, while lower numbers of aldehydes and the same number of alcohols in comparison to our study. Results of a study by Lima Neto et al. [25] did not show volatiles from the groups of alkanes, monoterpenes, disulfide, sulfoxide, benzenes, acids, furans, lactone, and pyrroles in pollen samples. On the other hand, terpenoids which were detected in the Brazilian bee pollen were not found in bee pollen from our study. A different number of volatiles detected may result from the botanical origin [20], time of harvest, and method of extraction [25].

Table 1. The profile of volatile compounds identified in bee products with their potential aroma descriptions and relative abundance presented.

No.	Compound	Aroma Description	LRI exp.	LRI lit.	Beeswax	Beebread	Bee Pollen	Honey
						Peak Areas [ppm]		
X1	2,4-dimethyl-heptane	-	789	797	26.07 ± 2.47 a B	12.29 ± 0.06 d C	36.73 ± 2.94 a A	-
X2	4-methyl-octane	-	822	823	12.56 ± 0.19 fg BC	6.42 ± 0.64 efg C	36.98 ± 4.42 a A	-
X3	nonane	Alkane *	899	900	7.60 ± 0.45 e A	4.50 ± 0.38 ghijk B	-	-
X4	3,8-dimethyl-decane	-	948	MS	-	5.35 ± 0.47 fghi A	5.02 ± 0.04 f A	-
X5	2-methyl-nonane	-	954	960	-	8.07 ± 0.57 ef B	6.59 ± 0.47 f B	-
X6	decane	-	1000	1000	9.75 ± 0.17 de A	22.55 ± 0.34 a A	23.49 ± 3.40 bc A	-
X7	4-methyl-decane	-	1001	1005	-	4.57 ± 0.35 ghij A	4.11 ± 0.59 f A	-
X8	2,6-dimethyl-nonane	-	1010	1014	3.99 ± 0.20 fg A	15.05 ± 0.82 bc A	6.50 ± 0.08 f B	-
X9	2,6,11-trimethyl-dodecane	-	1012	1015	-	23.92 ± 0.63 a A	17.95 ± 3.10 cd A	-
X10	α-pinene	pine, turpentine *	1029	1030	-	-	-	-
X11	5-methyl-decane	-	1059	MS	3.14 ± 0.32 fgh	2.95 ± 0.08 hijkl B	19.61 ± 0.47 bc A	-
X12	dimethyl disulfide	sulfurous *	1087	1094	-	2.92 ± 0.44 hijkl A	-	2.08 ± 0.24 h A
X13	hexanal	green, fruity **	1090	1093	-	13.71 ± 1.46 cd A	4.03 ± 0.00 f B	-
X14	benzyl methyl sulfoxide	-	1102	MS	-	-	24.94 ± 3.28 b	-
X15	1-(3,3-dimethylbutyl)benzene	-	1120	MS	-	1.83 ± 0.07 jkl A	2.48 ± 0.35 f A	-
X16	2,6,10-trimethyl-tetradecane	-	1190	MS	-	2.57 ± 0.05 jkl A	2.77 ± 0.07 f B	-
X17	dodecane	alkane *	1225	1227	10.14 ± 0.44 de B	2.79 ± 0.20 ijkl C	12.79 ± 0.16 de A	-
X18	nonadecane	alkane *	1265	1263	2.81 ± 0.24 fgh B	5.98 ± 0.66 efg A	-	-
X19	styrene	pungent **	1270	1273	-	2.75 ± 0.05 ijkl A	-	-
X20	o-cymene	-	1275	1276	-	-	2.72 ± 0.10 f A	-
X21	octanal	aldehydic *	1291	1291	18.60 ± 0.41 b A	1.90 ± 0.21 jkl B	0.70 ± 0.01 f C	2.39 ± 0.29 gh
X22	p-cymene	sweet	1293	1292	-	-	-	13.67 ± 2.10 d
X23	2-penten-1-ol	fruity	1315	1314	-	-	1.42 ± 0.01 f	-
X24	6-methyl-5-hepten-2-one	green, citrus	1342	1341	-	4.40 ± 0.18 ghijk A	2.38 ± 0.08 f B	-
X25	1-tridecene	pleasant	1344	1342	-	22.09 ± 0.45 a A	0.72 ± 0.02 f B	-
X26	2-propenyl 2-propenoic acid, ester	-	1456	MS	-	-	1.52 ± 0.02 f	-
X27	tetradecane	alkane *	1399	1400	1.91 ± 0.14 fgh A	2.51 ± 0.37 jkl A	1.76 ± 0.04 f A	-
X28	nonanal	aldehydic *	1416	1418	15.18 ± 1.75 c A	1.63 ± 0.18 l B	2.78 ± 0.04 f B	3.44 ± 0.12 gh B
X29	1-heptanol	green *	1458	1459	1.55 ± 0.01 fgh	-	-	-
X30	cis-linalool oxide	flower, woody	1462	1461	-	-	-	2.43 ± 0.10 gh
X31	acetic acid	sour, pungent	1468	1470	-	17.37 ± 1.34 b A	20.08 ± 3.41 bc A	3.57 ± 0.02 gh B
X32	furfural	bready *	1477	1476	2.08 ± 0.15 fgh C	24.07 ± 2.06 a B	3.87 ± 0.04 f C	37.19 ± 1.51 b A
X33	2-ethyl-1-hexanol	citrus *	1490	1493	1.54 ± 0.02 fgh B	-	-	2.02 ± 0.01 h A
X34	(E,E)-2,4-heptadienal	-	1497	MS	-	-	0.81 ± 0.06 f	-
X35	decanal	aldehydic *	1499	1500	13.18 ± 1.01 c	-	-	-
X36	(E,E)-3,5-octadien-2-one	-	1545	MS	-	1.92 ± 0.11 jkl B	2.19 ± 0.04 f A	-
X37	benzaldehyde	almond-like	1558	1562	4.32 ± 0.13 f B	5.55 ± 0.96 fgh B	6.05 ± 0.88 f B	52.39 ± 1.74 a A

Table 1. Cont.

No.	Compound	Aroma Description	LRI exp.	LRI lit.	Beeswax	Beebread	Bee Pollen	Honey
						Peak Areas [ppm]		
X38	1-octanol	waxy *	1561	1561	1.35 ± 0.04 fgh	-	-	-
X39	linalool	floral, sweet	1566	1565	-	-	-	28.79 ± 1.09 c
X40	1-ethyl-1H-pyrrole-2-carbaldehyde	burnt, roasted	1612	1616	-	1.58 ± 0.01 f A	0.93 ± 0.07 f B	-
X41	hotrienol	sweet tropical *	1633	1632	-	-	-	8.45 ± 0.14 ef
X42	dihydro-4-methyl-2(3H)-furanone		1649	1653	-	-	-	9.17 ± 0.09 e
X43	1- nonanol	floral *	1660	1668	0.85 ± 0.01 h	-	-	-
X44	butyrolactone	sweet, caramel **	1677	1673	-	1.81 ± 0.39 kl A	2.24 ± 0.26 f A	-
X45	2/3-methyl-butanoic acid	sweaty	1682	1684	-	1.20 ± 0.14 l B	-	14.91 ± 1.15 d A
X46	2-methyl-hexanoic acid	fruity	1755	1757	-	11.43 ± 1.05 d A	4.67 ± 0.23 f B	-
X47	verbenone (I)	minty, spicy **	1732	1733	-	-	0.66 ± 0.03 f	-
X48	3-methylvaleric acid	animal *	1823	1826	-	-	-	2.91 ± 0.01 gh
X49	hexanoic acid	fatty *	1844	1842	-	8.51 ± 0.13 e A	6.70 ± 0.02 ef B	5.27 ± 0.02 fg C
X50	p-cymen-8-ol	floral, sweet **	1879	1872	-	-	-	2.59 ± 0.06 gh
X51	benzyl alcohol	sweet, floral *	1900	1906	0.98 ± 0.01 gh B	-	-	2.09 ± 0.01 gh A
X52	phenylethyl alcohol	floral *	1923	1931	0.69 ± 0.01 h B	-	-	2.47 ± 0.01 gh A
X53	heptanoic acid	cheesy *	1991	1990	-	-	-	2.84 ± 0.01 gh
X54	2-propenoic acid, 3-phenyl-, methyl ester	fruity	2056	2065	-	-	0.92 ± 0.11 f	-
X55	nonanoic acid	waxy *	2200	2202	-	1.21 ± 0.25 l B	-	3.02 ± 0.02 gh A
	sum				138.29 ± 5.73 C	245.40 ± 1.70 A	267.13 ± 8.32 A	201.70 ± 3.39 B

* www.flavornet.org (accessed on 12 May 2021); ** pherobase; LRI: Linear retention indices (exp.: Experimental, lit.: Literature); MS: Compounds identified according to their mass spectrum. Values were presented as mean ± standard deviation. a–l Values followed by different letters in the same column are significantly different ($p < 0.05$) and A–C values followed by different letters in the same row are significantly different ($p < 0.05$) as determined by Tukey's multiple comparison test.

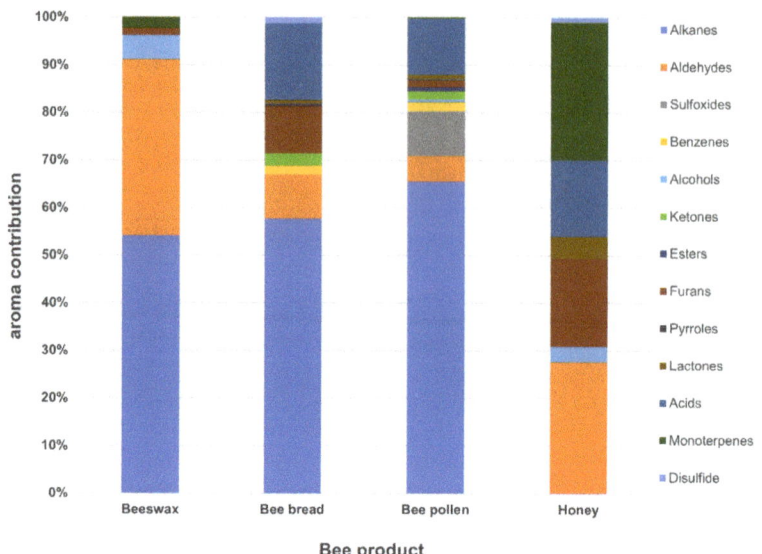

Figure 1. Chemical classes contribution of volatile compounds detected in bee products.

The next bee product with the high number of volatile compounds (32) was beebread. It contained 15 alkanes (X1, X2, X3, X4, X5, X6, X7, X8, X9, X11, X16, X17, X18, X25, and X27), four aldehydes (X13, X21, X28, and X37), five acids (X31, X45, X46, X49, and X55), two benzenes (X15 and X19), two ketones (X15 and X19), and others (disulfides, furans, pyrroles, and lactones). The results indicated a high contribution of alkanes in beebread, representing 47% of all identified compounds, and their 57.7% contribution to the aroma profile (Figure 1). Volatiles detected in the beebread belonged to nine out of the 13 chemical groups classified in the bee products. Moreover, it was the only sample from the tested bee products whose volatile profile was common with the profiles of the other samples, probably since it is a combination of honey and pollen [3,29]. Furthermore, our study showed that the volatiles present in bee pollen played a more important role in creating the profile of volatile compounds of beebread than these of honey.

To the best of our knowledge, there is paucity of data on the profile of volatile compounds in beebread. Results of a study by Kaškonienė et al. [30] also showed 32 volatiles in beebread from Lithuania. Comparing the beebread from Poland and Lithuania, the same 11 volatiles (2,4-dimethyl-heptane, nonane, decane, dimethyl disulfide, dodecane, octanal, nonanal, acetic acid, furfural, benzaldehyde, and hexanoic acid) were present in both samples. As mentioned above, the main compounds in beebread in our research were decane (X6), furfural (X32), 2,6,11-trimethyl-dodecane (X9), and 1-tridecene (X25), each contributing approximately 9% to the total volatiles (Table 1). However, the beebread from Lithuania had high contents of dimethyl sulphide, 1-heptadecene, acetic acid, nonane, and furfural, accounting for 20.0%, 13.9%, 13.4%, and 9.8% of the total volatiles, respectively [30]. Furfural was detected in the samples from Lithuania and Poland. Moreover, its content in both beebreads was the same (9.8% of the aroma profile). It is common knowledge that the furfural concentration is associated with the heat treatment [31], therefore, the high contribution of this compound may depend on the extraction method (SPME). The next compound with a high contribution in the beebread from Lithuania was acetic acid. However, its content in the sample from Poland was lower by 6.3%. The source of acetic acid in beebread is unknown, but there are two possibilities. As mentioned above, the beebread is a mixture of honey and bee pollen, which undergoes fermentation by enzymes of the bee's saliva and Lactobacillus [32], resulting in the formation of acetic acid. On the other hand, acetic acid—as a biological metabolic intermediate—occurs naturally in plant juices [33], which may also be collected by bees.

In honey, we identified 20 compounds in total, including: six acids (X31, X45, X48, X49, X53, and X55), six monoterpenes (X20, X22, X30, X39, X41, and X50), three volatiles from the group of alcohols (X33, X51, and X52), two aldehydes (X28 and X37), and one compound from each of the following groups: Disulfides (X12), furans (X32), and lactones (dihydro-4-methyl-2(3H)-furanone). These results concerning the detected chemical families are consistent with findings reported by other authors [20]. On the other hand, a recent study demonstrated a higher number of volatiles (42 compounds) in honey from China, which included 14 aldehydes, 12 ketones, seven alcohols, three acids, three pyrazines, one ether, one ester, and one terpene [34]. In Polish honey, we did not detect volatiles from ketone, pyrazine, ether, and terpene groups, but were identified compounds from groups of: Monoterpenes, disulfides, and furans, which were not detected in the Chinese samples. Moreover, the other research also showed more volatiles (60) detected in unifloral Chinese kinds of honey [35].

The contribution of the sum of aldehydes in the honey accounted for over 27%. However, the contribution of monoterpens was higher than that of aldehydes and was estimated at 29% (Figure 1). In the case of acids, the major compound was 2/3-methyl-butanoic acid accounting for 51.5% and 7.4% of the total acids and volatiles, respectively. The major monoterpene turned out to be linalool, which accounted for 49.4% and 14.3% of total monoterpenes and aroma profile, respectively. However, the results presented in Table 1 indicate that the major volatile in the honey was benzaldehyde. Its contribution in the tested sample was approximately 26% of the total volatiles. Moreover, it was present in all bee products, with the highest relative content detected in honey, being higher by 88.5%, 89.4%, and 91.8% than in bee pollen, beebread, and beeswax, respectively (Table 1). Furfural was also present in the honey in a relatively high amount (18.4% of peak areas). It was also detected in the bee pollen, beebread, and beeswax, while as in the case of benzaldehyde, the highest contribution of this compound was determined in honey.

The composition of volatiles of beeswax was significantly different than the profile of volatile compounds in honey, bee pollen, and beebread (Table 1). The HS-SPME-GC/MS analysis showed the presence of only five of the thirteen classes of chemical compounds detected in bee products. As mentioned above, a total of 20 compounds were identified in beeswax, including eight alkanes (X1, X2, X3, X5, X7, X17, X18, and X27), six alcohols (X29, X33, X38, X43, X51, and X52), four aldehydes (X21, X28, X35, and X37), and another two volatiles containing one monoterpene (X10) and one furan (X32). In the beeswax, the alkanes were the dominant group of compounds with 54.1% of contribution to the aroma profile (Figure 1). The next group of volatiles was aldehydes, which accounted for 37.1% of the aroma profile. Moreover, alcohols with the higher number of compounds (6) than aldehydes (4) were characterized by a lower contribution (5.0%) to the sum of volatiles detected in beeswax. The beeswax had the highest content of 2,4-dimethyl-heptane (18.9% of total peak area). Moreover, octanal and nonanal were the major contributors to the volatile composition in beeswax, accounting for 13.4% and 11.0%, respectively. On the other hand, the content of 4-methyl-decane (X7), α-pinene (X10), nonadecane (X18), tetradecane (X27), 1-heptanol (X29), furfural (X32), 2-ethyl-1-hexanol (X33), benzaldehyde (X37), 1-octanol (X38), 1-nonanol (X43), benzyl alcohol (X51), and phenylethyl alcohol (X52) were below 3.2%. Although beeswax is a precious product and its aroma is one of its incredibly privileged properties, unfortunately, only one team has undertaken an attempt of analyzing the profile of its volatile compounds so far. The study by Ferber and Nursten [36] showed the presence of 48 volatiles, divided into four different chemical classes. The larger number of volatile compounds identified in the above cited study compared to our research was due to the fact that the cited authors used two columns of different polarities. Moreover, the composition of beeswax is influenced by the genetic determinants of the bee colony and environmental factors [37].

2.2. Sugar Contents in Bee Products

The results of the sugar content analysis in bee bread, bee pollen, beeswax, and honey are provided in Table 2. Fructose, glucose, and sucrose contents were determined, and the total sugar content and fructose/glucose (F/G) ratio were calculated, as well. Glucose and fructose were the major components in the investigated bee products. The highest total sugar content was determined in honey (54.02 g/100 g), then pollen (25.63 g/100 g), and bee bread (16.81 g/100 g), and the lowest one in beeswax (2.09 g/100 g). Glucose and fructose contents in honey were lower than these noted by Szczęsna et al. [38] for rape honey, i.e., 37.6 and 37.3 g/100 g, respectively. Fructose, with the mean content of 35.91 g/100 g, followed by glucose, with the mean content of 35.00 g/100 g, were also detected in floral honey samples in the study by Kirs et al. [39]. The differences in sugar content could be related to the various floral/geographical origin of honey from the cited and our study. However, the range of glucose and fructose contents determined in our study were similar to those presented for multifloral light honey [40]. Moreover, the analyzed honey met the requirements set out in the European regulations concerning the sum of fructose and glucose. Furthermore, the honey sample did not exceed the permissible levels for sucrose [11,41]. Only a low content of sucrose was detected at 0.10 g/100 g for honey, at 0.11 for beebread, and at 0.01 for bee pollen. No sucrose was reported in beeswax in our study. This is in agreement with the data reported by Bertoncelj et al. [42], who determined the contents of sucrose in bee pollen only in the range of 0.05–0.28 g/100 g.

Table 2. The content of sugars (fructose, glucose, and sucrose at the level of g/100 g), total amount of sugars, and fructose/glucose ratio (F/G) determined in bee products.

Bee Products	Sugar [g/100 g]			TOTAL Amount of Sugars	F/G Ratio
	Fructose	Glucose	Sucrose		
Beeswax	0.06 ± 0.00 [d]	2.03 ± 0.14 [d]	-	2.09	0.03
Beebread	11.58 ± 0.12 [c]	5.12 ± 0.33 [c]	0.11 ± 0.00 [a]	16.81	2.26
Bee pollen	14.06 ± 0.01 [b]	11.56 ± 0.43 [b]	0.01 ± 0.00 [b]	25.63	1.22
Honey	27.60 ± 0.48 [a]	26.32 ± 0.49 [a]	0.10 ± 0.01 [a]	54.02	1.05

Values were presented as mean ± standard deviation. Values followed by different letters in the same column are significantly different ($p < 0.05$), as determined by Tukey's multiple comparison test.

The fructose/glucose (F/G) ratio is a standard parameter launched in many research dedicated to bee products. Therefore, the F/G ratio was also examined in our study and ranged from 2.26 (in beebread) to 0.03 (in beeswax) (Table 2). Its highest value was noted in beebread, followed by bee pollen (F/G = 1.22). This data is similar to those reported by Martins et al. [43] and Kalaycıoglu et al. [44], who calculated the F/G ratio in the range of 1.01–2.24 for bee pollen from Brazil, and in the range of 1.12–2.53 for bee pollen from Turkey, respectively. The F/G of honey analyzed in our study reached 1.05 and was similar to that noted by Szczęsna et al. [45] and Kirs et al. [39]. Moreover, Kirs et al. [39] emphasized that the F/G ratio could be a good indicator of honey origin. They reported that the mean F/G = 1.03 could be a marker of floral honey, which is correct in the context of our multiflorous honey. The lowest F/G value was determined for beeswax, since its major sugar is glucose, which is less sweet than fructose. Therefore, the content of sugars is important not only from the viewpoint of product authenticity, but also regarding consumer acceptance, as consumers definitely prefer more sweet products (in both taste and aroma).

2.3. QDA Profile of Bee Products in the Aspect of Color and Odor Quality

The QDA method of sensory profiling has already been used to determine the properties of honey samples [46,47]. The crucial points in this method include an appropriate training of the sensory panel and the choice of descriptors and standards related to them. Therefore, in the primary step, panelists prepare definitions of odor terms and choose reference materials, which help them identify the selected odors. In this case, honey-like was defined as an aromatic impression related to honey, and the standard used in this case was an essence on the tissue paper le nez du café no. 19 honeyed; sweet was described as a delicate sweet note characteristic of cotton candy (cotton candy); acid as an odor associated with organic acid, e.g., organic acids (lemon slice); pungent as an intensive odor of spices (black pepper); waxy as an odor characteristic of beeswax (beeswax); and plant-based as an odor related to plant material.

The results of QDA are presented in Figure 2. According to the scores given by the sensory panel to the color of bee products, beeswax was the less yellow product with a value of 3.53 arbitrary units (au), in comparison to beebread (9.00 au), honey (7.00 au), and bee pollen (5.00 au). Moreover, the sensory panel identified six odor descriptors for bee products: Honey-like, sweet, acid, pungent, waxy, and plant-based. Honey received the highest score for honey-like odor (2–3 times higher compared to beeswax and beebread), and had the sweetest aroma compared to the other bee products. This result is in agreement with findings from our previous study (data not published), indicating that multiflorous honey possesses the sweetest and honey-like odor compared to other types of honey. Castro-Vázquez et al. [46] defined the main honey descriptors as floral and fruity. However, they can be contributed to sweet and honey-like sensory markers. In the bee pollen, panelists did not note the honey-like aroma, but its aroma was described using attributes of sweet, acid, and plant-based. Moreover, Sipos et al. [48] characterized the aroma of bee pollen as sour/acid at the same level as in our study. Furthermore, in the present study, beebread exhibited the highest intensity for the attributes of acid aroma, whereas in beeswax acidity was not recognized at all. Other odor descriptors, including waxy, pungent, and plant-based were noted only in beeswax, honey, and pollen, respectively. The waxy in beeswax and plant-based in pollen received the high scores of 6.00 and 5.80 au, respectively, which might be deemed sensory markers of these two products.

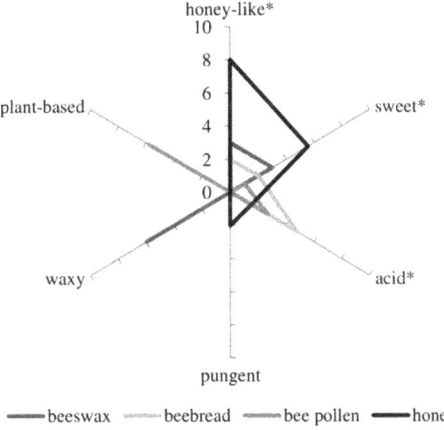

aroma descriptor	beeswax	beebread	bee pollen	honey
honey-like	3.00[b]	2.00[c]	0.00[d]	8.00[a]
sweet	3.00[b]	2.00[c]	1.02[d]	5.57[a]
acid	0.00[d]	4.80[a]	2.70[b]	1.50[c]
pungent	0.00[b]	0.00[b]	0.00[b]	2.02[a]
waxy	6.00[a]	0.00[b]	0.00[b]	0.00[b]
plant-based	0.00[b]	0.00[b]	5.80[a]	0.00[b]

Figure 2. Spider diagram of mean scores of aroma attributes of bee products. Different letters in the same row and "*" in the graph mean significantly different ($p < 0.001$).

In a future perspective, the sensory analysis could be helpful to choose bee products with the highest rate of aroma notes and with high consumer acceptability. Therefore, the sugar content and profile of volatile compounds could be correlated with sensory descriptors of bee products. The relationship between each analysis is described in the next step of this manuscript—the PLS analysis.

2.4. The PLS Analysis

The partial least squares (PLS) analysis was performed to evaluate the overall association between the bee products, aroma compounds, and sensory descriptors (Figure 3). The analysis performed explained 80.9% of the total variation. Beebread was located on the top left and was positively associated with one sensory descriptor (acid odor) and eleven volatiles (3,8-dimethyl-decane, 2,6,11-trimethyl-dodecane, hexanal, 1-(3,3-dimethylbutyl) benzene, 2,6,10-trimethyl-tetradecane, 6-methyl-5-hepten-2-one, acetic acid, (E,E)-3,5-octadien-2-one, 1-ethyl-1H-pyrrole-2-carbaldehyde, butyrolactone, and 2-methyl-hexanoic acid). Most of the compounds that correlated with beebread were characterized by green, fruity, citrus-like, and/or pungent aroma. Moreover, bee pollen was located in the upper left quadrant, and was moderately correlated with one volatile compound (5-methyl-decane). On the other hand, honey was located on the top right. It was positively correlated with twelve volatile compounds (o-cymene, p-cymene, cis-linalool oxide, furfural, benzaldehyde, linalool, hotrienol, dihydro-4-methyl-2(3H)-furanone, 2/3-methyl-butanoic acid, p-cymene-8-ol, 3-methylvaleric acid, and heptanoic acid), but mostly with nonanoic acid. Moreover, honey was positively correlated with the sensory descriptor of pungent odor, which may be positively influenced by the presence of o-cymene, p-cymene, cis-linalool oxide, benzaldehyde, linalool, hotrienol, dihydro-4-methyl-2(3H)-furanone, 2/3-methyl-butanoic acid, p-cymene-8-ol, 3-methylvaleric acid, and heptanoic acid, and positively correlated with the concentration of sugars, but mostly with glucose. The highest correlation was achieved between waxy odor and molecules of α-pinene, octanal, nonanal, 1-heptanol, 1-octanol, and 1-nonanol, while a lower correlation was observed in the case of waxy odor and nonane. Honey and beebread were positively correlated with furfural (they had the highest content of this compound). The aroma compounds, including 2-penten-1-ol, 2-propenyl 2-propenoic acid ester, (E,E)-2,4-heptadienal, verbenone (I), 2-propenoic acid, 3-phenyl-methyl ester, and 5-methyl-decane, were highly correlated with the plant-based note, whereas hexanoic acid was positively associated with acid odor. Moreover, a high correlation was observed between honey-like and sweet aroma with the phenylethyl acid, which has a characteristic floral note. Another correlation was observed between 2,6-dimethyl-nonane and yellow color. It can be concluded that the sugars were highly correlated with volatile compounds (Figure 3). Sucrose, fructose, and glucose were highly correlated with the 2,6-dimethyl-nonane, dimethyl disulfide, and nonanoic acid, respectively. Furthermore, a positive correlation was noted between sugar content and pungent, honey-like, and sweet odor.

In addition, some correlations were observed between individual aroma compounds A high correlation was demonstrated between 4-methyl-octane, 4-methyl-decane, and tetradecane, all being alkanes. The next group of positively correlated volatiles included 2,4-dimethyl-heptane, 2-methyl-nonane, and dodecane, which were also representatives of alkanes. It should be noticed that the sensory attributes can by no means be associated with an individual aroma compound. The unique sensory characteristics of the studied bee products were attributed to the superimposed and synergistic effects of volatile compounds [49], which in the case of bee products correspond mostly to the botanical and geographical origin [19,20].

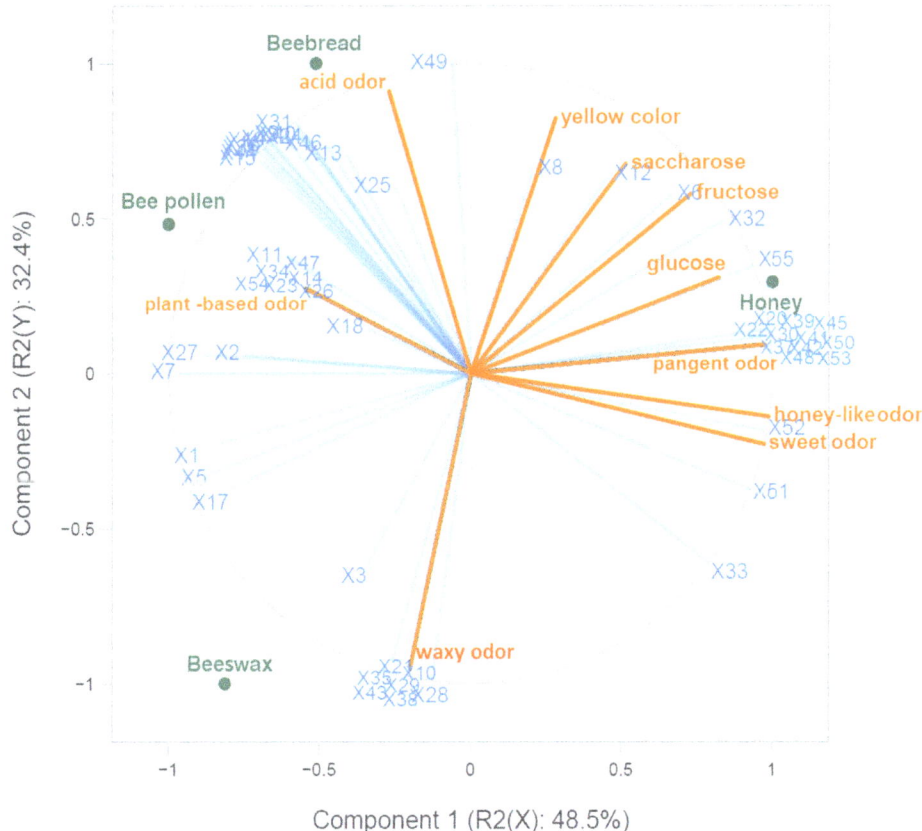

Figure 3. PLS correlation analysis of samples, between aroma compounds and sensory attributes; X1–X55 represent the number of compounds identified in the bee products (Table 1).

3. Materials and Methods

3.1. Chemicals and Reagents

The mix of C6-C30 *n*-alkanes, D-(+)-glucose, D-(−)-fructose, sucrose, methanol, acetonitrile, and water was purchased from Sigma-Aldrich (St. Louis, MO, USA).

3.2. Research Material

Beeswax, beebread, bee pollen, and multiflorous honey were obtained from the same apiary from the Kujawy region (central of Poland) by a professional beekeeper. The samples were collected in 2019, packed in polypropylene bags, and kept refrigerated at 4 °C before analyzed.

3.3. Determination of Volatile Composition of Bee Products

The volatile analysis of bee products was carried out by the headspace solid phase microextraction gas chromatography mass spectrometry (HS-SPME GC/MS) according to the method developed by Plutowska et al. [50] with some modifications. Therefore, 2 g of each bee product were weighed into a 20-mL vial. The vials were placed on an Eppendorf agitator/heater (Germany), shaken and heated (40 °C, 50 min, 600 rpm), and then the volatiles were allowed to absorb onto the SPME fiber for 15 min at 50 °C (without shaking). A 50/30 µm stable DVB/CAR/PDMS fiber (Supelco, Bellefonte, PA, USA) was used. The injection was done manually. Thus, the SPME fiber was introduced into the chromatograph injection port (splitless mode), where the analytes were desorbed at 250 °C for 5 min and transferred onto a capillary column (DB-WAX, 30 m, 0.25 mm × 0.50 μm). The analyses were performed using a gas chromatograph (Agilent Technologies 7890A GC system, Santa Clara, CA, USA) coupled to a mass spectrometer (Agilent Technologies 5975C VL MSD system, Santa Clara, CA, USA). The temperature was initially set to 40 °C and held for 5 min. Then, it was increased to 200 °C and held for 1 min. Finally, it was increased to 240 °C and held for 5 min. In the method, helium was used as the carrier gas with a constant flow rate of 1 mL/min.

The analyses were carried out in triplicate. The compounds were identified by comparing the obtained linear retention indices, retention times, and mass spectra with data from the Wiley Registry 7th Edition Mass Spectral Library (Wiley and Sons Inc., Weinheim, Germany) and the National Institute Standards and Technology (NIST) 2005 Mass Spectral Library. Linear retention indices (LRIs) were calculated using the C6-C30 n-alkanes mix (Sigma-Aldrich, MO, USA). The results were presented as the total peak area of the individual compounds identified in the database.

3.4. Determination of Sugar Content in Bee Products

About 0.5 g of bee product samples were dissolved in 5 mL of 70% methanol. Thus, the prepared solutions were boiled for 15 min. Afterwards, the samples were cooled to room temperature and next filtered through a PTFE syringe filter (pore size 0.45 μm). Additionally, before analysis by high performance liquid chromatography (HPLC), the sample was centrifuged (MPW-260R, Poland) for 10 min (7500 RPM) [51].

The contents of glucose, fructose, and sucrose were determined using a Sykam apparatus (Fürstenfeldbruck, Germany) equipped with a refractometric detector (RI) from the same manufacturer. The separation was carried out on a Cosmosil Sugar-D 250 × 4.6 column (Nacalai Tesque, INC; Kyoto, Japan) at 35 °C. The mobile phase consisted of acetonitrile and water (75:25, v/v), while the elution was carried out in an isocratic gradient by 15 min with a flow rate of 1 mL/min. Sugars were identified based on the retention times of the available standards, and their contents were calculated based on the concentration of the respective standard and expressed as g/100 g sample [52].

3.5. Quantitative Descriptive Analysis (QDA) of Bee Products

The sensory quality of honey products, including beeswax, bee bread, pollen, and honey, was evaluated by the QDA method, according to ISO standard [53]. The assessment was carried out by a sensory panel (six persons–five women and one man) previously selected, trained, and monitored according to the ISO guidelines [54]. The panel determined the list of all descriptors of aroma and appearance for the evaluated samples. There were six attributes of odor (honey-like, sweet, acid, pungent, waxy, and plant-based) and one attribute of appearance–color, which was defined as a yellow color intensity according to a color pattern RAL 075 80 60—scale value 5 with light–dark scale edges. The attributes of odor had non intensive–very intensive scale edges.

The assessments were carried out at a sensory laboratory room, which fulfils the requirements of the ISO standards [55]. The sensory laboratory was a temperature and humidity-controlled room (21 °C, 40%), equipped with 10 individual boxes, computers, and appropriate lighting. A computerized sensory program FIZZ (Biosystemes, Counternon, France) was used for the evaluations, and for collecting results, followed by their statistical analysis and graphical presentation.

3.6. Statistical Analysis

The data are presented as mean values ± standard deviations of triplicate measurements. The differences between samples were analyzed by a one-way ANOVA with Tukey's multiple comparison test ($p < 0.05$) using STATISTICA 13.0 (StatSoft Inc., Tulsa, OK, USA).

The data from the sensory analysis was evaluated using FIZZ (Biosystemes, Counternon, France). Then, the ANOVA was conducted to determine which sensory attributes were statistically significant, and Tukey's test was performed to show similarities and differences between the investigated products assessed by the sensory panel.

PLS was fitted with the use of "plsdepot" package in R (version 4.0.4, R Foundation for Statistical Computing, Vienna, Austria).

4. Conclusions

As already mentioned, this is the first study characterizing the composition of volatile compounds in four different bee products (beeswax, beebread, bee pollen, and honey) collected in the same batch. The conducted study showed that each bee product possessed its own unique profile of volatiles. However, the volatiles present in bee pollen played an important role in creating the volatile compound profile of beebread. The bee pollen was characterized by the highest number of volatiles, when compared to the other bee products. Alkanes were the main compounds detected in the bee products (except for honey). Three volatile compounds (nonanal, benzaldehyde, and furfural) were present in all samples of beeswax, beebread, bee pollen, and honey. Moreover, the major volatiles in beeswax were 2,4-dimethyl-heptane, octanal, nonanal, and 4-methyl-octane; in beebread: Decane, 1-tridecane, and furfural; in bee pollen: 2,4-dimethyl-heptane, 4-methyl-octane, decane, acetic acid, and 5-methyl-decane, whereas in honey: Benzaldehyde, furfural, and linalool. Thereby, the content of sugars, color, and odor descriptors such as honey-like, sweet, and pungent, were positively correlated only in honey. Beeswax was related to a waxy odor, bee pollen to a plant-based aroma, and beebread to an acid aroma. Therefore, it can be concluded that the volatile profile as well as sugar content, and furthermore sensory analysis depend on the stage of their processing by bees. Moreover, the characteristic of the volatile and sensory profiles and sugar content in the bee products can help the broader use of these products in the food industry to develop new functional foods.

Author Contributions: Conceptualization, M.S. and T.S.; methodology, M.S., P.H., and T.S.; formal analysis, M.S., P.H., G.L., and T.S.; investigation, M.S., P.H., G.L., and T.S.; writing—original draft preparation, M.S. and T.S.; writing—review and editing, M.S. and T.S.; visualization, M.S. and T.S. All authors have read and agreed to the published version of the manuscript.

Funding: This research received no external funding.

Institutional Review Board Statement: Not applicable.

Informed Consent Statement: Not applicable.

Data Availability Statement: Not applicable.

Acknowledgments: The authors would like to thank the panelists for their involvement in the sensory analysis of bee products.

Conflicts of Interest: The authors declare no conflict of interest.

References

1. Kacániová, M.; Vuković, N.; Chlebo, R.; Haščik, P.; Cubon, J.; Dżugan, M.; Pasternakiewicz, A. The antimicrobial activity of honey, bee pollen loads and beeswax from Slovakia. *Arch. Biol. Sci.* **2012**, *64*, 927–934. [CrossRef]
2. Denisow, B.; Denisow-Pietrzyk, M. Biological and therapeutic properties of bee pollen: A review. *J. Sci. Food Agric.* **2016**, *96*, 4303–4309. [CrossRef]
3. Kieliszek, M.; Piwowarek, K.; Kot, A.M.; Błażejak, S.; Chlebowska-Śmigiel, A.; Wolska, I. Pollen and bee bread as a new health-oriented products: A review. *Trends Food Sci. Technol.* **2018**, *71*, 170–180. [CrossRef]
4. Waheed, M.; Hussain, M.B.; Javed, A.; Mushtaq, Z.; Hassan, S.; Shariati, M.A.; Khan, M.U.; Majeed, M.; Nigam, M.; Mishra, A.P.; et al. Honey and cancer: A mechanistic review. *Clin. Nutr.* **2019**, *38*, 2499–2503. [CrossRef]
5. Kaškoniene, V.; Venskutonis, P.R. Floral markers in honey of various botanical and geographic origins: A review. *Compr. Rev. Food Sci. Food Saf.* **2010**, *9*, 620–634. [CrossRef]
6. Kopała, E.; Kuźnicka, E.; Balcerak, M. Survey of consumer preferences on the bee product market. Part 1. Honey. *Ann. Warsaw Univ. Life Sci. SGGW Anim. Sci.* **2019**, *58*, 153–158. [CrossRef]

7. Aparna, A.R.; Rajalakshmi, D. Honey—Its characteristics, sensory aspects, and applications. *Food Rev. Int.* **1999**, *15*, 455–471. [CrossRef]
8. Komosinska-Vassev, K.; Olczyk, P.; Kaźmierczak, J.; Mencner, L.; Olczyk, K. Bee pollen: Chemical composition and therapeutic application. *Evid. Based Complement. Alternat. Med.* **2015**, *6*, 297425. [CrossRef]
9. Bobis, O.; Marghitas, L.A.; Dezmirean, D.; Morar, O.; Bonta, V.; Chirila, F. Quality parameters and nutritional value of different commercial bee products. *Bul. Univ. Agric. Sci. Vet. Med. Cluj-Napoca Anim. Sci. Biotechnol.* **2010**, *67*, 1–2.
10. Bogdanov, S. Beeswax: Production, properties composition and control. In *Beeswax Book*; Muehlethurnen, Switzerland, 2009; Volume 1, pp. 1–17. Available online: citeseerx.ist.psu.edu/viewdoc/download?doi=10.1.1.567.4441&rep=rep1&type=pdf (accessed on 12 May 2021).
11. Regulation EC 1333/2008 (2008). Regulation (EC) No 1333/2008 of the European Parliament and of the Council of 16 December 2008 on food additives; Official Journal of the European Union: 16-33 (L 354). Available online: https://eur-lex.europa.eu/legal-content/EN/TXT/?uri=celex%3A32008R1333 (accessed on 12 May 2021).
12. EFSA. Beeswax (E901) as a glazing agent and a carrier for flavours. Scientific opinion of the pane on food additives, flavourings, processing aids and materials in contact with food (AFC). *EFSA J.* **2007**, *615*, 1–28.
13. Escudero, O.; Dobre, I.; Fernandez-Gonzalez, M.; Seijo, M.C. Contribution of botanical origin and sugar composition of honeys on the crystallization phenomenon. *Food Chem.* **2014**, *149*, 84–90.
14. Kaškoniene, V.; Venskutonis, P.R.; Čeksterytė, V. Carbohydrate composition and electrical conductivity of different origin honeys from Lithuania. *LWT Food Sci. Technol.* **2010**, *43*, 801–807. [CrossRef]
15. Tomczyk, M.; Zaguła, G.; Dżugan, M. A simple method of enrichment of honey powder with phytochemicals and its potential application in isotonic drink industry. *LWT Food Sci. Technol.* **2020**, *125*, 109204. [CrossRef]
16. Fratini, F.; Cilia, G.; Turchi, B.; Felicioli, A. Beeswax: A minireview of its antimicrobial activity and its application in medicine. *Asian Pac. J. Trop. Med.* **2016**, *9*, 839–843. [CrossRef]
17. Kostić, A.Ž.; Milinčić, D.D.; Barać, M.B.; Ali Shariati, M.; Tešić, Ž.L.; Pešić, M.B. The Application of Pollen as a Functional Food and Feed Ingredient—The Present and Perspectives. *Biomolecules* **2020**, *10*, 84. [CrossRef]
18. Amores-Arrocha, A.; Roldán, A.; Jiménez-Cantizano, A.; Caro, I.; Palacios, V. Evaluation of the use of multiflora bee pollen on the volatile compounds and sensorial profile of Palomino fino and Riesling white young wines. *Food Res. Int.* **2018**, *105*, 197–209. [CrossRef]
19. Rodriguez-Flores, M.S.; Falcão, S.I.; Escuredo, O.; Seijo, M.C.; Vilas-Boas, M. Description of volatile fraction of *Erica* honey from the northwest of the Iberian Peninsula. *Food Chem.* **2021**, *336*, 127758. [CrossRef]
20. Da Silva, P.M.; Gauche, C.; Gonzaga, L.V.; Costa, A.C.O.; Fett, R. Honey: Chemical composition, stability and authenticity. *Food Chem.* **2016**, *196*, 309–323. [CrossRef]
21. Conte, P.; Del Caro, A.; Urgeghe, P.P.; Petretto, G.L.; Montanari, L.; Piga, A.; Fedda, C. Nutritional and aroma improvement of gluten-free bread: Is bee pollen effective? *LWT Food Sci. Technol.* **2020**, *118*, 108711. [CrossRef]
22. Sancho-Galán, P.; Amores-Arrocha, A.; Jiménez-Cantizano, A.; Palacios, V. Use of multiflora bee pollen as a flor velum yeast growth activator in biological aging wines. *Molecules* **2019**, *24*, 1763. [CrossRef]
23. Keskin, M.; Özkök, A. Effects of drying techniques on chemical composition and volatile constituents of bee pollen. *Czech J. Food Sci.* **2020**, *38*, 203–208. [CrossRef]
24. Carpes, S.T.; de Alencar, S.M.; Cabral, I.S.R.; Oldoni, T.L.C.; Mourao, G.B.; Haminiuk, C.W.I.; da Luz, C.F.P.; Masson, M.L. Polyphenols and palynological origin of bee pollen of *Apis mellifera* L. from Brazil. Characterization of polyphenols of bee pollen. *CYTA J. Food* **2013**, *11*, 150–161. [CrossRef]
25. Lima Neto, J.S.; Lopes, J.A.D.; Moita Neto, J.M.; Lima, S.G.; Luz, C.F.P.; Citró, A.M.G.L. Volatile compounds and palynological analysis from pollen pots of stingless bees from the mid-north region of Brazil. *Braz. J. Pharm. Sci.* **2017**, *53*, e14093. [CrossRef]
26. Kaškoniene, V.; Kaškonas, P.; Maruška, A. Volatile compounds composition and antioxidant activity of bee pollen collected in Lithuania. *Chem. Pap.* **2015**, *69*, 291–299. [CrossRef]
27. Pilevar, Z.; Bahrami, A.; Beikzadeh, S.; Hosseini, H.; Jafari, S.M. Migration of styrene monomer from polystyrene packaging into foods: Characterization and safety evaluation. *Trends Food Sci. Technol.* **2019**, *91*, 248–261. [CrossRef]
28. Schwarz, K.J.; Boitz, L.I.; Methner, F.J. Enzymatic formation of styrene during wheat beer fermentation is dependent on pitching rate and cinnamic acid content. *J. Inst. Brew.* **2012**, *118*, 280–284. [CrossRef]
29. Sawicki, T.; Bączek, N.; Starowicz, M. Characterisation of the total phenolic, Vitamins C and E content and antioxidant properties of the beebread and honey from the same batch. *Czech J. Food Sci.* **2020**, *38*, 158–163. [CrossRef]
30. Kaškoniene, V.; Venskutonis, P.R.; Čeksterytė, V. Composition of volatile compounds of honey of various floral origin and beebread collected in Lithuania. *Food Chem.* **2008**, *111*, 988–997. [CrossRef]
31. Srivastava, R.; Bousquieres, J.; Cepeda-Vazquez, M.; Roux, S.; Bonazzi, C.; Rega, B. Kinetic study of furan and furfural generation during baking of cake models. *Food Chem.* **2018**, *267*, 329–336. [CrossRef]
32. Ivanišová, E.; Kačániová, M.; Frančáková, H.; Petrová, J.; Hutková, J.; Brovarskyi, V.; Velychko, S.; Adamchuk, L.; Schubertová, Z.; Musilová, J. Bee bread—Perspective source of bioactive compounds for future. *Potravináestvo* **2015**, *9*, 592–598. [CrossRef]
33. Kim, J.M.; To, T.K.; Matsui, A.; Tanoi, K.; Kobayashi, N.I.; Matsuda, F.; Seki, M. Acetate-mediated novel survival strategy against drought in plants. *Nat. Plants* **2017**, *3*, 17097. [CrossRef]

34. Wang, X.; Rogers, K.M.; Li, Y.; Yang, S.; Chen, L.; Zhou, J. Untargeted and targeted discrimination of honey collected by *Apis cerana* and *Apis mellifera* based on volatiles using HS-GC-IMS and HS-SPME-GC–MS. *J. Agric. Food Chem.* **2019**, *67*, 43, 12144–12152. [CrossRef] [PubMed]
35. Zhang, Y.-Z.; Si, J.-J.; Li, S.-S.; Zhang, G.-Z.; Wang, S.; Zheng, H.-Q.; Hu, F.-L. Chemical Analyses and Antimicrobial Activity of Nine Kinds of Unifloral Chinese Honeys Compared to Manuka Honey (12+ and 20+). *Molecules* **2021**, *26*, 2778. [CrossRef] [PubMed]
36. Ferber, C.E.M.; Nursten, H.E. The aroma of beeswax. *J. Sci. Food Agric.* **1977**, *28*, 511–518. [CrossRef]
37. Cornara, L.; Biagi, M.; Xiao, J.; Burlando, B. Therapeutic properties of bioactive compounds from different honeybee products. *Front. Pharmacol.* **2017**, *8*, 412. [CrossRef]
38. Szczęsna, T. Study on the sugar composition of honeybee collected pollen. *J. Apic. Sci.* **2007**, *51*, 15–21.
39. Kirs, E.; Pall, R.; Martverk, K.; Laos, K. Physicochemical and melissopalynological characterization of Estonian summer honeys. *Procedia Food Sci.* **2011**, *1*, 616–624. [CrossRef]
40. Borawska, M.; Arciuch, L.; Puścian-Jakubik, A.; Lewoc, D. Content of sugars (fructose, glucose, sucrose) and proline in different varieties of natural bee honey. *Probl. Hig. Epidemiol.* **2015**, *96*, 816–820.
41. European Union Directive (EU). Council Directive 2001/110/EC Relating to Honey; Official Journal of the European Communities: 2002. Available online: https://eur-lex.europa.eu/legal-content/EN/TXT/PDF/?uri=OJ:L:2014:164:FULL&from=IT (accessed on 12 May 2021).
42. Bertoncelj, J.; Polak, T.; Pucinar, T.; Lilek, N.; Borovšak, A.K.; Korošec, M. Carbohydrate composition of Slovenian bee pollens. *Int. J. Food Sci. Technol.* **2018**, *53*, 1880–1888. [CrossRef]
43. Martins, M.C.T.; Morgano, A.M.; Vicente, E.; Baggio, S.R.; Rodriguez-Amaya, D.B. Physicochemical composition of bee pollen from eleven Brazilian states. *J. Apic. Sci.* **2011**, *55*, 107–116.
44. Kalaycıoglu, Z.; Kaygusuz, H.; Doker, S.; Kolaylı, S.; Bedia Erim, F. Characterization of Turkish honeybee pollens by principal component analysis based on their individual organic acids, sugars, minerals, and antioxidant activities. *LWT Food Sci. Technol.* **2017**, *84*, 402–408. [CrossRef]
45. Szczęsna, T.; Rybak-Chmielewska, H.; Waś, E.; Kachaniuk, K.; Teper, D. Characteristics of polish unifloral honeys. I. Rape honey (*Brassica napus* L. var. *Oleifera* Metzger). *J. Apic. Sci.* **2011**, *55*, 111–119.
46. Castro-Vázquez, L.; Díaz-Maroto, M.C.; González-Viñas, M.A.; Pérez-Coello, M.S. Differentiation of monofloral citrus, rosemary, eucalyptus, lavender, thyme and heather honeys based on volatile composition and sensory descriptive analysis. *Food Chem.* **2009**, *112*, 1022–1030. [CrossRef]
47. Marcazzan, G.L.; Mucignal-Caretta, C.; Marchese, C.M.; Piano, M.L. A review of methods for honey sensory analysis. *J. Apic. Res.* **2018**, *57*, 75–87. [CrossRef]
48. Sipos, L.; Végh, R.; Bodor, Z.; Zinia Zaukuu, J.-L.; Hitka, G.; Bázár, G.; Kovacs, Z. Classification of bee pollen and prediction of sensory and colorimetric attributes—a sensometric fusion approach by e-Nose, e-Tongue and NIR. *Sensors* **2020**, *20*, 6768. [CrossRef] [PubMed]
49. Starowicz, M.; Lelujka, E.; Ciska, E.; Lamparski, G.; Sawicki, T.; Wronkowska, M. The application of *Lamiaceae* Lindl. promotes aroma compounds formation, sensory properties, and antioxidant activity of oat and buckwheat-based cookies. *Molecules* **2020**, *25*, 5626. [CrossRef]
50. Plutowska, B.; Chmiel, T.; Dymerski, T.; Wardencki, W. A headspace solid-phase microextraction method development and its application in the determination of volatiles in honeys by gas chromatography. *Food Chem.* **2011**, *126*, 1288–1298. [CrossRef]
51. Brummer, Y.; Cui, S.W. Understanding Carbohydrate Analysis. In *Food Carbohydrates*; Cui, S.W., Ed.; Taylor & Francis Group: Boca Raton, FL, USA, 2005; Volume 2, pp. 67–104.
52. Saeeduddin, M.; Abid, M.; Jabbar, S.; Hu, B.; Hashim, M.M.; Khan, M.A.; Xie, M.; Wu, T.; Zeng, X. Physicochemical parameters, bioactive compounds and microbial quality of sonicated pear juice. *Int. J. Food Sci. Technol.* **2016**, *51*, 1552–1559. [CrossRef]
53. ISO. PN-EN ISO 13299:2016-05. In *Sensory Analysis- Methodology- General Guidance for Establishing a Sensory Profile*; ISO: Warsaw, Poland, 2016.
54. ISO. PN-EN ISO 8586:2014. In *Sensory Analysis: General Guidelines for the Selection, Training and Monitoring of Selected Assessors and Expert Sensory Assessors*; ISO: Warsaw, Poland, 2014.
55. ISO. ISO 8589: 2010. In *Sensory Analysis: General Guidance for the Design of Test Rooms*; ISO: Warsaw, Poland, 2010.

Article

An Analytical Method for the Biomonitoring of Mercury in Bees and Beehive Products by Cold Vapor Atomic Fluorescence Spectrometry

Maria Luisa Astolfi [1,*], Marcelo Enrique Conti [2], Martina Ristorini [3], Maria Agostina Frezzini [4], Marco Papi [5], Lorenzo Massimi [4] and Silvia Canepari [4]

1. Department of Chemistry, Sapienza University of Rome, Piazzale Aldo Moro 5, 00185 Rome, Italy
2. Department of Management, Sapienza University of Rome, via del Castro Laurenziano 9, 00161 Rome, Italy; marcelo.conti@uniroma1.it
3. Department of Bioscience and Territory, University of Molise, 86090 Pesche, Italy; m.ristorini@studenti.unimol.it
4. Department of Environmental Biology, Sapienza University of Rome, Piazzale Aldo Moro 5, 00185 Rome, Italy; mariaagostina.frezzini@uniroma1.it (M.A.F.); l.massimi@uniroma1.it (L.M.); silvia.canepari@uniroma1.it (S.C.)
5. Association of Beekeepers of Rome and Province, via Albidona 20, 00118 Rome, Italy; buteo.betta@gmail.com
* Correspondence: marialuisa.astolfi@uniroma1.it; Tel./Fax: +39-06-4991-3384

Abstract: Bees and their products are useful bioindicators of anthropogenic activities and could overcome the deficiencies of air quality networks. Among the environmental contaminants, mercury (Hg) is a toxic metal that can accumulate in living organisms. The first aim of this study was to develop a simple analytical method to determine Hg in small mass samples of bees and beehive products by cold vapor atomic fluorescence spectrometry. The proposed method was optimized for about 0.02 g bee, pollen, propolis, and royal jelly, 0.05 g beeswax and honey, or 0.1 g honeydew with 0.5 mL HCl, 0.2 mL HNO_3, and 0.1 mL H_2O_2 in a water bath (95 °C, 30 min); samples were made up to a final volume of 5 mL deionized water. The method limits sample manipulation and the reagent mixture volume used. Detection limits were lower than 3 µg kg^{-1} for a sample mass of 0.02 g, and recoveries and precision were within 20% of the expected value and less than 10%, respectively, for many matrices. The second aim of the present study was to evaluate the proposed method's performances on real samples collected in six areas of the Lazio region in Italy.

Keywords: bees; beehive products; biomonitoring; cold vapor atomic fluorescence spectrometry; sample preparation; toxic metal

1. Introduction

Mercury (Hg) is an ubiquitous and toxic metal that continues to be a public health concern [1–3]. It is released into the environment from both natural and anthropogenic sources [4]. Mercury is present in the atmosphere as an elemental form (Hg^0) and it is accumulated through the terrestrial and aquatic food webs as an organic form (methylmercury) [5,6]. Although Hg is associated with several adverse human health effects [7], it is still widely used in the chloralkali industry, for gold mining, and the production of dental amalgam, batteries, pesticides, fungicides, disinfectants, and antiseptics [8]. Because of the mentioned toxic properties, Hg monitoring in food and environmental samples is essential in order to perform reliable risk assessments and take appropriate actions to protect human health and the environment [9]. According to the current air quality directives in Europe, industrial activities must reduce Hg emissions by implementing control programs and integrated pollution prevention and, at the same time, by improving air quality assessment and monitoring programs [10–13]. Mercury in the atmosphere is mainly assessed by making punctual measurements with manual or automated air quality monitoring stations [14] and applying standardized methodologies, based on current legislation [10]. However, due to the high costs, the monitoring networks for Hg pollution assessment are still characterized by low temporal (generally on an annual basis) and spatial coverage [15]. For the above reasons, there is growing interest in alternative air monitoring techniques such as plant, insects, lichens, and mosses that can provide reliable time-integrated estimates of air pollution in a given area at low cost [16–22]. In particular, bees and their products such as honey, propolis, and pollen have been proposed as bioindicators of environmental Hg contamination [23–25]. The assessment of Hg levels in bee products is important not only for their use as possible bioindicators for environmental contamination purposes, but also for the potential human exposure due to their dietary, pharmaceutical, and cosmetic use [26–30].

Mercury has been studied in honey samples by several authors [24,25,31–42], whereas limited literature data are available regarding the Hg determination in beeswax [42], pollen [24,25,39,41,43,44], propolis [24,45–47], and bees [24,25,31,39,47–49]. Microwave-assisted digestion is the most commonly used technique for preparing bee samples and hive products [32,35,42–44]. However, the microwave-assisted digestion method requires certain sample masses and reagent volumes, often leading to high final dilution factors and a consequent increase in the method detection limits [25,50]. In contrast, some authors have miniaturized digestion of honey, pollen, and/or bees by heating them in a heat block (80–100 °C) and using very small reagent volumes [25,33]. Throughout the literature, many studies have quantified Hg concentrations in bees and beehive product matrices with atomic absorption spectroscopy [38] and inductively coupled plasma–mass or optical emission spectrometry (ICP-MS or ICP-OES, respectively) [25,32,40–43,49], often coupled to cold vapor generation (CV) for matrix separation [24,33–35,37,44,45], electrothermal atomic absorption spectrophotometry [47], and direct Hg analysis using automated commercial instruments such as the advanced mercury analyzer (AMA) [36,39,46] or direct mercury analyzer (DMA) [37,48]. CV atomic fluorescence spectrometry (AFS) is a good alternative for total Hg determination, and has been commonly employed for the analysis of several biological and environmental matrices [51], food [52], and human bodily fluids and tissues [53–55]. Despite this, CV-AFS has rarely been applied for the determination of Hg in honey [35], and, to the best of our knowledge, this technique has not been applied in bees and other beehive products.

This study aimed to miniaturize the sample digestion of bees and beehive products to achieve accurate and reproducible results with low detection limits for Hg determination by CV-AFS. The proposed analytical method was applied to commercial honeydew and royal jelly samples and bees, honey, beeswax, pollen, and propolis samples collected from six central Italy areas were characterized by different exposure to environmental pollution.

2. Results and Discussion

2.1. Comparison with Previous Methods

The analytical characteristics comparison of the method proposed in the present study with others already developed for Hg determination in bee and beehive product samples is shown in Table S1. In this study, the sample digestion was miniaturized by reducing all volumes and masses to allow sample preparation in one disposable test tube. This prevented sample loss due to the transfer in different tubes and minimized possible contamination. In addition, the use of smaller volumes of reagents allows for a lower final dilution factor (5×), and lower method detection limit (DL) and decreases the consumable and chemical waste generated, meeting the ever-increasing demand to comply with green chemistry requirements. The dilution factor, together with the sample mass, the reagent purity, and the chosen instrument, can affect the Hg DL in bees and beehive products, where this metal is generally present in low concentrations. To decrease the method DL, the sample mass can be increased, but sometimes this is not possible (such as for bees or specific pollen). Furthermore, if the analytical method requires sample digestion, the increase in sample mass must necessarily be accompanied by an appropriate volume of reagents to ensure complete sample digestion. Even for methods that do not require sample pre-treatment such as AMA or DMA (Table S1), the sample mass cannot be randomly increased to ensure complete drying, ashing, and atomization of the sample. In addition, amounts larger than 100 mg of sample can produce a build-up of combustion gases, resulting in a rapid increase of pressure in the furnace [37]. In the literature (see Table S1), some studies have used a large final dilution of bees and beehive product samples (25–50×) [24,33,35,36,40,41,44,47,49] to reduce the acidity of the final digest, sometimes compromising the Hg determination. For this purpose, in this study, various sample aliquots (0.05–1 g for honey and honeydew, 0.02–0.2 g for bees, beeswax, pollen, propolis, royal jelly) were digested at the maximum temperature of 95 °C, considering two different times (30 or 60 min) and using the smallest amount of reagent mixture (0.5 mL HCl, 0.2 mL HNO_3, and 0.1 mL H_2O_2) to employ the smallest dilution factor final (5×). The choice of acid and oxidizing agent (HNO_3 and H_2O_2, respectively) is widely agreed by most of the literature for the selected matrices [24,25,33,35,36,38,41,42,44,47–49], while HCl was selected according to the manufacturer's recommendations.

The proposed sample preparation also appears to be the fastest procedure (digestion time, 30 min for 120 samples or more) compared to the other sample treatments reported in the literature [25,33,40–42,44,47] (Table S1), resulting in being suitable for routine analysis with high sample throughput and biomonitoring. However, it should be noted that possible volatile Hg species such as organometallic compounds or metal nanoparticles could be lost during digestion due to their volatilization.

The analytical characteristics of the proposed method are detailed in the following sections.

2.2. Linearity and Selectivity

The linearity and selectivity of the proposed method were evaluated by preparing calibration curves in aqueous [3% (v/v) HCl and HNO_3] standards and using the standard addition method at Hg concentrations of 0.00, 0.02, 0.04, 0.1, 0.2, 0.4, 0.8, 1.0, and 1.5 µg L^{-1}. Digested samples of each matrix (20 mg) were diluted to reach the same acid ratio as the aqueous standard solutions and used to create calibration curves with the standard addition method. The linearity ranges from 0.02 to 1.5 µg L^{-1} were checked through the linear regression coefficient (R^2) and verified by the Mandel fitting test. Calibration curve points with percent relative deviation ≥10% from calculated concentrations were tested and removed using the instrument software. The parameters of the calibration curves after the outliers' removal are presented in Table 1. Data of the calibration curves using aqueous standards were obtained by nine independent replicates. The dynamic range was compared with that of other previously published methods (Table S1). In particular, CV-AFS allows for the determination of Hg in a wide range of concentrations, showing a dynamic range greater than that possible with other techniques such as CV-AAS, ICP-OES, and DMA. The matrix effect was evaluated by comparing the slopes of the calibration curves obtained from aqueous standards and the standard addition method (Table 1). Most of the results showed good data dispersion; however, some standard deviation values were of the same order of magnitude as the intercept data, generating a large statistical uncertainty on these data. The t-test at a 95% confidence level was used to evaluate possible significant differences between the angular coefficients of the calibration curves, in accordance with previous studies [25,37,56]. There were no apparent matrix effects between the aqueous and standard addition calibration curves. Thus, these results agree with those obtained for bees, honey, and pollen by other authors [25,37].

Table 1. Comparison of calibration curve parameters for Hg determination by cold vapor atomic fluorescence spectrometry (CV-AFS).

Calibration Standards	Parameter [a]				
	a	s(a)	b	s(b)	R^2
Aqueous standards	9.68×10	9.68×10	1.86×10^4	1.16×10^3	0.999
Bee-addition standards	1.32×10^2	3.14×10	1.64×10^4	2.30×10^3	0.999
Beeswax-addition standards	7.65×10	1.05×10	1.71×10^4	2.22×10^3	0.999
Honey-addition standards	1.07×10^2	1.34×10^2	1.61×10^4	2.05×10^3	0.999
Honeydew-addition standards	1.38×10^2	8.58×10	1.65×10^4	1.78×10^3	0.999
Pollen-addition standards	1.30×10^2	3.73×10	1.62×10^4	2.07×10^3	0.998
Propolis-addition standards	3.85×10^2	2.50×10^2	1.76×10^4	1.40×10^3	0.999
Royal jelly-addition standards	7.65×10	1.05×10	1.80×10^4	9.19×10^2	0.998

[a] a, intercept; s(a), standard deviation of intercept; b, slope; s(b), standard deviation of slope; R^2, correlation coefficient.

2.3. Detection and Quantification Limits

The DL was calculated based on the calibration curve using software prepared by the Regional Agency for Environmental Protection [57]. Therefore, the DL can be expressed as DL = 3.3 σ/b; where the coefficient 3.3 is called the expansion factor and is obtained assuming a 95% confidence level; σ is the standard deviation of the response of the curve; and b is the calibration curve slope. The reached DL of 0.01 µg L^{-1} for aqueous calibration confirmed the excellent sensitivity of the proposed method. The QL was set at the lowest standard curve points of calibration, which was 0.02 µg L^{-1}. The DL and QL varied depending on the mass of the analyzed matrix and dilution required before analysis (in this study, 5×). In particular, for a mass of 0.02, 0.05, 0.1, 0.2, and 1 g, the DL was 3, 1, 0.5, 0.3, and 0.05 µg kg^{-1}, and the QL was 5, 2, 1, 0.5, 0.1 µg kg^{-1}, respectively. As shown in Table S1, the obtained DLs are comparable to previously reported AMA or DMA analysis [36,37,39] and ICP-MS analysis [25] and lower than CV-AAS or CV-ICP-OES analysis [24,33].

2.4. Accuracy and Precision

Due to the lack of certified reference material of bees and beehive products, the accuracy and precision (as repeatability and intermediate precision) of the proposed method were evaluated by recovery tests in agreement with other authors [25,36,37] and as indicated by Commission Decision no. 657/2002 [58]. Samples of each matrix (0.05–1 g for honey and honeydew, 0.02–0.2 g for bees, beeswax, pollen, propolis, royal jelly) were spiked with Hg at low (0.02 µg L^{-1}), intermediate (0.2 µg L^{-1}), and high (1 µg L^{-1}) concentration and then digested. The method performance at levels near the QL was evaluated considering the smallest mass of each matrix and the shortest digestion time (30 min). The same solutions were again analyzed on two separate days to assess intermediate precision. The recovery and precision (such as repeatability) data are shown in Table 2 and Tables S1 and S2. Intermediate precision data (not shown) were very similar to the repeatability values.

In summary, the digestion time and mass for each matrix suitable for obtaining recoveries and precision within 20% of the expected value and less than 10%, respectively, were tabulated (Table 3). In this study, the major sources of uncertainty were the recovery of the procedure, instrumental calibration, and repeatability of the measurements. In contrast, the samples' weights were the lowest contribution to the Hg uncertainty, in agreement with a previous study [59].

Table 2. Recovery and precision data for Hg in bees and beehive products ($n = 3$) by water bath digestion (95 °C, 30 min).

Matrix	Mass (g)	Low Level Spike (0.02 µg L^{-1})		Intermediate Level Spike (0.2 µg L^{-1})		High Level Spike (1 µg L^{-1})	
		R%	CV%	R%	CV%	R%	CV%
Honey	0.05	86	5.4	116	9.3	96	7.8
Honeydew	0.1	89	0.4	113	10	91	8.4
Pollen	0.02	92	9.6	90	3.7	95	3.6
Propolis	0.02	104	9.8	98	8.6	91	2.5
Beeswax	0.05	92	10	111	8.5	99	2.0
Royal Jelly	0.02	117	9.3	108	4.4	110	0.9
Bees	0.02	95	8.9	97	4.5	91	10

Table 3. Summary of mass and digestion time that can be used in bees and beehive products.

Matrix	Mass [a] (g)	Digestion Time [a] (min)
Bees	0.02–0.2	30 or 60
Beeswax	0.02	60
	0.05–0.1	30 or 60
	0.2	60
Honey	0.05	30 or 60
	0.1	60
Honeydew	0.05	60
	0.1	30 or 60
	0.2	60
Pollen	0.02–0.2	30 or 60
Propolis	0.02	30 or 60
Royal Jelly	0.02–0.2	30 or 60

[a] The method performance at levels near the QL (0.02 µg L^{-1}) was evaluated considering the smallest mass of each matrix and the shortest digestion time (30 min).

2.5. Hg Concentrations in Real Samples

Bees and beehive products (honey, beeswax, pollen, and propolis) from various geographical areas in central Italy (Figure 1) and commercial samples of both honeydew and royal jelly were analyzed to demonstrate the applicability of the proposed method for routine analysis and biomonitoring.

Mercury pollution is an important environmental and public health issue. Elemental Hg can be emitted into the atmosphere by both anthropogenic (mainly artisanal gold mining, fossil fuel combustion, and cement production) and natural (such as a geothermal activity) sources [39]. Subsequently, Hg is transported to land and surface waters through wet and dry deposition, where it can undergo a bioconversion into more volatile or soluble forms such as methylmercury and return into the atmosphere or bioaccumulate in food chains [39]. Additionally, bees are continuously exposed to contaminants including Hg. Every day during foraging activities, bees gather nectar, plant resins, and water in the border of 7 km^2 around their beehive and may come into contact with chemicals [47,49]. Therefore, the bee was proposed as a multi-sample contaminant collector because of its high mobility, contact with possible chemicals through inhalation, digestion, and hairs covering its body [40,47,49]. Contaminants adhered to the hairs such as particles of soil and dust can be carried into the beehive, thus affecting the composition of the beehive products [42,49]. In addition, Hg captured by the leaves of plants or absorbed from the soil through the plant root system can influence the nectar and pollen composition, which are brought back into the beehive [42,44]. Furthermore, propolis, produced with plant resin and mixed with salivary secretions and wax, due to the sticky nature of gum, might be used as a bioindicator of atmospheric pollution [46,47].

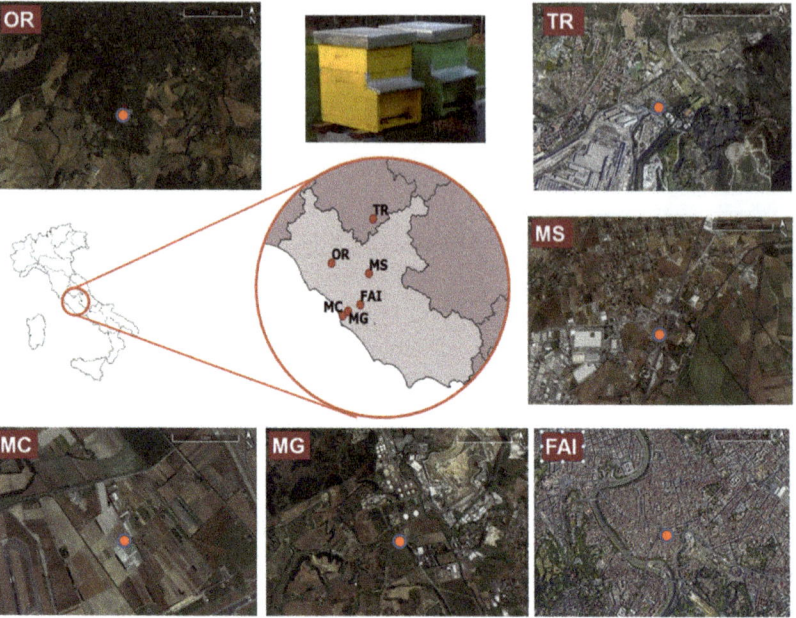

Figure 1. Geographic location of sampled apiaries (central Italy).

These considerations form the basis with which bees and their products have been proposed as reliable bioindicators of the environment including the atmosphere and pollution [40,42,43,46–49,60].

The Hg levels in all royal jelly samples were lower than DL, while in honeydew samples, they were 0.83 ± 0.34 µg kg^{-1}. As shown in Table 4, the Hg concentrations were above the DL for many matrices, showing that the proposed method can be used to determine the Hg level in bees and beehive products. Although variation in Hg level across different areas for each matrix indicates the possibility of using the proposed method for biomonitoring, alternative and parallel measurements of the contamination of the environmental compartment of interest are necessary.

For honey, the mean Hg concentrations in this study (0.91–3.37 µg kg^{-1}) were in agreement with the mean contents found in Croatia (0.47–0.52 µg kg^{-1}) by Bilandžić et al. [36] and China (0.34–4.00 µg kg^{-1}) by Ru et al. [35]. In another Italian study, Hg levels were lower than the quantification limit of 2 µg kg^{-1} [32]. The mean Hg levels in pollen (3.2–12.8 µg kg^{-1}) were similar to the concentrations reported in Poland (3.6–6.6 µg kg^{-1}) by Roman [43] and in Brazil (0.4–6.8 µg kg^{-1}) by Morgano et al. [44]. According to our mean data in propolis (4.6–14.8 µg kg^{-1}), studies from Croatia by Cvek et al. [45] and Spain by Bonvehí and Bermejo [46] reported Hg concentrations as a median of 12 µg kg^{-1} and mean of 8.0 ± 2.5 µg kg^{-1}, respectively. For bees, there are few available data in the literature because of the limited amount of this matrix and consequently high DL values [24,25,40,47,48]. A study by Toth et al. [39] reported Hg concentrations of 39.892 ± 0.035 µg kg^{-1} and 8.224 ± 0.028 µg kg^{-1} in bees from two locations in eastern Slovakia. These results are in accordance with our data ranging from 0.53 to 31 µg kg^{-1}. For beeswax, only one study by Bommuraj et al. [42] reports a concentration value of Hg equal to 62 µg kg^{-1}, while our data fell in the range of <1–12.7 µg kg^{-1}. Unfortunately, it was not possible to make a comparison with the literature for honeydew and royal jelly.

Table 4. Comparison of mercury occurrence (µg kg^{-1}) in six sampled apiaries (central Italy).

Matrix	Statistics	OR	FAI	MC	MG	MS	TR
Honey	N	14	6	10	10	10	4
	Mean	**0.91** [a,b,c,d,e]	**2.26** [a]	**2.68** [b]	**2.66** [c]	**2.43** [d]	**3.37** [e]
	SD	0.23	0.69	0.75	0.36	0.60	0.60
	Median	0.78	1.92	2.97	2.53	2.07	3.37
	Minimum	0.66	1.80	1.35	2.30	1.95	2.95
	Maximum	1.25	3.06	3.17	3.23	3.20	3.80
Pollen	N	14	NS	12	10	4	4
	Mean	**3.2** [a,b]	-	7.6	7.5	**12.8** [a]	**10.4** [b]
	SD	1.4	-	2.0	1.9	8.0	2.5
	Median	3.0	-	7.2	7.0	12.8	10.4
	Minimum	<3	-	5.1	5.9	7.2	8.7
	Maximum	5.6	-	10.2	10.6	18.5	12.2
Propolis	N	10	NS	NS	4	6	NS
	Mean	**4.6** [a]	-	-	**7.54** [b]	**14.8** [a,b]	-
	SD	1.2	-	-	0.65	2.1	-
	Median	4.8	-	-	7.54	15.7	-
	Minimum	<3	-	-	7.08	12.4	-
	Maximum	5.7	-	-	8.00	16.4	-
Beeswax	N	14	8	14	12	12	4
	Mean	**2.8** [a]	5.9	6.4	4.8	4.9	**11.5** [a]
	SD	1.6	2.8	3.0	1.7	3.1	1.8
	Median	2.8	5.2	4.5	4.2	3.9	11.5
	Minimum	<1	3.5	3.3	3.5	3.0	10.2
	Maximum	5.7	9.5	10.6	8.1	11.2	12.7
Bees	N	14	6	14	14	10	4
	Mean	**1.76** [a,b,c]	**16.2** [a]	11.0	**16.2** [b]	**17.1** [c]	14.5
	SD	0.85	2.7	5.8	5.6	8.5	5.2
	Median	2.09	15.1	10.4	14.5	17.0	14.5
	Minimum	0.53	14.4	1.6	8.3	9.1	10.9
	Maximum	2.65	19.3	20.3	25.3	31.0	18.2

N, samples number; SD, standard deviation; NS, not sampled. [a,b,c,d,e] The data in bold with the same superscript letters within rows were significantly different ($p < 0.01$; ANOVA test).

In this study, the Hg levels showed a typical distribution related to anthropogenic development of the areas. Furthermore, in agreement with the observations of other authors on the biological barrier capacity of bees for the contamination of honey by Cd and Tl [40], bees also seem to work as biofilters for Hg. In fact, the Hg levels were generally lower in the rural site (OR) and honey samples and higher in the sites with greater anthropogenic impact and bee samples. In particular, bees showed approximately ten times higher mean concentrations in the FAI (16.2 ± 2.7 µg kg^{-1}), MG (16.2 ± 5.6 µg kg^{-1}), and MS (17.8 ± 8.5 µg kg^{-1}) areas than in the OR site (1.76 ± 0.85 µg kg^{-1}). The lowest Hg level was detected in honey samples from OR (0.91 ± 0.23 µg kg^{-1}). The principal anthropogenic sources of Hg pollution are industrial and urban discharge and combustion [35,38].

Our results agree with numerous other studies [60,61]. In Toth et al. [39], a statistically significant relationship was described between the locality and Hg content in bees and bee pollen. Moreover, in the study by Dżugan et al. [40], the sampling area and its related emission sources influenced toxic metal concentration in both bee bodies and honey. However, the Hg content in bees may also depend on other factors such as method of rearing bee colonies (including supplemental feeding), age of worker bees, and physiological and health status of bee specimens and bee colonies [62]. Due to its physical feature (sticky) and its chemical composition (mainly polyphenols, amino acids, terpenes, and steroids), propolis can absorb Hg and other metals [47,63], thus it can also be used as a bioindicator of air pollution [63,64].

Especially for honey, the assessment of Hg levels is important not only for environmental protection but also for food quality and consumer health [38]. Currently, the Hg presence in honey must not follow specific regulations. However, the Codex Alimentarius states that honey shall be free from metals in amounts that may result in a hazard to human health [65]. A provisional tolerable weekly intake (PTWI) of 0.3 mg (0.042 mg/day) for a 70-kg person (0.004 mg/kg body weight/week) was designed for Hg [66]. Considering the highest Hg concentration of the whole campaign (3.80 µg kg^{-1}), a 20-g daily honey consumption represents a weekly intake of circa 0.2% of the PTWI for Hg. This Hg intake is well below the recommended dose, and the consumption of honey is not considered dangerous for human health.

3. Materials and Methods

3.1. Study Areas and Sample Collection

Samples of royal jelly (n = 2) and honeydew (n = 2) of different brands were purchased in duplicate from the Italian market, while samples of bees, beeswax, honey, pollen, and propolis were collected from six different apiaries across central Italy from April 2018 to June 2019 (Figure 1). Two beehives were selected at each apiary, and the beekeepers sampled their bee colonies and beehive products into polyethylene screw-cap containers once every two months in the late morning. The six study locations were chosen to represent sites with different human activities and environmental impacts. Terni (TR) was selected as an industrial area affected by the steel mill industry. Rome [city center on the roof of the Apicultural Italian Federation (FAI), Anagnina (MS), Malagrotta (MG), and Maccarese (MC)] was chosen as an urban area influenced by different emission sources such as traffic pollution in FAI and MS; biomass burning in FAI; various industrial plants such as refinery, gasifier, hospital waste incinerator, landfill of municipal waste, and quarries for the extraction of building materials in MG; and intense air and ship traffic in MC (located next to Fiumicino airport). Finally, Oriolo Romano (OR) in Viterbo province was selected as a rural area.

After sampling, the samples were transported to the laboratory. For each beehive, bees (n = 20) were dried in a freeze drier (at least 48 h for constant weight) and then were ground in a ceramic mortar coated with parafilm. Beeswax samples were separated from honey, washed in deionized water until all of the residual honey was removed, and dried using a freeze dryer (at least 24 h for constant weight). All of the obtained samples were thoroughly mixed to have a homogeneous sample and were stored at −18 °C until analysis.

3.2. Materials and Reagents

Certified Hg standard solution of 1002 ± 7 mg L^{-1} in 10% HNO$_3$ was obtained from SCP Science (Baie D'Urfé, Quebec, Canada) and was used for further dilutions in order to prepare eight calibration standard solutions in the range from 0.02 to 1.5 µg L^{-1}. HNO$_3$ (67%, suprapure), HCl (30%, suprapure), NaOH (98%, anhydrous pellets, and RPE for analysis, ACS–ISO) were purchased from Carlo Erba Reagents (Milan, Italy) and H$_2$O$_2$ (30%, suprapure) and NaBH$_4$ were obtained from Merck KgaA (Darmstadt, Germany). Deionized H$_2$O (resistivity, ≤18.2 MΩ cm) from an Arioso Power I RO-UP Scholar UV system (Human Corporation, Songpa-Ku, Seoul, Korea) was used throughout the study.

Graduated tubes (2.5, 5, and 10 mL in polypropylene) were purchased from Artiglass S.R.L. (Due Carrare, PD, Italy), and syringe filters (0.45-µm pore size and cellulose nitrate membrane) were obtained from GVS Filter Technology (Indianapolis, IN, USA).

3.3. Sample Preparation and Analysis

Preliminary experiments were conducted to optimize the sample digestion using a water bath (WB12, Argo Lab, Modena, Italy) at 95 °C and ~1 bar. A freeze dryer (Heto Power Dry LL1500, Thermo Electron Corporation, Waltham, USA) was employed with a vacuum of 10^{-3} mbar and a condensing plate temperature of −40 °C to dry the beeswax and bee samples. An aliquot of the samples (0.05–1 g for honey and honeydew, 0.02–0.2 g for bees, beeswax, pollen, propolis, royal jelly) was treated with 0.5 mL HCl, 0.2 mL HNO$_3$, and 0.1 mL H$_2$O$_2$ into open graduated tubes for 30 or 60 min under a fume hood. Digestion blanks (n = 10) were carried out in the same way. All solutions of the digested samples were colorless and without suspended solid particles except for the honey, honeydew, and propolis solutions obtained from digestion of the largest mass. Thus, digested samples were diluted to a final volume of 5 mL with deionized water, filtered, and then analyzed with an

AFS 8220 (Beijing Titan Instrumental Co. Ltd., Beijing, China) with Ar (99.999% purity, SOL Spa, Monza, Italy) as a carrier gas. HCl (5%, v/v) was used as a carrier liquid, and 2% (w/v) $NaBH_4$ in 0.5% (w/v) NaOH was used as a reducing agent. The instrumental optimized parameters were previously described [59]. Duplicate analyses were performed for each sample. Blanks and control standards (at 0.4 µg L^{-1}) were run every 20 determinations to evaluate instrument drift.

3.4. Quality Assurance

The analytical performance parameters of selectivity, linearity, detection and quantification limit (DL and QL, respectively), precision, and accuracy were evaluated. The validation process was performed using spiked real sample assays. Method blanks, in-house quality control samples, and spiked and non-spiked real samples (three replicates each) were prepared along with every digested sample batch. Hg standard solution at 2, 20, or 100 µg L^{-1} was made for spikes; 0.05 mL of the spike solution was added to appropriate tubes 30 min before reagents and digestion. At the instrument, the concentration was 0.02, 0.2, or 1 µg L^{-1}. For the recovery determination, the non-spiked real sample concentration was subtracted from that measured in the spiked real sample.

An eight-point calibration curve consisting of Hg concentrations between 0.02 and 1.5 µg L^{-1} was prepared using aqueous standards and the standard addition method for each matrix. The DL was defined as the Hg concentration corresponding to three times the standard deviation of the digestion blanks ($n = 10$).

3.5. Statistical Analysis

Statistical analysis was performed using the SPSS 25.0 program (IBM Corp., Armonk, NY, USA). All data were normally distributed as confirmed by the Kolmogorov–Smirnov test. One-way ANOVA, followed by Bonferroni post-hoc test, was used to determine the significant differences among the Hg concentrations for each matrix in different geographical areas. The probability level of $p < 0.05$ was considered statistically significant. For statistical analysis, in samples where the Hg concentration was below DL, the used values were one half of DL.

4. Conclusions

Coupling water bath digestion with CV-AFS analysis proved to be a good analytical tool for evaluating Hg contamination in bees and beehive products (beeswax, honey, honeydew, pollen, propolis, and royal jelly). Due to the possibility of preparing the sample using same single autosampler tubes, the optimized digestion procedure allows for the prevention of sample loss, minimize manipulation, and reduce both the reagent volumes and final dilution. The proposed method is suitable for small masses (down to 0.02 g) of all selected matrices and can be used for biomonitoring and food quality control. In particular, the results from the application in the field of the proposed method showed a higher Hg concentration in bees than the other matrices considered and in areas with a higher anthropogenic impact than the background site. In the future, considering alternative and parallel measurements of the contamination of the environmental compartment of interest, it will be interesting to evaluate whether bees and hive products can indeed be used to assess environmental spatial changes in Hg levels. However, the determination of Hg concentrations in beehive products is also important for potential human dietary exposure. The Hg concentrations in the analyzed samples of honey, honeydew, and royal jelly are not a cause for concern for consumer health effects. Furthermore, the data in this study can be used as a reference for comparing Hg concentrations to other countries in the world.

Supplementary Materials: The following are available online. Table S1: Summary of the analytical characteristics of the proposed method and comparison with some of the previous methods published during the last decade (2010–2020), Table S2: Recovery and precision data for Hg in bees and beehive products by water bath digestion (95 °C, 30 min or 60 min).

Author Contributions: Conceptualization, M.L.A.; Methodology, M.L.A.; Validation, M.L.A.; Formal analysis, M.L.A. and L.M.; Investigation, M.L.A., M.R., and M.A.F.; Resources, M.L.A., M.E.C., M.P., and S.C.; Data curation, M.L.A.; Writing—original draft preparation, M.L.A.; Writing—review and editing, M.L.A., M.E.C., and S.C.; Visualization, M.L.A.; Supervision, M.L.A.; Funding acquisition, M.E.C. All authors have read and agreed to the published version of the manuscript.

Funding: This research was partially funded by Sapienza University of Rome, project 2018, grant number RG11816432851FA6.

Data Availability Statement: The data presented in this study are available on request from the corresponding author.

Acknowledgments: The authors wish to thank Massimo Marcolini (Association of Beekeepers of Rome and Province), and Fabrizio Piacentini (Italian Beekeeping Federation—FAI) for their kind support in the bee and beehive product sampling. We also wish to thank Giulia Vitiello and Elisabetta Marconi for their excellent support in the treatment and classification of the samples and Helga Liselotte for her support during our stays in Oriolo Romano.

Conflicts of Interest: The authors declare no conflict of interest.

Sample Availability: Some samples used in this study are available from the authors.

References

1. Langford, N.J.; Ferner, R.E. Toxicity of mercury. *J. Hum. Hypertens.* **1999**, *13*, 651–656. [CrossRef]
2. Clarkson, T.W.; Magos, L. The toxicology of mercury and its chemical compounds. *Crit. Rev. Toxicol.* **2006**, *36*, 609–662. [CrossRef]
3. Selin, N.E. Global biogeochemical cycling of mercury: A review. *Annu. Rev. Environ. Resour.* **2009**, *34*, 43–63. [CrossRef]
4. United Nations Environment Programme (UNEP). *Global Mercury Assessment 2013: Sources, Emissions, Releases and Environmental Transport*; UNEP Chemicals Branch: Geneva, Switzerland, 2013; Available online: https://wedocs.unep.org/handle/20.500.11822/7984 (accessed on 2 May 2021).
5. Agency for Toxic Substances and Disease Registry. *Toxicological Profile for Mercury*; Department of Health and Human Services: Atlanta, GA, USA, 1999. Available online: http://www.atsdr.cdc.gov/toxprofiles/tp46.pdf (accessed on 2 May 2021).
6. Berglund, M.; Lind, B.; Björnberg, K.A.; Palm, B.; Einarsson, O.; Vahter, M. Inter-individual variations of human mercury exposure biomarkers: A cross-sectional assessment. *Environ. Health* **2005**, *4*, 1–11. [CrossRef]
7. Mercury and Health Fact Sheet. Available online: https://www.who.int/news-room/fact-sheets/detail/mercury-and-health (accessed on 2 May 2021).
8. European Union. European Food Safety Authority. Mercury as undesirable substance in animal feed. Scientific opinion of the Panel on Contaminants in the Food Chain. *EFSA J.* **2008**, *654*, 1–74.
9. European Union. European Food Safety Authority (EFSA). Scientific Opinion on the risk for public health related to the presence of mercury and methylmercury in food. *EFSA J.* **2012**, *10*, 2985.
10. European Union. European Parliament. Directive 2004/107/EC of the European Parliament and of the Council of 15 December 2004 relating to arsenic, cadmium, mercury, nickel and polycyclic aromatic hydrocarbons in ambient air. *Off. J. Eur. Communities* **2004**, *23*, 3–16.
11. European Union. European Parliament. Directive 2008/50/EC of the European Parliament and of the Council of 21 May 2008 on ambient air quality and cleaner air for Europe. *Off. J. Eur. Union* **2008**, *152*, 1–44.
12. European Union. European Parliament. Directive 2010/75/EU of the European Parliament and of the Council of 24 November 2010 on industrial emissions (integrated pollution prevention and control). *Off. J. Eur. Union* **2010**, *334*, 17–119.
13. European Union. European Parliament. Regulation (EU) 2017/852 of the European Parliament and of the Council of 17 May 2017 on mercury, and repealing Regulation (EC) No 1102/2008. *Off. J. Eur. Union* **2017**, *137*, 1–21.
14. Pirrone, N.; Aas, W.; Cinnirella, S.; Ebinghaus, R.; Hedgecock, I.M.; Pacyna, J.; Sprovieri, F.; Sunderland, E.M. Toward the next generation of air quality monitoring: Mercury. *Atmos. Environ.* **2013**, *80*, 599–611. [CrossRef]
15. Sprovieri, F.; Pirrone, N.; Ebinghaus, R.; Kock, H.; Dommergue, A. A review of worldwide atmospheric mercury measurements. *Atmos. Chem. Phys. Discuss.* **2010**, *10*, 8245–8265. [CrossRef]
16. Conti, M.E.; Cecchetti, G. Biological monitoring: Lichens as bioindicators of air pollution assessment—A review. *Environ. Pollut.* **2001**, *114*, 471–492. [CrossRef]
17. Wolterbeek, H.T. Biomonitoring of trace element air pollution: Principles, possibilities and perspectives. *Environ. Pollut.* **2002**, *120*, 11–21. [CrossRef]
18. Bargagli, R. Moss and lichen biomonitoring of atmospheric mercury: A review. *Sci. Total Environ.* **2016**, *572*, 216–231. [CrossRef] [PubMed]
19. Fortuna, L.; Candotto Carniel, F.; Capozzi, F.; Tretiach, M. Congruence evaluation of mercury pollution patterns around a waste incinerator over a 16-year-long period using different biomonitors. *Atmosphere* **2019**, *10*, 183. [CrossRef]

20. Massimi, L.; Conti, M.E.; Mele, G.; Ristorini, M.; Astolfi, M.L.; Canepari, S. Lichen transplants as indicators of atmospheric element concentrations: A high spatial resolution comparison with PM 10 samples in a polluted area (Central Italy). *Ecol. Indic.* **2019**, *101*, 759–769. [CrossRef]
21. Vitali, M.; Antonucci, A.; Owczarek, M.; Guidotti, M.; Astolfi, M.L.; Manigrasso, M.; Avino, P.; Bhattacharya, B.; Protano, C. Air quality assessment in different environmental scenarios by the determination of typical heavy metals and Persistent Organic Pollutants in native lichen Xanthoria parietina. *Environ. Pollut.* **2019**, *254*, 113013. [CrossRef] [PubMed]
22. Ristorini, M.; Astolfi, M.L.; Frezzini, M.A.; Canepari, S.; Massimi, L. Evaluation of the efficiency of Arundo donax L. Leaves as biomonitors for atmospheric element concentrations in an urban and industrial area of central Italy. *Atmosphere* **2020**, *11*, 226. [CrossRef]
23. Bargańska, Z.; Ślebioda, M.; Namieśnik, J. Honey bees and their products: Bioindicators of environmental contamination. *Crit. Rev. Environ. Sci. Technol.* **2016**, *46*, 235–248. [CrossRef]
24. Maragou, N.C.; Pavlidis, G.; Karasali, H.; Hatjina, F. Cold vapor atomic absorption and microwave digestion for the determination of mercury in honey, pollen, propolis and bees of greek origin. *Glob. NEST J.* **2016**, *18*, 690–696.
25. Grainger, M.N.C.; Hewitt, N.; French, A.D. Optimised approach for small mass sample preparation and elemental analysis of bees and bee products by inductively coupled plasma mass spectrometry. *Talanta* **2020**, *214*, 120858. [CrossRef]
26. Burdock, G.A. Review of the Biological Properties and Toxicity of Bee Propolis. *Food Chem. Toxicol.* **1998**, *36*, 347–363. [CrossRef]
27. Kalogeropoulos, N.; Konteles, S.J.; Troullidou, E.; Mourtzinos, I.; Karathanos, V.T. Chemical composition, antioxidant activity and antimicrobial properties of propolis extracts from Greece and Cyprus. *Food Chem.* **2009**, *116*, 452–461. [CrossRef]
28. Tsiapara, A.V.; Jaakkola, M.; Chinou, I.; Graikou, K.; Tolonen, T.; Virtanen, V.; Moutsatsou, P. Bioactivity of Greek honey extracts on breast cancer (MCF-7), prostate cancer (PC-3) and endometrial cancer (Ishikawa) cells: Profile analysis of extracts. *Food Chem.* **2009**, *116*, 702–708. [CrossRef]
29. Melliou, E.; Chinou, I. Chemical constituents of selected unifloral Greek bee-honeys with antimicrobial activity. *Food Chem.* **2011**, *129*, 284–290. [CrossRef] [PubMed]
30. Burlando, B.; Cornara, L. Honey in dermatology and skin care: A review. *J. Cosmet. Dermatol.* **2013**, *12*, 306–313. [CrossRef]
31. Toporcák, J.; Legáth, J.; Kul'ková, J. Levels of mercury in samples of bees and honey from areas with and without industrial contamination. *Vet. Med.* **1992**, *37*, 405–412.
32. Pisani, A.; Protano, G.; Riccobono, F. Minor and trace elements in different honey types produced in Siena County (Italy). *Food Chem.* **2008**, *107*, 1553–1560. [CrossRef]
33. Dos Santos Depoi, F.; Bentlin, F.R.S.; Pozebon, D. Methodology for Hg determination in honey using cloud point extraction and cold vapour-inductively coupled plasma optical emission spectrometry. *Anal. Methods* **2010**, *2*, 180–185. [CrossRef]
34. Domínguez, M.A.; Grünhut, M.; Pistonesi, M.F.; Di Nezio, M.S.; Centurión, M.E. Automatic flow-batch system for cold vapor atomic absorption spectroscopy determination of mercury in honey from argentina using online sample treatment. *J. Agric. Food Chem.* **2012**, *60*, 4812–4817. [CrossRef] [PubMed]
35. Ru, Q.-M.; Feng, Q.; He, J.-Z. Risk assessment of heavy metals in honey consumed in Zhejiang province, southeastern China. *Food Chem. Toxicol.* **2013**, *53*, 256–262. [CrossRef]
36. Bilandžić, N.; Gačić, M.; Đokić, M.; Sedak, M.; Šipušić, Đ.I.; Končurat, A.; Gajger, I.T. Major and trace elements levels in multifloral and unifloral honeys in Croatia. *J. Food Compos. Anal.* **2014**, *33*, 132–138. [CrossRef]
37. Vieira, H.P.; Nascentes, C.C.; Windmöller, C.C. Development and comparison of two analytical methods to quantify the mercury content in honey. *J. Food Compos. Anal.* **2014**, *34*, 1–6. [CrossRef]
38. Meli, M.A.; Desideri, D.; Roselli, C.; Benedetti, C.; Feduzi, L. Essential and toxic elements in honeys from a region of central Italy. *J. Toxicol. Environ. Health* **2015**, *78*, 617–627. [CrossRef]
39. Toth, T.; Kopernicka, M.; Sabo, R.; Kopernicka, T. The evaluation of mercury in honey bees and their products from eastern Slovakia. *Sci. Pap. Anim. Sci. Biotechnol.* **2016**, *49*, 257–260.
40. Dżugan, M.; Wesołowska, M.; Zaguła, G.; Kaczmarski, M.; Czernicka, M.; Puchalski, C. Honeybees (Apis mellifera) as a biological barrier for contamination of honey by environmental toxic metals. *Environ. Monit. Assess.* **2018**, *190*, 101. [CrossRef]
41. Jovetić, M.S.; Redžepović, A.S.; Nedić, N.M.; Vojt, D.; Đurđić, S.Z.; Brčeski, I.D.; Milojković-Opsenica, D.M. Urban honey—The aspects of its safety. *Arh. Hig. Rada Toksikol.* **2018**, *69*, 264–274. [CrossRef]
42. Bommuraj, V.; Chen, Y.; Klein, H.; Sperling, R.; Barel, S.; Shimshoni, J.A. Pesticide and trace element residues in honey and beeswax combs from Israel in association with human risk assessment and honey adulteration. *Food Chem.* **2019**, *299*, 125123. [CrossRef] [PubMed]
43. Roman, A. Concentration of chosen trace elements of toxic properties in bee pollen loads. *Pol. J. Environ. Stud.* **2009**, *18*, 265–272.
44. Morgano, M.A.; Martins, M.C.T.; Rabonato, L.C.; Milani, R.F.; Yotsuyanagi, K.; Rodriguez-Amaya, D.B. Inorganic contaminants in bee pollen from southeastern Brazil. *J. Agric. Food Chem.* **2010**, *58*, 6876–6883. [CrossRef] [PubMed]
45. Cvek, J.; Medić-Šarić, M.; Vitali, D.; Vedrina-Dragojević, I.; Šmit, Z.; Tomić, S. The content of essential and toxic elements in Croatian propolis samples and their tinctures. *J. Apicult. Res.* **2008**, *47*, 35–45. [CrossRef]
46. Bonvehí, J.S.; Bermejo, F.J.O. Element content of propolis collected from different areas of South Spain. *Environ. Monit. Assess.* **2013**, *185*, 6035–6047. [CrossRef]
47. Matin, G.; Kargar, N.; Buyukisik, H.B. Bio-monitoring of cadmium, lead, arsenic and mercury in industrial districts of Izmir, Turkey by using honey bees, propolis and pine tree leaves. *Ecol. Eng.* **2016**, *90*, 331–335. [CrossRef]

48. Perugini, M.; Manera, M.; Grotta, L.; Abete, M.C.; Tarasco, R.; Amorena, M. Heavy metal (Hg, Cr, Cd, and Pb) contamination in urban areas and wildlife reserves: Honeybees as bioindicators. *Biol. Trace Elem. Res.* **2011**, *140*, 170–176. [CrossRef]
49. Zaric, N.M.; Deljanin, I.; Ilijević, K.; Stanisavljević, L.; Ristić, M.; Gržetić, I. Assessment of spatial and temporal variations in trace element concentrations using honeybees (Apis mellifera) as bioindicators. *PeerJ* **2018**, *6*, e5197. [CrossRef] [PubMed]
50. Astolfi, M.L.; Conti, M.E.; Marconi, E.; Massimi, L.; Canepari, S. Effectiveness of different sample treatments for the elemental characterization of bees and beehive products. *Molecules* **2020**, *25*, 4263. [CrossRef]
51. Melaku, S.; Gelaude, I.; Vanhaecke, F.; Moens, L.; Dams, R. Comparison of pyrolysis and microwave acid digestion techniques for the determination of mercury in biological and environmental materials. *Microchim. Acta* **2003**, *142*, 7–12. [CrossRef]
52. Astolfi, M.L.; Marconi, E.; Protano, C.; Canepari, S. Comparative elemental analysis of dairy milk and plant-based milk alternatives. *Food Control* **2020**, *116*, 107327. [CrossRef]
53. Schlathauer, M.; Reitsam, V.; Schierl, R.; Leopold, K. A new method for quasi-reagent-free biomonitoring of mercury in human urine. *Anal. Chim. Acta* **2017**, *965*, 63–71. [CrossRef]
54. Astolfi, M.L.; Protano, C.; Schiavi, E.; Marconi, E.; Capobianco, D.; Massimi, L.; Ristorini, M.; Baldassarre, M.E.; Laforgia, N.; Vitali, M.; et al. A prophylactic multi-strain probiotic treatment to reduce the absorption of toxic elements: In-vitro study and biomonitoring of breast milk and infant stools. *Environ. Int.* **2019**, *130*, 104818. [CrossRef]
55. Astolfi, M.L.; Pietris, G.; Mazzei, C.; Marconi, E.; Canepari, S. Element levels and predictors of exposure in the hair of Ethiopian children. *Int. J. Environ. Res. Public Health* **2020**, *17*, 8652. [CrossRef] [PubMed]
56. Oliveira, S.S.; Alves, C.N.; Morte, E.S.B.; Júnior, A.D.F.S.; Araujo, R.G.O.; Santos, D.C.M.B. Determination of essential and potentially toxic elements and their estimation of bioaccessibility in honeys. *Microchem. J.* **2019**, *151*, 104221. [CrossRef]
57. Tenaglia, H.; Venturini, E.; Raffaelli, R. Linee guida per la validazione dei metodi analitici e per il calcolo dell'incertezza di misura. In *Manuali ARPA*; Agenzia Regionale Prevenzione e Ambiente dell'Emilia Romagna: Bologna, Italy, 2003.
58. European Commission. Commission Decision 2002/657/EC of 12 August 2002 implementing Council Directive 96/23/EC concerning the performance of analytical methods and the interpretation of results (notified under document number C(2002) 3044) (Text with EEA relevance). *Off. J. Eur. Communities* **2002**, *221*, 8–36. Available online: https://eur-lex.europa.eu/legal-content/EN/TXT/PDF/?uri=CELEX:32002D0657&rid=13 (accessed on 3 August 2021).
59. Astolfi, M.L.; Protano, C.; Marconi, E.; Piamonti, D.; Massimi, L.; Brunori, M.; Vitali, M.; Canepari, S. Simple and rapid method for the determination of mercury in human hair by cold vapour generation atomic fluorescence spectrometry. *Microchem. J.* **2019**, *150*, 104186. [CrossRef]
60. Herrero-Latorre, C.; Barciela-García, J.; García-Martín, S.; Peña-Crecente, R.M. The use of honeybees and honey as environmental bioindicators for metals and radionuclides: A review. *Environ. Rev.* **2017**, *25*, 463–480. [CrossRef]
61. AL-Alam, J.; Chbani, A.; Faljoun, Z.; Millet, M. The use of vegetation, bees, and snails as important tools for the biomonitoring of atmospheric pollution—A review. *Environ. Sci. Pollut. Res.* **2019**, *26*, 9391–9408. [CrossRef]
62. Zhelyazkova, I. Honeybees—Bioindicators for environmental quality. *Bulg. J. Agric. Sci.* **2012**, *18*, 435–442.
63. Finger, D.; Filho, I.K.; Torres, Y.R.; Quináia, S.P. Propolis as an indicator of environmental contamination by metals. *Bull. Environ. Contam. Toxicol.* **2014**, *92*, 259–264. [CrossRef]
64. Conti, M.E.; Botrè, F. Honeybees and their products as potential bioindicators of heavy metals contamination. *Environ. Monit. Assess.* **2001**, *69*, 267–282. [CrossRef]
65. Codex Alimentarius. Standard for Honey (CXS 12-1981). Available online: http://www.fao.org/fao-who-codexalimentarius/sh-proxy/en/?lnk=1&url=https%253A%252F%252Fworkspace.fao.org%252Fsites%252Fcodex%252FStandards%252FCXS%2B12-1981%252FCXS_012e.pdf (accessed on 2 May 2021).
66. Joint FAO; World Health Organization; WHO Expert Committee on Food Additives. *Evaluation of Certain Food Additives and Contaminants: Seventy-Second Report of the Joint FAO/WHO Expert Committee on Food Additives*; World Health Organization: Geneva, Switzerland, 2011.

MDPI
St. Alban-Anlage 66
4052 Basel
Switzerland
Tel. +41 61 683 77 34
Fax +41 61 302 89 18
www.mdpi.com

Molecules Editorial Office
E-mail: molecules@mdpi.com
www.mdpi.com/journal/molecules

www.ingramcontent.com/pod-product-compliance
Lightning Source LLC
LaVergne TN
LVHW070432100526
838202LV00014B/1579